Archives
of
Virology
Supplementum 6

P. P. Liberski

The Enigma of Slow Viruses
Facts and Artefacts

Springer-Verlag Wien New York

Dr. Pawel P. Liberski
Electron Microscopic Laboratory
Department of Oncology
Medical Academy Lodz
Łodz, Poland

ISSN 0939-1983
ISBN-13: 978-3-211-82427-6   e-ISBN-13: 978-3-7091-9270-2
DOI: 10.1007/978-3-7091-9270-2

This book is dedicated to

Dr. Carleton Gajdusek
on the occasion of his 70th birthday

Preface

After my return to Poland having spent several years in the Laboratory of Central Nervous System Studies of the National Institute of Health, Bethesda, USA I moved after a period of time to one of the very few islands in Poland not devastated by the communist mentality, the Department of Oncology, Medical Academy, Lodz created by Prof. Leszek Wozniak. At the same time I suggested this review to Prof. H.-D. Klenk, the editor of the special issue of *Archives of Virology*.

Such a work was not possible without the invaluable assistance of many investigators who read chapters or generously provided scientific materials or illustrations: Prof. Jan Albrecht, Polish Academy of Sciences, Warsaw, Poland; Dr. David M. Asher, Laboratory of Central Nervous System Studies, NIH, Bethesda, USA; Dr. Jan Boellaard, University of Tübingen, Federal Republic of Germany; Dr. Paul Brown, Laboratory of Central Nervous System Studies, NIH, Bethesda, USA; Prof. Herbert Budka, Neurological Institute of Vienna University, Austria; Dr. D. Carleton Gajdusek, Laboratory of Central Nervous System Studies, NIH, Bethesda, USA; Dr. Clarence J. Gibbs, Jr, Laboratory of Central Nervous System Studies, NIH, Bethesda, USA; Dr. Peter Gibson, formerly with the Neuropathogenesis Unit, Edinburgh, Scotland; Dr. Lev Goldfarb, Laboratory of Central Nervous System Studies, NIH, Bethesda, USA; Dr. Hugh Fraser, Neuropathogenesis Unit, Edinburgh, Scotland; Dr. James Hope, Neuropathogenesis Unit, Edinburgh, Scotland; Dr. Richard H. Kimberlin, SARDAS, Edinburgh, Scotland; Prof. Laura Manuelidis, Yale University, USA; Prof. Stanley B. Prusiner, University of California San Francisco, USA; Dr. Gerald A.H. Wells, Central Veterinary Laboratory, Ministry of Agriculture, Fisheries and Food, U.K., Dr. Richard Yanagihara, Laboratory of Central Nervous System Studies, NIH, Bethesda, USA. A few of them, however, preferred to stay anonymous. This work could not have been accomplished without generous support from the Fogarty International Center and the National Institutes of Health while in USA, and the Neurological Institute, University of Vienna, Austria. Prof. Miroslaw J. Mossakowski and the Polish Academy of Sciences, Warsaw, is acknowledged for support while in Poland. Last but not least, I acknowledge the continuous support and encouragement from my collaborators in Poland: Prof. Janusz Alwasiak, Department of Oncology, Lodz, Poland; Dr. Maria Barcikowska, Polish Academy of Sciences, Warsaw; Prof. Hubert Kwiecinski, Department of Neurology, Warsaw; Dr. Barbara Mirecka, Department of Oncology, Lodz; Prof. Wielislaw Papierz, Department of Pathological Anatomy, Lodz; and the skilful technical assistance of Ryszard Kurczewski, Elzbieta Naganska, Leokadia Romanska and Kazimierz Smoktunowicz.

Pawel P. Liberski

Contents

Abbreviations

An index of abbreviations used in this review. The most commonly used abbreviations such as DNA, and abbreviations of restrictions enzymes and mice strains are not listed.

A – agouti locus on chromosome 2 in mice
22A – a strain of scrapie virus
79A – a strain of scrapie virus
139A – a strain of scrapie virus
A2B5 – a specific astrocyte surface antigen
AchE – acetylcholinesterase
ACPMs – abnormal configurations of plasma membranes
AD – Alzheimer's disease
ALS – amyotrophic lateral sclerosis
ARIA – acetylocholine receptor-inducing activity
ASF – ammonium sulphate fraction
AV – autophagic vacuoles
BBB – blood brain barier
B2m – beta-2-microglobulin locus on chromosome 2 in mice
BSE – bovine spongiform encephalopathy
C506 – the "american" strain of scrapie virus
22C – a strain of scrapie virus
CH 1641 – a strain of scrapie virus
ChAT – choline acetyltransferase
ch-PrLP – chicken prion-like protein
CJD – Creutzfeldt-Jakob disease
C-J virus – Creutzfeldt-Jakob disease virus
CLB – cerebellar lamellar bodies
CRD – cross-reactive determinant
C_0t curve – DNA reassociation curve
CWD – chronic wasting disease
2D – two dimensional
7D – a strain of scrapie virus, probably the same as ME7
D_{37} – a dose which reduces survival of viruses (cells) to 37%; a measure of the exponential portion of the survival curve
DEP – diethylpyrocarbonate –
Dh – dominant hemimelia gene in mouse
DHPG – 3,4-dihydroxyphenylethyleneglycol
D-loop – displacement loop of mitochondrial DNA
DN – dystrophic neurites
DOC – sodium deoxycholate
DOPAC – dihydroxyphenylacetic acid
ds – double stranded

DS 500 – dextran sulphate 500
EAE – experimental allergic encephalomyelitis
EMC – encephalomyocarditis virus
endo H – endoglycosidase H
GABA – gamma aminobutyric acid
GFAP – glial fibrillary acidic protein
Gp34 – glycoprotein of apparent molecular weight 34 kDa (the same as PrP 33–35^{sc})
GPI – glycoinositol phospholipid
GSSD – Gerstmann-Straussler-Scheinker (disease)
GSSS – Gerstmann-Straussler-Scheinker (syndrome)
Ha – suffix which designates hamster
HaPrPcDNA – hamster PrP cDNA
HD – L-histidine decarboxylase
HF – hydrogen fluoride
5-HIAA – 5-hydroxyindoloacetic acid
HNAD – human neuroaxonal dystrophy
HPA – Helix pomatia agglutinin
HPLC – high performance liquid chromatography
5-HTP – L-5-hydroxytryptophan
Hu – a suffix which designates human
HuPrPcDNA – human PrP cDNA
IBR – the New York State Institute of Basic Research in Developmental Disabilities
i.c. – intracerebral route of inoculation
IFN – interferon gamma
INH – isonicotinic hydrazide
i.p. – intraperitoneal route of inoculation
Itp – ITPase locus on chromosome 2 in mice
i.v. – intravenous route of inoculation
263K – a strain of scrapie virus
kb – kilobase
KPC – kuru plaque core fraction from GSS brain
22L – a strain of scrapie virus
ME7 – a strain of scrapie virus
MER – methanol extraction residue of BCG
MHC – major histocompatibility class I antigen
MS – multiple sclerosis
NAD – neuroaxonal dystrophy
NEPHGE – nonequilibrium pH gradient electrophoresis
NFP – neurofilament protein
NFT – neurofibrillary tangles
NGF – nerve grow factor
o.e.r. – oxygen enhancement ratio
ORF – open redaing frame, gene coding sequence
P1 – an antibody against a synthetic peptide encompassing codons 90–102 of PrP
    cDNA
P2 – an antibody against a synthetic peptide encompassing codons 15–40 of PrP cDNA
P3 – an antibody against a synthetic peptide encompassing codons 220–232 of PrP
    cDNA
P5 – an antibody against a synthetic peptide encompassing codons 140–172 of PrP
    cDNA
PAGE – polyacrylamide gele electrophoresis

PCR – polymerase chain reaction

PGA – peanut agglutinin

pGp135 – a clone homologous to the hamster D-loop

PHA – phytohaemagglutynin

PID – prion incubation determinant gene located in the D-subregion of the H-2
complex on chromosome 17 in mice

PIPL – phosphatidylinositol specific phospholipase C

PK – proteinase K

PNGase F – peptide-N^4-(N-acetyl-beta-glucosaminyl) asparaginase F

Prn – prion gene complex on chromosome 2 in mouse

Prn-i – a gene which controls scrapie incubation time, tightly linked to or the same as
Prn-p

Prn-p – PrP gene on chromosome 2 in mouse

Prn-p^a – an allele of Prn-p gene which control scrapie shorter incubation time in mice

Prn-p^h – an allele of Prn-p gene which control scrapie longer incubation time in mice

PRNP – PrP gene on chromosome 20 in humans (The PRP designation for the human
PrP gene could not be used as it has already been used to designate the proline rich
protein (PRP) gene [867]).

PrP – any prion protein (without specification of molecular weight)

PrP followed by any number – PrP of apparent molecular weight specified by this
number

PrP 27–30 – prion protein of apparent molecular weight 27–30 kDa

PrP 33–35 prion protein of apparent molecular weight 33–35 kDa

PrP_o – PrP-precursor protein *in vitro*

PrP_1 – unglycosylated form of PrP *in vitro*

PrP_2 – glycosylatef=d form of PrP *in vitro*

PrP^c – cellular isoform of PrP 33–35

PrP^{CJD} – CJD isoform of PrP 33–35

PrP^{GSS} – GSS isoform of PrP 33–35

PrP^{sc} – scrapie isoform of PrP 33–35

PrP MHM2$_{Ala182/198}$ – recombinant gene containing the entire mouse PrP open reading
frame in which Leu^{108} and Val^{111} were replaced with methionines while two
asparagine-linked glycosylation sites were abolished [996].

PrPs – any prion proteins (without specification of molecular weight)

PSTV – potato spindle tuber viroid

QNB – [^{3}H]quinuclidynylbenzilate

RCA – Ricinus communis agglutinin

RRL – rabbit reticulum lysate

SAF – scrapie associated fibrils

s.c. – subcutaneous route of inoculatio

Sc237 – a strain of scrapie virus, the same as 263K

Scr-1 – a clone from cDNA library which hybridized preferentially to scrapie-infected
brains; later shown to represent 3′ non coding region of GFAP

SDAT – senile dementia of ALzheimer type

SDS-PAGE – sodium dodecyl sulphate polyacrylamide gele electrophoresis

Sinc – scrapie incubation period gene in mice

Sinc^{p7} – an allele of Sinc gene which controls prolonged incubation period in VM mice
following infection with the ME7 strain of scrapie virus

Sinc^{s7} – an allele of Sinc gene which controls shorter incubation period in C57Bl mice
following infection with the ME7 strain of scrapie virus

SIP – scrapie incubation period gene in sheep, probably the same as Prn-p (i)

SP54 – polysulphated polyanion (M_r 3500 – 5000)

SprP-8 – the 3 kb insert of chromosomal PrP gene

ss – single stranded

SSBP/1 – scrapie sheep brain pool/1 – scrapie inoculum containing a mixture of strains (22A, 22C and 22L) of scrapie virus

SSVE – the subacute spongiform virus encephalopathies

STE – stop transfer efector

TEM – transmissible mink encephalopathy

TFMSA – trifluoromethane sulfonic acid

Tg – transgenic mice

Tg (GSS MoPrP) – transgenic mice constructed with PrP gene with Pro to Leu substitution at codon 101

Tg(Prn-p^b) – transgenic mice constructed with Prn-p^b allele of the Prn-p gene

Tg(SHa PrP) – transgenic mice constructed with Syrian hamster PrP gene

TM1 – the first membrane spanning region of transmembrane variant of PrP

TNF-alpha – tumor necrosis factor alpha

TVS – tubulovesicular structures

87V – a strain of scrapie virus

VSGs – variant surface glycoproteins of Trypanosoma brucei

VSV G a glycoprotein of the vesicular stomamtitis virus

WG – wheat germ

WGA – wheat germ agglutinin

1. Introduction: subacute spongiform virus encephalopathies from the perspective of a neuroscientist

"The most important works [. . .] admit that we did not learn everything: but the amount of space devoted to the achievements, with occasional footnote mentions of what remained unknown – those very proportions suggested that we had mastered the Labyrinth, with the exception of few corridors – dead ends, no doubt, probably burried with rubble – whereas in fact we did not get as far as the entrance. Doomed forever to conjecture, having chipped a few flecks from the lock that sealed the gate, we delighted in the glitter that gilded our fingertips. But of what was locked we know nothing. And yet, surely, one of the first duties of a scientist is to determine the extent not of the acquired knowledge, for that knowledge will explain itself, but, rather, of the ignorance, which is the invisible Atlas beneath that knowledge."

Stanislaw Lem "His Master's Voice",
translated from the Polish by M. Kandel,
A Helen and Kurt Wolff Book, Harcourt
Brace Jovanovich, Publ, San Diego – New York –
London, 1983

Scrapie, a naturally occuring neurodegenerative disease of sheep and sometimes goats [25, 86, 907, 1050], is a prototypic disease for the whole group of the subacute spongiform virus encephalopathies. Kuru (Figs. 1–4) was the first human disease of this type to be discovered in 1957 by Gajdusek and Zigas (Fig. 1) [374, 381–382, 389–390, 508–509], and its discovery opened the whole field in the human biomedical sciences by the very realization of the fact that viruses may induce disease months or even decades after infections, and that these slow virus diseases are more compatible with classical degenerations of the nervous system than with inflammatory disorders of the brain. Incidentally, this viewpoint was a common knowledge in the veterinarian sciences perhaps from the time of transmission experiments of Cuille and Chelle [239–241] and seminal work by Sigurdsson [964].

More than a quarter of a century since discovery of kuru, and more than half a century following the first transmission of scrapie, the very

Fig. 1. The Nobel laureate, Dr. D. Carleton Gajdusek (arrow), during his field studies in Papua New Guinea. Courtesy of Dr. D.C. Gajdusek, National Institutes of Health, Bethesda, USA

Fig. 2. Kuru victims. Courtesy of Dr. D.C. Gajdusek, National Institutes of Health, Bethesda, USA

Fig. 3. Two persons in the early stages of kuru. Courtesy of Dr. D.C. Gajdusek, National Institutes of Health, Bethesda, USA

nature of the infectious virus remains unknown. However, two major discoveries lead to the recent developments which have been achieved by use of the sophisticated *armamentarium* of modern molecular biology. First, in 1981 Patricia Mertz, New York, discovered scrapie associated fibrils (SAF) [764] and in 1982 Stanley B. Prusiner, San Francisco, discovered a unique protein associated with scrapie infectivity, designated PrP (from prion protein) [109]. It has been shown soon that PrP is a component of SAF [296, 886]. The discovery of PrP enabled the cloning and sequencing of its gene, which appeared to be a cellular gene [208, 820], and by a series of elegant experiments using transgenic mice technologies, it was shown that PrP is crucial for the pathogenesis of scrapie (and related disorders) (*vide infra*). Even more interesting, anti-PrP sequences (an open reading frame situated on the anti-sens DNA strand of the PrP gene) was reported [494]. By the same token, however, all discoveries made in the last decade did not prove beyond the point of a common acceptance that PrP is the only component of the scrapie virus [333, 561, 563, 566, 908]. Some investigators still believe that it is not

Fig. 4. A person in the end stage of kuru. Courtesy of Dr. D.C. Gajdusek, National
Institutes of Health, Bethesda, USA

Fig. 5. A sheep affected with terminal scrapie. Note skin areas completely devoid of
wool. Courtesy of Dr. Richard Kimberlin, SARDAS, Edinburgh, U.K.

even a part of it [122, 716]. This current level of understanding is surprisingly not much different from that of the past. Indeed, it is most fascinating, and on the other hand ridiculous, that almost everything which is being said currently, has been said many times previously, but based on much narrower experimental basis. Thus, scrapie virus was regarded as self-replicating protein [457, 657], plasma membrane [423, 518] or small oligonucleotide linked by the yet to be discovered "linkage substance" to the plasma membrane [2]. Recently, even the spurious link between scrapie virus and HIV-1 has been reported [484]. Few of the most unsubstantiated hypotheses were, however, completely ruled out – the sarcosporidia [746], viroids [287–289, 720] and spiroplasmas [56, 452, 654] as the cause of diseases are only fingerprints of the imagination of its inventors.

It seems that there are several reasons for a current situation in which most of the investigators believe in the most unorthodox "protein only" hypothesis of the scrapie virus (agent), a hypothesis which would be probably unacceptable as the framework of thinking for "conventional" virologists. Indeed, even before their full characterization, nobody invoked self-replicating proteins for such elusive agents as hepatitis delta or Borna viruses. Two major obstacles to perform breakthrough experiments are enormous difficulties of experimentation with a "sticky" agent and the ignorance of the current generation of molecular biologists of natural disorders. Scrapie is the best example of it. Every hypothesis concerning the causative agent (virus) must explain the very fact that scrapie is the natural infectious disease spreading laterally from sheep to sheep and vertically from sheep to lambs [128, 266, 844]. An epidemic of scrapie among sheep which had received vaccination against louping ill is one of the best known example of such a virus-like infection [449]. Analogously, the hypothesis must explain only too well known iatrogenic cases of CJD [133, 139, 222, 615, 1012] and recent epidemic of bovine spongiform encephalopathy, regarded by most as scrapie passaged from sheep to cattle by dietary practices [252, 1041]. Furthermore, it must explain different passageable characteristics of strains of scrapie virus [267, 284]. Last but not least, it must explain the virus-like properties of virus spreding and targeting [30, 245–246, 328]. At the present time, the "protein-only" hypothesis may easily explain the conversion of PrP^c into $PrP^{sc(CJD)}$ but it still cannot explain the other data and mostly it ignores the very existence of them.

The purpose of this large review is to summarize almost all existing data on scrapie and related infections, asking a question whether all these data may fit one complete pattern. Having written this review, the author is convinced even less than before, that such a task is not possible to be accomplished at the present time. The emerging picture is not only incomplete, what is a normal situation in any fast moving field, but even worse, there is no consensus concerning many fundamental questions. Even some of the most simple problems, as the morphogenesis of spongiform vacuoles for instance, are still waiting to be solved. The forefront of scrapie research, the search for the infectious virus is as incomplete now as it was almost a decade ago and it is still possible that the virus is, even in a part, unknown. However, the major proportion of this review covers the disease itself at many levels and these data form a framework against which any hypothesis concerning the virus itself must be tested.

2. The molecular biology of the slow viruses

"After more than twenty years of unrewarded effort in numerous laboratories to detect an infection-specific nucleic acid, and despite the logical impossibility of precluding the existence of something that has not been found, it would seem appropriate as a purely practical matter to accept what has been discovered and move ahead with it, letting whatever theoretical objections that attend the idea of a "replicating protein" work themselves out in the course of time"

Brown P, Liberski PP, Wolff A, Gajdusek D.C.
Proc Natl Acad Sci USA 1990; 87:7240–7244

2.1. The search for the virus-specific nucleic acid

"The Fourth Wonder on my list is an infectious agent known as the scrapie virus, which causes a fatal disease of the brain in sheep, goats and several laboratory animals. A close cousin of scrapie is the C-J virus, the cause of some cases of presenile dementia in human beings. These are called "slow viruses", for the excellent reason that an animal exposed to infection today will not become ill until a year and a half or two years after today. The agent, whatever it is, can propagate itself in abundance from a few infectious units today to more than a billion next year. I used a phrase "whatever it is" advisedly. Nobody has yet been able to find any DNA or RNA in the scrapie or C-J viruses. It may be there, but if so it exists in amounts too small to detect. Meanwhile, there is a plenty of protein, leading to a serious proposal that the virus may indeed be *all* protein. But protein, so far as we know, does not replicate itself, not on this planet anyway. Looked at this way, the scrapie agent seems to be the strangest thing in all biology and, until someone in some laboratory figures out what it is, a candidate for a Modern Wonder.

Lewis Thomas – Late night thoughts listening to Mahler's Ninth
Symphony. Bantam Books, Toronto 1983

All viruses, and there is no exception to this claim, contain nucleic acid – DNA or RNA enveloped with several proteins. The smallest infectious agents, plant viroids contain only RNA and not proteins [287,

288]. The scrapie virus (and analogously, viruses of CJD and the other spongiform encephalopathies) seems to be that exception (at least several investigators believe it) in so far that the infection-specific nucleic acid has not been found. In this chapter I will review all efforts to isolate infection-specific nucleic acid from scrapie- and CJD-affected brains and spleens, unsuccessful to date and more than often misleading.

2.1.1. Radiation experiments

Perhaps most of the ideas, reverberating over a quater of this century, that scrapie virus may be devoid of nucleic acid stem from early irradiation experiments. Alper, Haig and Clarke [19] were the first to use ionizing radiation from a linear accelerator to calculate the target size of the scrapie virus. The scrapie material was irradiated dry to prevent the potential indirect effects of radiation. The inactivation curve was exponential, suggesting a single-hit process, yielding a D_{37} (a dose leaving residual activity 37% or e^{-1}) of 43 k.grays (1 gray = 100 rads). The target size calulated according to Lea's method was equal to 150 000 (subsequenty corroborated by Field et al. [329] and Alper et al. [17]) and, if the genome of the scrapie virus consisted of nucleic acid, it could be no larger than 800 nucleotides. Furthermore, as u.v. irradiation at a wavelength of 254 nm did not produce any significant decrease of scrapie infectivity at a dose of $2.4 \, kJ/m^2$, Alper et al. [19] concluded that "the agent may be able to increase in quantity without itself containing nucleic acid. This possibility is supported by the data from electron irradiation, since these yield a target size which is implausible small as a nucleic acid core". Latarjet [647–648] reported a similar exponential inactivation curve to yield a minimum target size of approximately 64 000. Noteworthy, an initial increase of infectivity titer with a single dose of approximately 3 k. grays was observed. Such a phenomenon of increased infectivity titer following a small irradiation dose has been observed for "conventional" viruses and is explained on the basis of disagregation [647].

Using u.v. irradiation of different wavelengths Alper et al. [19] did not obtain any significant inactivation of the scrapie material, suggesting once more that the replication of the agent may not be dependent on an intrinsic nucleic acid. In a subsequent experiment Latarjet et al. [647–648] using near monochromatic u.v. irradiated scrapie material at different wavelengths (237, 250, 254 and 280 nm) to yield an exponential inactivation curve with a D_{37} of $22.4 \, kJ/m^2$ or $21.5 \, kJ/m^2$ [647]. In another experiment, Prusiner reported the D_{37} of $42 \, kJ/m^2$ [863]. The inactivation at 250 and 280 nm was virtually the same while at 237 it was six-fold more effective [863]. The authors stressed the striking difference between many biological systems that are dependent on nucleic acids for their replication (a maximum inactivation at 250 to 270 nm and a minimum at 235 to 245 nm, for viruses) and scrapie virus with a maximum inactivation at 237 nm. In further experiments, Latarjet and co-workers [647] used a giant monochromator to irradiate scrapie preparations at 250, 237, 225 and 210 nm. As internal controls, T_2 bacteriophage, RNase, peroxidase, and the endotoxin from E.coli were used. The monochromatic efficiency e_{lambda} for scrapie was equal to 34 at 210, nm and 1 at 250 nm, while for DNA containing T_2 phage e_{lambda} was 1 at both wavelengths). Remarkably, peroxidase showed inactivation values comparable to those of scrapie virus (e_{lambda} = 1 at 250 nm and 33 at 210 nm). While these data strongly support the hypothesis that the scrapie virus may lack a large nucleic acid, a small oligonucleotide, perhaps linked to a glycoprotein might also fit the data (vide infra in the context of PrP).

It was shown subsequently, that infectivity of the Eiru isolate of kuru virus was not reduced by a radiation dose of 200 k. grays [420, 648]. Similarly, 150 k. grays did not decrease the infectivity of three Creutzfeldt-Jakob disease (CJD) isolates [420, 648]. However, as kuru and CJD viruses were propagated in nonhuman primates, including chimpanzees, it was not possible to perform end-point titrations and obtain dose-response curves (for obvious reasons). However, D_{37} was estimated to be higher than 45 k.grays. Thus, the viruses of kuru and CJD appeared to be as resistant to ionizing radiation as the scrapie virus.

For most complex biological systems (cells, tissues and organisms), the presence of oxygen during irradiation is detrimental, while for simple systems, such as nucleic acids, it is of no significance or either protective [16, 648]. The *oxygen enhancement ratio* (o.e.r) is defined as the ratio of the doses necessary to produce the same effect in the presence and absence of oxygen [18]. This ratio is approximately 3 for most living systems and <1 for nucleic acids or viruses. The results of the first series of experiments were inconclusive [17]. In subsequent experiments, Alper and co-workers [18] reported that, in contrast to nucleic acid containing viruses, the presence of oxygen during the irradiation of scrapie-affected brain suspensions caused marked radiosensitization with an o.e.r of approximately 12.5. These investigators noted that such an effect of oxygen resembled that encountered during the irradiation of membranes, and thus supported the membrane hypothesis suggesting that the scrapie agent is an abnormal membrane structure (vide infra).

Recently, these early data were corroborated using more sophisticated equipment and using highly purified scrapie material containing mostly, if not exclusively, the glycoprotein, PrP (vide infra) [70] (Fig. 6). To

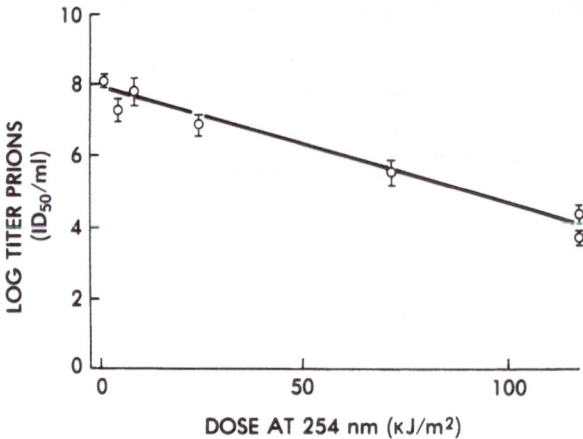

Fig. 6. UV irradiation of scrapie infectivity [70]. Courtesy of Prof. Stanley B. Prusiner, University of California, San Francisco and the editor of Journal of Virology

measure the amounts of u.v. radiation, dosimetry experiments were performed simultaneously using physical methods and the estimation of thymine dimer formation in both eucaryotic and procaryotic DNA [70]. As an internal control, the filamentous bacteriophage M13, RNase A and PSTV (potato spindle tuber viroid) were used. Following doses of less than $300 \, J/m^2$ the M13 titer decreased by a factor of 10^{10} to yield a D_{37} of $6.5 \, J/m^2$ (Fig. 6). PSTV was more resistant to u.v. irradiation yielding a D_{37} of $4800 \, J/m^2$. Irradiation of RNase A yielded a D_{37} of 11 800. In agreement with earlier data, the scrapie virus revealed the highest D_{37} of approximately $20\,000 \, J/m^2$ which is closer to that of some proteins (like DNase A) than to viruses or viroids. The D_{37} of scrapie virus was used to estimate the size of a putative scrapie nucleic acid. If the genome of the scrapie virus were single stranded (ss) then its estimated size would be 4 bases; and it would be in a range of 30 to 45 bases if the genome were double stranded (ds) [70]. This small size of the putative nucleic acid genome of the scrapie virus would make it difficult to detected by biochemical methods and force investigators to use molecular cloning techniques (vide infra).

Recent experiments have corroborated the small target size of the scrapie genome [70]. Microsomal fractions isolated from scrapie-affected hamster brains, scrapie-associated fibrils (prion rods; see below) and liposomes [363, 365] containing scrapie infectivity were irradiated. The inactivation curves for all preparations were exponential suggesting a single-hit process. The target size was about 55 kDa regardless of the preparation, in good agreement with previously published data [70]. However, as Latarjet pointed out [648], if there is a system which can repair the putative scrapie genome, the size of it may be underestimated.

For example, the size of the polyoma virus genome estimated by irradiation is 125 000 while the actual size equals 2 400 000. As scrapie infectivity is tightly associated with membranes, an efficient repair mechanism may be possible.

However, estimation of the size of putative scrapie genome by irradiation has been subjected to strong criticism [911, 914]. Instead of using the "target" theory to calculate the size of scrapie genome, Rohwer [911] and Rohwer and Gajdusek [914] made a direct comparison of scrapie resistance, represented by the D_{37} value, with that of viruses with genomes of known molecular weights. In this way the putative scrapie genome was estimated to be 0.75×10^6 for ss DNA or 1.6×10^6 for ds DNA. Such sizes, albeit small, fit at one end of the continous spectrum of "conventional" viruses and are 10–20 times larger than that estimated on the basis of target theory. This challenge by Rohwer [911, 914] was later questioned by Alper [15], but primary criticism remains. At the very least, it must be allowed that the target theory is not the only basis on which to interpret data from irradiation studies, and other options should be borne in mind when building theoretical models of the scrapie virus.

2.1.2. Fruitless attempts to isolate scrapie-specific infectious nucleic acid

The earliest attempts to identify scrapie-specific DNA or RNA from scrapie-infected brains by means of the simple biochemical techniques of that time were completely unsuccessful (for a review see, [517, 573, 577, 774]). At the beginning, the increased incorporation of [3H]- and [14C] thymidine into DNA extracted from scrapie-affected brains was discovered, particularly within a nuclear fraction. However, this change was not associated with the synthesis of scrapie virus-specific DNA but with abnormal metabolism of scrapie-affected mouse brain [560]. An increased incorporation of [3H] thymidine was found in both labile and relatively stable DNA fractions [577]. As neurons do not divide in adult brain, the only cells putatively responsible for such increased DNA turnover were glial and subependymal cells. However, as the turnover of labelled DNA was the same for all brain regions, the subependymal cells could not account entirely for observed increases in turnover of DNA.

Subsequently, Adams and co-workers [2–4] found an increased incorporation of [3H] thymidine, [14C] glucosamine and [14C] uridine diphosphoglucose (UDPG) into a post ribosomal pellet fraction of density 1.34 g/ml in a CsCl gradient. Chromatography on Sepharose 4B yielded two peaks of u.v. absorbtion at 260 nm and two corresponding peaks of radioactivity. The peak of [3H] thymidine and [14C] glucosamine

specific activity was located between these two u.v. absorption peaks.
The first peak was higher in scrapie-affected animals than in controls,
while the reverse was true for the second peak. The authors interpreted
those data in accord with their "linkage substance hypothesis" [2–4]
suggesting that the scrapie agent is a nucleic acid-polysaccharide complex
(because of incorporation of labelled precursors of both nucleic acid and
polysaccharides) linked to cellular membrane by a hypothetical "linkage
substance" resulting in an infectious entity. The chromatography of
nuclear DNA on hydroxyapatite yielded two absorption peaks [2]. The
second peak, consisting mostly of ds DNA, was approximately 3-fold
higher than the corresponding peak in control animals. In contrast,
when the large granular fraction of DNA was chromatographed on
hydroxyapatite, an additional small peak melting off at 50°C was seen in
scrapie-affected but not control animals. It is noteworthy that an analysis
of temperature elution of normal DNA from a hydroxyapatite column
suggested that the small peak could represent ss DNA. This information
was of potential significance as abnormal mitochondrial ss D-loop was
found recently to be overrepresented in purified scrapie material ([9–10];
see below).

Additional impetus for the search for the scrapie-specific nucleic acid
was provided by Diener's hypothesis that the scrapie agent, or at least its
genome, may be similar to that of plant viroids, small "naked" RNAs
that had just been discovered at the time [287, 564]. In particular,
Diener stressed the long incubation period of both groups of diseases,
similar resistance of both scrapie agent and viroids to inactivation by
ionizing radiation, similar "target" size calculated on the basis of inactivation curves (see above) and the failure to recognize either viroids
or scrapie virus by means of thin-section electron microscopy [287].
Several important differences have been stressed, however. In contrast
to viroids, scrapie virus is insensitive to nucleases and is markedly
unstable following phenol extraction. Diener's hypothesis was directly
tested by several groups of investigators, not the least being Diener
himself [289]. Marsh et al. [725] failed to detect infectious scrapie-specific
nucleic acid extracted with phenol. To monitor their extraction procedures, these investigators used encephalomyocarditis (EMC) virus,
recovering 7% of the infectious EMC RNA. Analogously, Ward et al.
[1034] using mengovirus as an internal control, could not detect infectious scrapie-specific RNA after phenol extraction. Both groups of
investigators concluded that the hypothesis of scrapie virus being similar
to plant viroids is highly unlikely. Manuelidis and Manuelidis [713, 715]
used the highly efficient nick translation method to label DNA extracted
from CJD-affected brains, followed by separation by PAGE with special
emphasis to those bands within a range of viroid size (<2 kb). Overall,

there were no differences between bands separated from CJD-affected and control animals. In one preparation, ^{32}P-labeled DNA from a rough endoplasmic reticulum fraction of CJD-affected hamsters yielded a smear in a range of 400 to 500 bp which was absent from control preparations. However, since this smear was absent from the synaptosomal microsomal CJD fraction, containing 3 to 8 times more CJD infectivity, it was unlikely that it represented CJD-specific DNA. Furthermore, a restriction analysis of synaptosomal microsomal, rough endoplasmic reticulum and cytosol brain fractions did not reveal any differences between CJD-affected and control animals. In conclusion, no ds or partially ds CJD-specific DNA fragments within a viroid size range were detected. It must be stressed, however, that if a putative CJD-specific DNA is completely single stranded (without "folding back") it would not have been detected by the nick translation technique. Malone et al. [703, 720, 723, 724] reported copurification of scrapie infectivity with low molecular weight nucleic acids in PAGE (RNA molecules of 23 kDa–70 kDa; ds DNA of 48 kDa–130 kDa). As scrapie infectivity was reported to be sensitive to DNAse, this low molecular-weight material was presumably DNA [724]. However, the application of procedures previously used to extract citrus exocortis viroid RNA to scrapie-infected brain resulted in complete loss of infectivity [725]. In contrast, no inactivation of scrapie infectivity was observed following digestion with either DNase I and II or RNase A and T_1 [876–877]. Furthermore, only a small amount of the scrapie infectivity entered the composite 2.5% polyacrylamide – 0.5% agarose gel, and confirmation of low molecular weight DNA copurifying with scrapie infectivity was not achieved [876–877].

The fundamental differences between scrapie virus and viroids have been elaborated by Diener and co-workers [289]. First, diethylpyrocarbonate completely inactivated scrapie virus leaving potato spindle tuber viroid (PSTV) intact. Secondly, while the loss of scrapie infectivity following exposure to diethylpyrocarbonate could be restored by treatment with hydroxylamine, the PSTV infectivity was dramatically diminished under similar experimental conditions. Thirdly, psoralens, linear tricyclic furocoumarin derivatives, that inactivate viruses by forming photoadducts with nucleic acids [753], readily inactivated PSTV while the scrapie virus remained completely resistant. Fourthly, Zn^{2+} ions completely inactivated PSTV but left scrapie infectivity unaltered. Fifthly, alkaline pH reduced the scrapie infectivity by a factor of 100 000 but PSTV infectivity was reduced by a factor of only 10. Overall, as Diener et al. [289] stressed "properties of PSTV are exactly opposite to those of the scrapie agent".

Several different approaches have been used to address the problem of putative nucleic acid within the scrapie agent. Manning and Millson

[707] attempted to enhance the infectivity of scrapie-associated nucleic acid by means of encapsulation in liposomes. The results with scrapie virus were completely negative, while the infectivity of encephalo-myocarditis (EMC) virus RNA encapsulated into liposomes, and used as an internal control, was enhanced 12-fold by intraperitoneal (but not intracerebral) inoculation. Borras and Gibbs [115–116] analysed the kinetics of DNA reassociation (C_0t curves) attempting to detect the presence of scrapie-specific nucleic acid. ^{125}I-labeled DNA enriched from an ammonium sulphate fraction (ASF) purified from scrapie-affected brains was one of the hybridization probe used. Further probes were constructed as derivatives of the primary one. No significant differences in C_0t curves were obtained with six different probes. However, the mass of putative scrapie nucleic acid present in ASF was estimated to be in a range of 1.6×10^{-4} to 1.6×10^{-3}. This would not be enough to be detected by biochemical methods. The same quantitative problem was addressed recently by Oesch and co-workers [818]. If the purified scrapie fractions contained 10^8 infectious units per ml, and the molecular weight of the nucleic acid was about 100 000, and the particles to infectivity ratio was one, the total amount of nucleic acid would be within the range of 0.2–0.4 fmole or 20 pg. Such an amount is at the limit of the level of detection using biochemical methods and cloning techniques would be necessary for detection. The strategy applied by Oesch an co-workers [818] consisted of cDNA synthesis using random oligo-nucleotide primers. The beta-globin gene RNA was used as an internal control. Two inserts of 900 bp and 200 bp hybridized to repetitive sequences of the hamster genome on Southern blots suggesting that small amounts of hamster nucleic acid was present in highly purified scrapie preparations reported to be free of nucleic acids on the basis of conventional methods [818]. However, no scrapie-specific clone was found. A similar approach has recently been used by other investigators. Wietgrefe and co-workers [1040] constructed cDNA from oligo(dT)-cellulose chromatographically purified poly(A^+)RNA from scrapie-infected mouse brain. For differential hybridization, this cDNA library was screen by ^{32}P-labeled cDNA reverse-transcribed from poly-(A^+)RNA of scrapie-infected and control brains. One clone (Scr-1) hybridized preferentially to scrapie-infected brains. However, in dot-blot experiments, Scr-1 was shown also to hybridize to control material although the extent was 20-fold less. On Northern blots, Scr-1 hybridized to the 3.3 kb RNA species. In *in situ* hybridization experiments, Scr-1 was located to neurons, mostly in scrapie-affected brains. Furthermore, Scr-1 hybridized to dystrophic neurites within neuritic plaques in human brains with Alzheimer's disease and rare senile plaques of multi-infarct dementia brains [1040]. Noteworthy, Scr-1 was also nuclease resistant as

the putative scrapie genome should be. While the significance of Scr-1 gene was unknown at the time of its discovery, it was subsequently established that Scr-1 clone represented 3′ non coding region of GFAP [290, 717]. The Scr-1 cDNA sequence is 98% homologous to the 3′ untranslated region of the mouse GFAP cDNA. Indeed, Scr-1 was further used as a probe to examine the expression of GFAP mRNA in CJD-infected hamsters [717].

Duguid and co-workers [300–301] used subtractive hybridization of a cDNA library to find 5 clones differently expressed in scrapie. None of these was a unique scrapie-associated genome but 3 genes were over-represented in scrapie-affected brains, coding for the following proteins: 1) glial fibrillary acidic protein (GFAP), presumably related to the reactive dense fibrillary gliosis typical in scrapie-affected brains; 2) metyllothionein II, and 3) the B chain of hamster alpha-crystallin. The last two proteins are homologous, in part, to heat-shock proteins, but their role in scrapie pathogenesis is unknown. As scrapie infectivity is associated with membranes [518] and cytoskeletal preparations [9–10], two cDNA libraries were constructed from nucleic acids found in these fractions and screened by means of differential hybridization [9–10]. Four clones hybridized preferentially with cDNA from cytoskeletal fractions of scrapie-infected brains. Although these clones hybridized strongly to scrapie cytoskeletal preparations in dot blots, none showed increased hybridization with total nucleic acids of scrapie-infected tissue. On Southern blots, all four clones hybridized to cytoplasmic (mitochondrial) DNA. Sequence analysis of the two larger clones (of 23 kb and 16 kb) showed 77 and 79% homology to the mouse mitochondrial genome and mitochondrial ATPase-6 gene, respectively. In addition, two smaller clones revealed strong sequence homology to the displacement-loop region (D-loop) of the mouse mitochondrial DNA. Furthermore, when the mitochondrial fraction was purified from scrapie-infected brains the titer of infectivity of these fractions was 10^8 LD_{50} [10]. As the mitochondrial fraction was enriched for scrapie infectivity and there is evidence that prion protein (PrP; vide infra) is also associated with scrapie infectivity, the subsequent approach of the Wisconsin group was to search for mitochondrial DNA within PrP-enriched fractions [10]. The probe used was one of two clones (pGp135) homologous to the hamster D-loop (an unusual triple-stranded region of mammalian mitochondrial DNA). Two different PrP preparations hybridized to the D-loop on slot-blots. On Southern blots, pGp135 hybridized to a 450 nucleotide band that was identical to the D-loop ss DNA fragment. To confirm the D-loop specificity of the 450 nucleotide band, strand-specific probes were further used. Only an RNA probe synthesized from T7 RNA polymerase promotor homologous to the D-loop band hybridized to the 450-

nucleotide band on Southern blot. If the abnormal ss D-loop DNA is indeed the scrapie virus genome, then it replicates when the mito-chondrial genome replicates. Furthermore, as the D-loop fragment was found in prion protein enriched preparations, this DNA may be in a nuclease-resistant form; an argument previously used as evidence that scrapie virus is putatively devoid of nucleic acids.

Possible RNA involvement in scrapie virus is similarly a controversial and inclonclusive issue. Recently, Dees and co-workers [261] reported finding a unique small RNA in membrane vesicles purified from scrapie-infected hamster brains. While no differences were detected in protein and lipid contents of these fractions [259–260] or in RNAs labeled by 3′-labeling and oligo(dA-dT)-cellulose chromatography, the 5′-labeling revealed a 100 nucleotide (4.3 S) RNA overrepresented in scrapie fractions. When two-dimensional RNA fingerprints were subjected to computer analysis, approximately half the oligonucleotides in scrapie

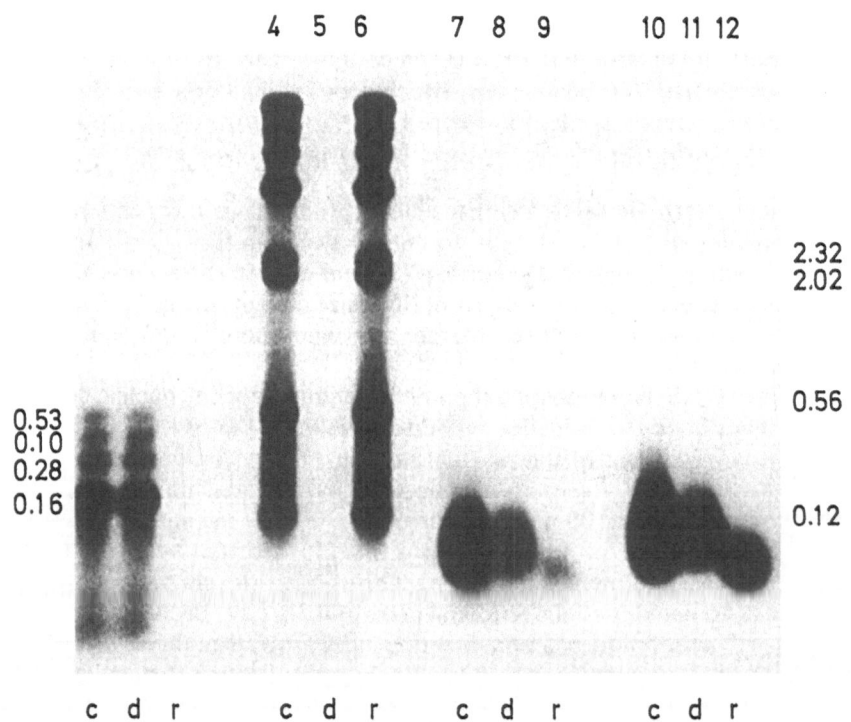

Fig. 7. Nucleic acid from CJD and control preparations, 5′-end labeled with poly-nucleotide kinase. Digests were resolved on a 1.4% denaturing glyoxal agarose for autoradiography. Identical brain equivalents were digested with: no nucleae (*o*); DNAse I (*d*); RNAse (*r*) [12]. Size of control markers are seen on the right in kb. Lanes: (*1–3*), known amounts of RNA; (*4–6*), known amount of DNA; (*7–9*), CJD brains; (*10–12*), uninoculated animals. Courtesy of Dr. Laura Manuelidis and the editor of Microbial Pathogenesis

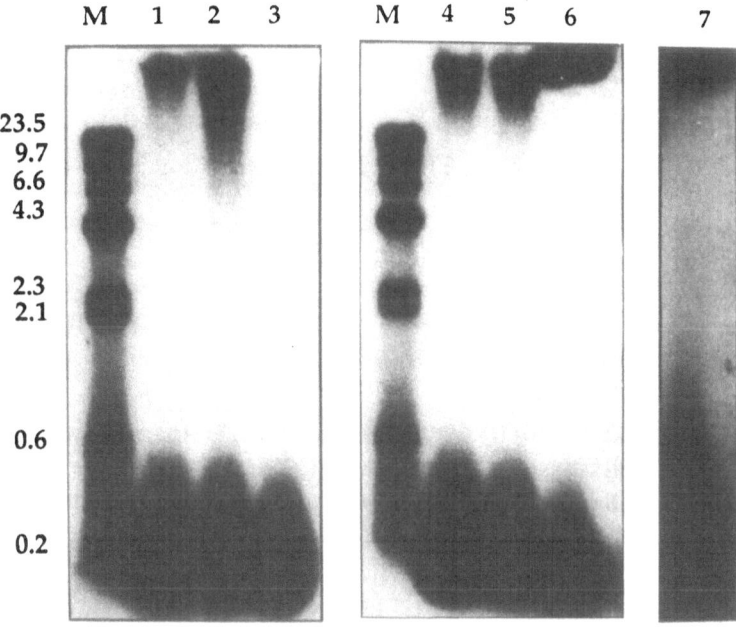

Fig. 8. An analysis of nucleic acids isolated from uninfected (*1–3*) and CJD (*4–7*) preparations 5′ end labeled with ^{32}P [967]. Note that all preparations show smears of species that migrate from 500 to >2 kb. Courtesy of Dr. Laura Manuelidis and the editor of Archives of Virology

RNA preparations differed from those of control preparations [405]. The biological significance of these findings, if any, is unknown.

Claims that scrapie and CJD infectivity is not associated with any kind of disease-specific nucleic acid have been challenged recently by showing that several species of nucleic acid, from 100 bp up to 2000 bp survived micrococcal nuclease digestion in purified CJD preparations (Fig. 7–8) [12, 967]. Furthermore, equilibrium centrifugation revealed a peak of CJD infectivity at 1.28 g/cm^3, separated from the bulk of PrP33–35^{CJD} (designated by the Yale group Gp34). This buoyant density corresponds to that of nucleoprotein complexes rather than to nucleic acid-free protein. As the amount of DNA and RNA in infectious CJD fractions was small (less than 2 ng of DNA and less than 6 ng of RNA), the polymerase chain reaction (PCR) was applied to amplify cDNA reverse-transcribed from potential total RNA of CJD-infected brain tissues [12] (Fig. 7). To amplify these unknown sequences, specific adaptors were blunt-end-ligated to cDNA. Although most if not all the amplified nucleic acid was host-derived, it clearly resisted RNAse treatment, affirming that nuclease-resistance of infectivity in purified scrapie fractions does not provide a sufficient experimental basis for claiming that the scrapie (CJD) virus cannot contain nucleic acid. Furthermore, such a strategy might yet to amplify and clone a scrapie (CJD)-specific nucleic acid.

2.2. The prion protein

2.2.1. The purification of the PrP 27–30, PrP 33–35^{sc} and PrP 33–35^c

2.2.1.1. Strategies used to purify the scrapie virus and scrapie specific protein

Numerous attempts to purify the scrapie virus to homogeneity have been unsuccessful, but its association with cellular membranes has been well established. The distribution of scrapie infectivity closely paralleled the activity of 5′-nucleotidase, a marker for cellular membranes, but not that of NADPH: cytochrome c reductase, a marker of endoplasmic reticulum [221]. Those data were interpreted within a framework of the "membrane hypothesis". Such an intimate association with membranes was challenged when Malone et al. [704] reported membrane-free scrapie infectivity in preparations treated by prolonged centrifugation. These early results were used as a starting point for a development of a purification scheme which eventually led to the discovery of PrP 27–30.

Differential centrifugation proved to be of little value to those investigators hoping that the scrapie virus would behave as a discrete homogenous particle like typical "conventional" viruses. Siakatos et al. [960] using centrifugation in sucrose gradients, found that scrapie infectivity was spread throughout the entire gradient with peaks at 1.19, 1.26 and 1.29 g/ml. Such heterogeneity was further confirmed in Gajdusek's laboratory by sedimentation to equilibrium in CsCl, sucrose and metrizamide gradients [913].

The early protocol developed in Prusiner's laboratory to purify the scrapie virus from murine spleen and brain used fixed-angle rotors [868, 869, 874, 875, 877, 883, 884]. Scrapie infectivity sedimented over a relatively narrow range of w^{2}t values between 10^9 to 10^{11} rad^2/sec, regardless of whether sonication or treatment with sodium deoxycholate (DOC) were applied. The second phenomenon indicated that scrapie infectivity is, contrary to the "membrane hypothesis", independent of cellular membranes that have been solubilized by means of DOC. Subsequent purification attempts used detergents.

All purification strategies suffered from imprecision and difficulty in the estimation of amounts of infectivity by end-point titration. The development of an bioassay based on the measurement of the lenght of incubation period allowed, although less accurate then end-point titrations, use many fewer animals and allowed faster progress in attempts to purify the scrapie virus [870, 871, 880]. Differential centrifugation followed by DOC treatment yielded a fraction (P_5) enriched for scrapie infectivity 20-fold relative to protein. When P_5 was subjected to near equilibrium centrifugation in a sucrose gradients, scrapie infectivity was distributed over the entire gradient between densities 1.08 to 1.30 g/cm^3 [874–875]. Rate-zonal centrifugation yielded particle sedimentation coefficients ranging from 40S to >600s. While 40S values were smaller than those of the smallest viruses known at that time, the sedimentation coefficients of >600S indicated a strong tendency of the virus to aggregate. The stickiness of the virus has been the "curse" of all those who tried (and still try) to purify it. In subsequent protocols differential and gradient centrifugation were suplemented with preparative electrophoresis [881] and then replaced by polyethylene glycol 8000 precipitation and rate-zonal centrifugation [868]. The extensive digestion with micrococcal nuclease and proteinase K improved the purification of the scrapie virus to between 100 to 1000-fold.

2.2.1.2. Purification of the PrP27–30

Prion protein (PrP), the protein tightly associated with scrapie-infectivity and regarded by most [75, 288, 367–369, 377–380, 385, 481, 861, 862, 864, 865, 866, 872, 873, 885, 1045], but not all investigators [294, 561, 564–566, 653, 714, 716, 857], to be a part of the infectious scrapie virus, was discovered at the beginning of the 1980s first as filamentous structures by negative-stained electron microscopy [761–765] and later as a unique band in PAGE [109]. However, the idea that protein is involved in scrapie infectivity predated this discovery by almost 15 years. Indeed, Griffith [457] and Gibbons and Hunter [423] were among the first to suggest the importance of protein. Analysing the possible role of glycoproteins in scrapie, Gibbons and Hunter stated that "a foreign oligosaccharide or polysaccharide or *glycoprotein* may induce the synthesis of polysaccharides which would also be able to function as a transferase; in this way a polysaccharide or oligosaccharide would appear to be self-replicating". That scrapie infectivity is sensitive to protease digestion had been known for a while but was largely ignored [774]. As early as 1969, Marsh and Hanson [721] working on the physicochemical characteristics of transmissible mink encephalopathy virus (TME, regarded as scrapie passaged to mink) reported a complete loss of infectivity following digestion with pronase. Millson, Hunter and Kimberlin [774] reported a more than 90% loss of scrapie infectivity following digestion with different proteases, in contrast to nuclease digestion which had no effect.

Cho [209] found markedly reduced scrapie infectivity titers following digestion with pronase and concluded that the infectious agent probably contained an essential protein.

Several studies performed in Prusiner's laboratory at the University of California, San Francisco concluded with the discovery of PrP 27–30 opened up new avenues in scrapie research and new horizons in the field of neurodegenerative diseases. Using an elaborate purification protocol, a fraction designated E_6 was obtained after preparative electrophoresis with subsequent electroelution or gel pulverizing. Treatment of these fractions with proteinase K at 100 ug/ml, a concentration previously reported to be ineffective [703], reduced titers of scrapie infectivity by 10 to 10^3-fold [109, 748, 868, 884]. Such effects of proteinase K were corroborated by Lax et al. [649]. Furthermore, diethylpyrocarbonate (DEP), a compound known to inactivate proteins by forming etoxyformylated adducts of amino acids, also markedly diminished scrapie infectivity titers [754]. It was also found that hydroxylamine restored the infectivity that was lost by the DEP treatment [754]. Treatment with $KSCN^-$, followed by freezing to $-20°C$, dramatically reduced scrapie infectivity titer [881]. The inactivation of more than 97% of infectivity over a narrow range of $KSCN^-$ concentrations suggested a highly cooperative process. Substitution of Na^+, Li^+ or guanidinum ions for K^+ potentiated the effects of SCN^- while dialysis of KSCN but not guanidinium SCN restored it [881]. Furthermore, infectivity was markedly reduced by exposure to alkali while it was stable in acidic conditions. All these data clearly implicated a protein as the target for inactivation procedures. A candidate for the protein associated with scrapie infectivity was identified by Bolton, McKinley and Prusiner in 1982 [109]. Their purification protocol included low-speed centrifugation, polyethylene glycol-8000 precipitation, digestion with micrococcal nuclease and proteinase K, ammonium sulphate precipitation and rate-zonal centrifugation on a discontinuous sucrose gradient. Most of the infectivity was found near the bottom of the gradient, and these fractions were highly enriched in a protein that appeared as a diffuse (heterogeneous) band of molecular weight of 27–30 kDa in PAGE [109]. This protein was later designated PrP for *p*rotease *r*esistant *p*rotein or *p*rion protein. PrP was resistant to proteinase K digestion provided it had not been denatured by heating to 100°C in the presence of SDS, a procedure which rendered it proteinase K-sensitive [748]. Thus, digestion with proteinase K, prior to denaturation, removed most proteins from both control and scrapie preparations leaving a proteinase K-resistant protein in the scrapie samples [748]. Subsequently, PrP 27–30 was found in scrapie-affected mouse brains [110].

2.2.1.3. PrP proteins are disease-specific for the whole group of unconventional slow virus disorders

Protein analogous to that purified from scrapie-affected brains was detected in naturally occurring and experimentally induced CJD. Bockman et al. [97] were the first to detect this proteins using antiserum raised against PrP 27–30, in fractions purified from brains of 3 CJD patients. Furthermore, PrP immunoreactive proteins were also found in Swiss mice inoculated with the Fujisaki strain of CJD virus. PrP from murine CJD exhibited apparent molecular weights 44 to 46 kDa, 27.5–30, 25–27, 22 to 24, and 18–20 kDa, while PrP purified from human cases of CJD showed a similar but not identical distribution of bands. Brown et al. [138] found PrP immunoreactive proteins (designated "marker" proteins) in 25 of 31 human CJD cases, 3 of 4 cases of kuru, 3 of 4 cases with Gerstmann-Sträussler-Scheinker (GSS) syndrome and none of 32 cases with Alzheimer disease, ALS-dementia complex, AIDS encephalopathy and several other cases of dementia. Sporadic and familial cases of CJD did not present any differences in the proportion of PrP immunopositive cases. It is noteworthy that PrP proteins were found in all cases with a long duration of the clinical disease but in only 19 of 25 CJD cases with durations shorter than 2 years. Such temporal distribution of PrP immunoreactivity resembles appearance of PrP-immunoreactive plaques among CJD cases with longer duration of the clinical disease [605]. In a subsequent study, Bockman et al. [98] found PrP in brain extracts of 14 patients with CJD; 6 of these cases proved transmissible. While human and mouse PrP shared some epitopes they also exhibited different ones. Western blots revealed that PrP purified from human CJD-affected brains reacted with antiserum raised against human PrP^{CJD} but not with serum against the PrP purified from CJD affected mouse brains at first passage [98]. Furthermore, Gibbs Jr et al. [421] detected PrP 27–30 in a brain of a CJD patient who was a recipient of human growth hormone purified from pituitaries. These studies have been extended to experimental CJD in hamsters and guinea pigs [716, 718, 968]. Sarcosyl treatment and proteinase K digestion followed by SDS-PAGE revealed a diffuse band of silver staining with an apparent molecular weight of approximately 29 kDa. This protein was clearly immunoreactive with anti-PrP antiserum on a Western blot. When proteinase K was omitted from the preparation method, two bands of apparent molecular weights 34–36 kDa and 29 kDa were found. It is clear that the 34–36 kDa protein detected in these CJD models is homologous to the PrP33–35^{sc} and the 29 kDa protein to the PrP 27–30, respectively. PrP proteins have been also found in brains of scrapie-affected sheep [921]. A major PrP band of apparent molecular weight 26–29 kDa was detected in all cases of experimentally transmitted scrapie in sheep but in only 25% of those with natural scrapie. In experimental sheep scrapie, PrP26–29 was accompanied by several other bands: 23–24 kDa, 21–22 kDa and 7-1-kDa. Interestingly, PrP was not detected by Western blot in some cases even when SAF (Scrapie associated fibrils) (vide infra) were easily visualized by negative stain electron microscopy. Furthermore, PrP 33–35 accompanied by a minor variant, PrP 26–29 has been recently associated with a bovine spongiform encephalopathy (BSE) discovered in British cattle [505–506, 1041].

Different models of scrapie and CJD are associated with different band patterns of PrP on SDS-PAGE and Western blots. For example, PrP purified from hamsters infected with the 263K strain of scrapie virus revealed a single diffuse band of molecular weight 26 kDa to 28 kDa on SDS-PAGE [544] which resolved, on Western blots, into proteins of molecular weights 26 to 28 kDa, 23 to 24 kDa and 19 to 20 kDa, respectively [543, 920]. In contrast, ME7 and 139A SAF revealed three polypeptides of molecular weight of 26 to 28 kDa, 23 to 24 kDa and 21 to 22 kDa irrespective of whether silver staining [544] or Western blotting [543] were used for their detection. Furthermore, the Western blot pattern of PrP purified from mice infected with the 87V strain of scrapie virus was more similar to that associated with the 263K strain than to the ME7 or 139A strains. While the intensities of silver-stained bands of PrP associated with the ME7 and 139A strains of scrapie virus were virtually identical, Western blots revealed reproducible differences which could be further used to differentiate between these strains. A different band pattern was observed with mice infected with the Obihiro strain of scrapie virus [993]. SDS-PAGE purified PrP yielded polypeptides of apparent molecular weight: 24.5 kDa, 21 kDa, 17 kDa and 11 kDa, respectively, following proteinase K treatment. Interestingly, similar polypeptides were detected in sheep affected with natural scrapie. However, additional bands of molecular weight greater than 24.5 kDa were visualized, while the 11 kDa peptide was not seen when polyclonal antibodies against the fraction containing PrP was used on Western blots. These PrP peptides were further analysed with an antiserum raised against a synthetic peptide encompassing the 15 N-terminal amino acids of PrP27–30 (Gly-Gln-Gly-Gly-Gly-Thr-His-Asn-Gln-Tyr-Asn-Lys-Pro-Ser-Lys) [962]. All three major PrP polypeptides were detected with this antiserum. When proteinase K was omitted from a purification protocol, additional PrP peptides of molecular weight 30 kDa and 34 kDa, respectively, were detected. Since the major polypeptides are recognized by an antiserum raised against N-terminal peptide and thus have the same N-terminal sequence, it is justifiable to suggest that the differences between them are brought about by posttranslational modification – probably in glycosylation (vide infra). Similar studies have been used to evaluate different CJD models [96]. Mice, guinea pigs and humans infected with different isolates of CJD virus exhibited PrP proteins of different molecular weights (18–20, 22–24, 25–27, and 27.5–30 kDa for mouse; 25–27, 29.5–31 kDa for guinea pigs and 15–15.5, 21–22 and 23–25 for humans, respectively). As both the hosts and the CJD isolates were different in this experiments, the question whether the differences in PrP variants were host- or isolate-directed could not be addressed. To attempt do this directly, PrP variants from a human CJD case and from mice infected with the analogous human isolate were analysed and proved to be different. Such a distinction of mouse and human PrP variants was further strengthened by a demonstration that a mouse antiserum raised against human PrP^{CJD} recognized the homologous human PrP variant but not the PrP variant purified from CJD-infected mouse brains [96, 98]. However, as the authors did not know whether the same strain of CJD virus was present in both humans and mice (strain selection is a well known process accurring during the primary isolation) the problem of host versus strain of the virus-associated PrP characteristics is yet to be settled. Furthermore, the data of Kascak et al. [543] of C57Bl mice infected with the ME7, 139A, 79A, 22C and 22L strains of scrapie virus yielded identical band patterns on Western blot when host was constant. Analogously, three different hamster species infected with the 263K strain of scrapie virus yielded different variants of PrP^{sc} as evidenced by different antigenicity [694].

2.2.1.4. PrP 33–35^{sc} and PrP 33–35^{c}

As first predicted by an analysis of PrP cDNA and Northern blot experiments [820], PrP of apparent molecular weight 33–35 kDa is present in both scrapie-affected (PrP33–35^{sc}) and control (PrP33–35^{c}) brains. Antisera raised against a synthetic peptide encompassing the first 13 amino acids of the N-terminus of the PrP 27–30 reacted with the PrP33–35^{sc} which, following digestion with proteinase K, yielded a core protein PrP 27–30 [48, 49, 50]. In contrast, PrP33–35^{c} was completely digested with proteinase K [820]. Furthermore, both PrP^{sc} and PrP^{c} were tested against three antisera raised against synthetic peptides encompassing the N-terminal 13 amino acids of PrP 27–30 (internal stretch of PrP^{sc}), presumed N-terminus of the PrP^{sc} (containing also the last 8 amino acids of its signal peptide) and 14 amino acids at the C-terminus of PrP^{sc}, respectively [51]. As predicted, PrP27–30 reacted only with antisera against the N- and C-termini of PrP27–30 but not against PrP^{sc}

N-terminus. This is because PrP27–30 is generated from PrP^{sc} by a limited proteolysis at the PrP^{sc} N-terminus [51]. While both PrP^{sc} and PrP^c react with the same antisera on Western blots, they differ in their physicochemical properties. Following ultracentrifugation of the detergent-extracted microsomal fraction, PrP^c remains in the supernatant. In contrast, PrP^{sc} is located in the pellet. Ultrastructurally, the supernatant containing the bulk of the PrP^c contained only membrane vesicles while detergent-extracted pellets containing PrP^{sc} revealed numerous "prion rods" (SAF) [766; vide infra]. The identity of PrP^{sc} and PrP^c has been established by direct amino acid sequencing [1023]. The amino acid sequences of both PrP^{sc} and PrP^c are identical and correspond to that predicted by cDNA analysis.

Both PrP^{sc} and PrP^c contain a C-terminal glycolipid anchor which can be cleaved by phosphatidylinositol specific phospholipase C (PIPLC) ([982]; see below). Differential sensitivities to PIPLC have been used to differentiate between PrP^c and PrP^{sc} [983]. Using dissociated cells from normal and scrapie-affected brains, Stahl et al. [983] found that PrP^c was released into the medium while, in contrast, PrP^{sc} remained mostly cell-bound. Furthermore, proteinase K hydrolyzed almost all PrP^c, consistent with its surface localization. PrP^{sc} was converted into PrP 27–30 suggesting that PrP^{sc} is accessible to proteinase K and, therefore, it should also be accessible, at least in part, to PIPLC. There is no good explanation for the differential release of PrP^{sc} and PrP^c by PIPLC. The self-aggregation of PrP^{sc}, the existence of transmembrane domains of PrP^{sc} (already detected *in vitro*) or an interaction with other proteins have been proposed as possible explanations for PrP^{sc} resistance to PIPLC digestion. However, it is difficult to reconcile the presence of the glycolipid anchor and the presence of transmembrane domains as evidenced by *in vitro* studies.

Several cell lines persistently infected with scrapie virus were used in further differentiation between the two PrP isoforms [997]. Both scrapie-infected and uninfected cells reacted strongly with anti-PrP27–30 antibodies in an ELISA test. Such immunoreactivity was much diminished following proteinase K digestion indicating that at least part of the immunoreactivity was associated with the presence of a protease sensitive variant of PrP namely PrP^c. However, PrP immunoreactivity of scrapie-infected cells but not uninfected cells was restored by denaturation with guanidine hydrochloride, and such restoration of immunoreactivity indicated the presence of PrP^{sc}. The presence of PrP^{sc} and PrP^c in cultured cells was further substantiated by differential release of PrP isoforms by PIPLC [983]. Furthermore, scrapie-infected cultured cells treated with guanidine hydrochloride revealed strong intracellular PrP immunoreactivity by immunofluorescence. PrP was resistant to both proteinase K and PIPLC indicating it was generated by PrP^{sc}. The requirement for treatment with guanidine hydrochloride suggested that the denaturation of PrP^{sc} is mandatory for detection with antibodies.

The intracellular localization of PrP^{sc} was established *in vitro* by means of confocal microscopy [997]. PrP^{sc} was localized roughly along the cisternae of the Golgi apparatus and the *trans*-Golgi network, as revealed by a comparison of PrP immunofluorescence and the WGA binding. In contrast, PrP^{c} remained at the cell surface [997]. The different topologies of PrP^{sc} and PrP^{c} was further evaluated by means of streptavidin shift PAGE/immunoblot analysis following exposure to sulfo-NHS-biotin reagent that covalently binds biotin to free NH$_2$ groups of cell surface proteins (biotinylated proteins bind streptavidin which causes retarded migration on SDS-PAGE) [113]. PrP^{c} released from cultured cells by PIPLC following biotinylation revealed marked retardation in the presence of streptavidin. In contrast, PrP^{sc} extracted from biotinylated cultures did not bind streptavidin demonstrating that most of the PrP^{sc} was not accessible to sulfo-NHS-biotin. This means that most of the PrP^{sc} did not accumulate on the cell surface. The kinetics of PrP^{sc} and PrP^{c} synthesis is also different [113]. PrP^{c} appeared immediately following [^{35}S]methionine pulse labeling and then declined steadily during the chase period. In contrast, no PrP^{sc} could be found immediately after labelling. After increasing the chase interval, proteinase K resistant PrP^{sc} became apparent. Thus, it seems, that PrP^{sc} is converted to its proteinase K resistant form by processing of PrP^{c}.

2.2.2. *The structure of the PrP proteins*

2.2.2.1. N-terminal sequence of the PrP

The amino-acid sequence of the N-terminus of PrP has been elucidated in several laboratories. First, PrP 27–30 has been purified from scrapie-affected hamster brains (paragraph: 2.2.1.). Microheterogeneous PrP 27–30 purified by size-exclusion HPLC did not reveal any covalently linked oligonucleotides as evidenced by its UV absorbtion spectrum [878]. Amino acid analysis of PrP 27–30 showed that the most common residues were glycine, glutamate/glutamine and aspartate/asparagine. The similarity of this amino acid composition to that encountered in A4 amyloid of Alzheimer disease has been noted [878]. The N-terminal sequence determined from the major signals was: N-Gly-Gln-Gly-Gly-Gly-Thr-His-Asn-Gln-Trp-Asn-Lys-Pro-Ser-Lys. Minor signals indicated that the protein had "ragged" ends – namely free N-termini started at different amino acids (X-X-X-Thr-His-Asn-X-Trp-X-Lys-Pro or X-X-

Pro-Trp-X-Gln-X-X-X-Thr-His-X-Gln-Trp, where X stands for amino acid not determined at this cycle). This sequence is unique as revealed by a search of computerized data bases. This N-terminal sequence enabled oligonucleotide probes to be made to identify the PrP cDNA [820]. Nucleotide sequencing of cDNA, and later the PrP gene, gave a complete amino acid sequence of the protein. Western blot analysis of control and scrapie-affected hamster brains revealed a presence of larger protein of apparent molecular weight 33–35 kDa (PrP33–35^{sc}) [820]. Following proteinase K digestion, the molecular weight of PrP 33–35^{sc} was shifted to that of PrP 27–30 (27–30 kDa) while the similar protein detected in control brains (PrP 33–35^{c}) was completely digested by proteinase K. This PrP 27–30 sequence has been confirmed in several different laboratories. Multhaup, Diringer and co-workers [792] used a more efficient method to purify PrP 27–30 [296, 496]. They purified a 26 kDa protein (analogous to the PrP 27–30) from hamster brains infected with the 263K strain of scrapie virus. Extraction of purified material with formic acid yielded another protein of apparent molecular weight 6–8 kDa. Hydrofluoric acid treatment also produced a single deglycosylated protein of apparent molecular weight 7 kDa. The amino acid sequence of the 7 kDa protein – N-Ser-Gly-Pro-Trp-Gly-Gln-Gly-Gly-Gly-Thr matched in part that of a minor signal sequence described already by Prusiner et al. [878], and its much lower molecular weight was probably the result of extensive degradation. Hope et al. [503] purified PrP33–35^{sc} accompanied by minor components of apparent molecular weight 26–29 kDa and 23–25 kDa, respectively, from hamster brains infected with the 263K strain of scrapie virus. The purification method did not use proteinase K, and protease inhibitors were added to stop endogenous enzyme activity. The sequence of PrP 33–35^{sc} purified by gel-filtration chromatography was: N-Lys-Lys-Arg-Pro-Lys-Pro-Gly-Gly-Trp-Gly-Gly-Ser-Arg-Tyr-Pro-Gly-Gly-Gln-Gly. This sequence matched that of amino acids 12 to 31 deduced from the cDNA sequence [820]. Both PrP^{sc} and PrP^{c} have been purified, without proteinase K, from scrapie-infected and control hamster brains, respectively [74, 108, 1023]. PrP33–35^{sc} (designated HaSp33–37; PrP33–35^{c} was designated HaCp33–37) was accompanied by a protein of apparent molecular weight 29–32 kDa [108]. When digested with proteinase K, HaSp33–37 (PrP^{sc}) was converted into PrP 27–30. The N-terminal sequence of this protein has been established: Lys-Lys-Arg-Pro-Lys-Pro-Gly-Gly-Trp-Asn-Thr-Gly-Gly-Ser-Arg-Tyr-Pro-Gly-Gln-Gly-Ser-Pro. This sequence corresponded to positions 1 to 22 deduced from the cDNA sequence [820]. The sequence of HaSp29–32 (Gln-Pro-His-Gly-Gly-Gly-Trp-Gly-Gln-Pro-X-Gly-Gly-Gly) corresponded to positions 36–50. It seems that HaSp29–32 is a product of a partial proteolytic cleavage of HaSp33–37

(PrP33–35^{sc}). Recently, Safar and co-workers in Gajdusek's laboratory [925] purified PrP^{sc} from scrapie-affected hamster brains by alternative procedures (also without a proteinase K digestion). The sequence of this protein (Ser-Lys-Arg-Pro-Lys-Pro-Gly-Gly-Trp-Asn) matched in part those previously found. In contrast to microheterogeneous hamster PrP33–35^{sc}, PrP purified from scrapie-affected mouse brains exhibited apparent molecular weights of 33–35 kDa, 26–29 kDa (PrP26–29) and 23–25 kDa (PrP23–25) [504]. As the purification protocol did not contain a proteinase K step [503], and several potent proteinase inhibitors were used to prevent a proteolysis, the apparent molecular weigth diversity presumably resulted from *in vivo* and not from *in vitro* proteolysis. Furthermore, when these PrP variants were purified in the presence of proteinase K, proteins of apparent molecular weight 27–30 kDa (PrP 27–30), 22–25 kDa (PrP 22–25) and 20 kDa (PrP 20), respectively, were obtained. Two dimensional gel analysis showed that proteinase K removed the more basic components of PrPs and, furthermore, PrP26–29 is different from PrP27–30, PrP23–25 from PrP22–25 and PrP20–21 from PrP20. The sequencing of PrPs purified by gel-filtration chromatography yielded one major sequence (Lys-Lys-Arg-Pro-Lys-Pro-Gly-Gly) irrespective of how many PrPs of different molecular weight were also present. This sequence matched exactly that deduced from PrP cDNA sequencing [208]. The minor signals yielded a sequence Phe^{152}-Val-His-Asp-X-Val-X-Ile-Thr; the two undetermined amino acids were predicted to be Cys and Asn, respectively. The Asn is a potential N-glycosylation site, and the absence of a signal in this position was the first direct evidence that PrP33–35^{sc} is indeed glycosylated. The PrPs sequenced after proteinase K digestion yielded multiple "ragged" ends and a major signal (Gly^{59}-Gln-Gly-Pro-His-Gly-Gly-Gly) similar but not identical to the sequence of minor signals of PrP26–29 and PrP23–25.

Recently, PrP was purified from brains with GSSS of Indiana kindred [992]. This Indiana GSSS family is characterized by tau-positive neurofibrillary tangles in addition to typical PrP plaques [409, 411] and the lack of typical for GSSS codon 102 mutation [511]. Proteins of amyloid plaque cores were purified, fractionated on a Sephadex G-100 column and gel filtrated to yield a peak of 11 kDa on SDS-PAGE. This protein reacted strongly with antibodies against PrP 27–30 and anti-synthetic peptide encompassing codons 90–102 (P1) of PrP cDNA while weakly with antibodies encompassing codons 140–172 (P5). It was unreactive with antibodies against synthetic peptides encompassing codons 15–40 (P2) and 220–232 (P3). This fraction was further HPLC-purified and the N-terminus sequenced to yield 29 residues beginning at codon 58 (Gly) of human PrP cDNA. These data of GSSS PrP are at variance with those of Hope et al. [503–506] who reported full length PrP purified from hamster, mice and cattle brains when proteases were omitted from a purification procedure.

As noted previously, PrP33–35 has also been detected in cattle brains infected with the virus of bovine spongiform encephalopathy (BSE) which seems to be scrapie passaged into British cattle [505–506]. The sequence of its 12 N-terminal amino acids (Lys-Lys-Arg-Pro-Lys-Pro-Gly-Gly-Gly-Trp-Asn-Thr) differed from that of rodents and human PrP sequences by one amino acid [505]. The bovine PrP has an extra Gly residue inserted between amino acids at positions 9 and 10 of rodent and human PrP [443].

2.2.2.2. The secondary structure of the PrP

The secondary structure and membrane topology of the PrP has been studied extensively by computer modelling [58]. Several regions have been identified within a sequence of PrP. These included: 1, an N-terminal putative signal peptide region from position -22 to -1, containing 12 hydrophobic amino acids at positions -22 to -18; 2, Pro- and Gly-rich series of hexa- (Gly-Gly-(Asn/Ser)-Arg-Tyr-Pro) and imperfect octapeptides (Pro-(His/Gln)-Gly-Gly-Gly-(-/Thr)-Trp-Gly-Gln) encompassing residues at positions 7–69 (it is noteworthy, that these Pro- and Gly-rich motifs are typical for proteins, like collagen, which are characterized by the ability of a polymerization into fibrillar structures, consistent with the ability of PrP to polymerize into SAF (prion rods)); 3, a short stretch encompassing residues 70–89, likely served to fix the first transmembrane region; 4, an Ala- and Gly-rich internal hydrophobic domain, which is a potential membrane-spanning region, encompassing positions 90–109; 5, the region between two hydrophobic domains encompassing residues 114–208 of alpha helical or beta strand structure where the beta strand structure is flanked by two alpha helices; 6, two Cys residues at positions 152 and 192; 7, two N-glycosylation sites at positions 159 and 173; and 8, C-terminal hydrophobic domain (positions 219–232).

The putative N-terminal signal peptide region of PrP is reminiscent of those signal peptides that, by an interaction with signal recognition particles (SRP) or docking proteins, direct the co-translational insertion into endoplasmic reticulum membranes. The presence of a signal peptide and two hydrophobic putative transmembrane domains is a prerequisite for the membrane topology of PrP. First, the signal peptide brings about the co-translational translocation into or across the membrane of endoplasmic reticulum. Then, a stop-transfer sequence (positions 90 to 113) determines further translocation of a nascent PrP chain into the endoplasmic reticulum lumen. As the two N-glycosylation sites are thought to be luminally located, another membrane-spanning domain between two hydrophobic regions has been postulated. This domain is hydrophilic and contains three Pro residues. On this model, the mature protein would have both the N- and the C-termini exposed at the extra cytoplasmic surface.

The predicted membrane topology of PrP has been investigated experimentally *in vitro* in a wheat germ (WG) cell-free system transfected with an SP6 transcript of PrP cDNA or mRNA isolated from control- or scrapie-affected hamster brains [486]. The results obtained with either a cloned PrP gene or with native mRNAs were identical. When the N-terminal signal peptide was cleaved in the presence of co-translationally added microsomal membranes, PrP-precursor protein (PrP_o) was transformed into a glycosylated form (PrP_2), and an unglycosylated (PrP_1) form of PrP as evidenced by endoglycosidase H digestion. PrP_2 and PrP_1 variants showed apparent molecular weights of 30 kDa and 25 kDa, respectively.

The topology of PrP_1 and PrP_2 in membranes was further studied by *in vitro* PrP synthesis with or without microsomal membranes [486–487]. PrP stretches that were inserted into membranes would have been protected. After treatment with proteinase K, PrP fragments were precipitated by N-terminus- or C-terminus-specific antisera, suggesting that both N- and C-termini are luminally located. The proteinase K protection was abolished by using of nondenaturing detergents to solubilize microsomal membranes. This suggested that at least portions of PrP are protected by and are presumably inserted into microsomal membranes. Furthermore, the proteinase K sensitivity suggested that the PrP^c and not proteinase K resistant PrP^{sc} isoform was synthesized *in vitro*. The differences between PrP^{sc} and PrP^c probably depend on postranslational modifications that were obviously not reproduced *in vitro*. However, when another *in vitro* system, (rabbit reticulum lysate: RRL) was used to translate PrP, the secretory form of PrP was unexpectedly found [487]. When PrP was translated in RRL in the presence of membranes, the same PrP forms found were similar to those found in the WG system. However, in contrast to the WG system where translated PrP was sensitive to proteinase K digestion, approximately two thirds of PrP translated in the RRL was resistant to proteinase K, suggesting either a protease-resistant form of PrP or a completely translocated protein. The second possibility was supported when the secretory variant of PrP was finally detected in Xenopus oocytes transfected with PrP transcripts and pulse-chased with [^{35}S] methionine [487]. In this system, the labelled PrP was released into the medium in a proteinase K sensitive form that would be similar to PrP^c. Furthermore, extraction of membranes with sodium carbonate, known to release secreted proteins, yielded approximately half of PrP_2 but not PrP_1 and thus suggested that PrP_2 exists in two forms: secretory, (released by sodium carbonate) and membrane-bound. The pulse-chase experiments suggested that both secretory and membrane-bound forms exist from the earliest time that the nascent PrP chain is synthesized.

The possible transmembrane orientation of PrP with two signal sequences and one putative stop transfer sequence, was further studied *in vitro* by means of chimeric proteins [690]. As the transmembrane variant of the PrP spans the membrane twice and the first membrane-spanning region (designated TM1) is located approximately 90 amino acids from the N-terminus, the different topology of the PrP in WG versus RRL systems may reflect the ability of the putative stop transfer sequence to halt the synthesized PrP chain at the TM1 domain. To uncouple translocation from PrP synthesis, a PrP cDNA truncated at a HicII restriction site (PrP/HcII) was constructed. This resulted in the synthesis of a PrP chain from the N-terminus through the TM1 domain. In the WG system, the N-terminal PrP fragment predominated when PrP synthesis was in the presence of intact membranes. If stop transfer did not occur the translation product would be fully protected. In contrast, the same experimental approach used in the RRL system produced a fully protected PrP fragment. Thus, the two *in vitro* systems contributed unknown factors which influenced the PrP topology independently of chain elongation. Furthermore, since the stop transfer sequence was present on the truncated PrP/HcII cDNA, it must be located upstream of the HcII restriction site. To evaluate this topogenic sequence further, a chimeric construct was made in which codons 74 to 114 of PrP (R_{74-114}) were inserted within codon 346 of the glycoprotein of vesicular stomatitis virus (VSV G). The topology of the VSV G was found to be directed by the topogenic sequence of PrP. The finding of such an unusual topogenic sequence may add to the unresolved problem of the basis of the physicochemical differences between PrP^{sc} and PrP^c. To determine the location of the topogenic sequence, a set of deletions within the PrP gene was constructed and expressed in the WG system in the presence of microsomal membranes [1070]. When both transmembrane domains were deleted, the product was completely translocated. However, deletion of 24 codons upstream to the TM1 region produced either many completely protected (translocated) chains, or a few chains displaying characteristic transmembrane topology. The deleted domain (designated stop transfer effector, STE) contains four Lys and two His residues resulting in a highly positive charge. Furthermore, when codons from the alpha-globin gene were inserted as spacers within the coding regions of PrP, the stop transfer reaction was disrupted only when the spacer was inserted between the STE and the TM1 domains. Thus, it has been demonstrated for the first time that the extracytoplasmic domain plays a role in stop transfer and this mechanism may participate in the generation PrP^{sc} during scrapie infection.

2.2.3. Postranslational modifications of the PrP

PrP is a glycoprotein. An analysis of the PrP cDNA and its chromosomal gene sequence revealed two potential N-linked glycosylation sites (Asn 181 and 197) [820]. Direct evidence that the Asn residue is indeed glycosylated came from the sequencing of murine PrP in which the absence of a signal in a predicted Asn position was found [504]. Eight PrP 27–30 charge variants of isoelectric points ranging from pH 4.6 to 7.9 have been detected by nonequilibrium pH gradient electrophoresis (NEPHGE) [111]. Analogously, NEPHEGE of PrP proteins purified from CJD-infected hamster brains (see below) yielded two clusters of proteins having isoelectric points of 6.8–7.5 and 5.5–7.5, respectively [968]. Digestion with neuraminidase resulted in an apparent shift of the isoelectric points of PrP charge isomers to those between pH 6.5 and 7.5 with a simultaneous reduction in their number to four. A similar shift has been observed following digestion with endoglycosidase H (endo H), resulting in six charge isomers having isoelectric points between 6.5 to 7.9. Furthermore, SDS-PAGE purified PrP was stained with PAS. All these data suggested that PrP 27–30 is a sialoglycoprotein. PrP 27–30 was resistant to endo H digestion. However, following peptide-N^4-(N-acetyl-beta-glucosaminyl) asparaginase F (PNGase F) digestion, the apparent molecular weight shifted from 27–30 kDa to 20–22 kDa [480]. Such a decrease suggests the removal of both Asn-linked oligo-saccharides from a polypeptide chain. Chemical deglycosylation with hydrogen fluoride (HF) which cleaves phosphodiester bonds or tri-fluoromethane sulfonic acid (TFMSA) resulted in proteins having apparent molecular weights of 19–21 kDa and 20–22 kDa, respectively. This further decrease of apparent molecular weight is consistent with the removal of the glycolipid anchor (see below) by HF treatment. Similarly, both PrP^{sc} and PrP^{c} digested with PNGase F yielded two variants of apparent molecular weights 28 kDa and 26 kDa, respectively. Notably the ratio of PrP28^{sc} to PrP26^{sc} was nearly one, while the ratio of PrP28^{c} to PrP26^{c} was >10. When lectin chromatography was used, approximately 20% of PrP 27–30 bound to Lens culinaris (lentil) lectin or wheat germ agglutinin (WGA), while smaller amounts bound to Ricinus communis agglutinin (RCA) or concanavalin A (Con A). The binding to peanut agglutinin (PGA) and Helix pomatia agglutinin (HPA) was negligible. PrP binding to WGA was decreased and that to RCA was increased by digestion with sialidase. The last finding suggested the presence of galactose in a penultimate position, while the binding of approximately 20% of PrP to the lentil lectin column suggested that a proportion of the N-linked oligosacharides may be a of moderately branched complex residue containing fucose. These data have been extended recently by Somerville and Ritchie [974–975]. On two-dimensional (2D) gels, PrP resolved into three relatively more acidic components (of molecular weight 29 kDa, 25 kDa and 20 kDa, designated A29, A25 and A20, respectively) and three more basic regions (34 kDa, 29 kDa and 25 kDa, designated B34, B29 and B20, respectively). The acidic PrP variants are proteolytic cleavage products of the more basic PrP peptides and it seems that the proteolytic cleavage arises endogenously. Studies with several lectins revealed four major patterns of binding: 1, no detectable binding; 2, more binding to B25 than to B29 and B34; 3, the reverse situation – more binding to B29 and B34 than to B25; 4, binding exclusively to B29 and B34 but not to B25. Furthermore, when PNGase F was used, the B34 PrP variant shifted to that of undigested B29 and then to a sharp doublet band of molecular weight 25 kDa. PNGase digestion altered the molecular weigth of B29 to that of B25. These findings can be interpreted to suggest that B25 has neither N-glycosylation site occupied, B29 has one and B34 has both [974–975]. Similar results have been obtained for CJD PrP [968]. A protein analogous to the PrP33–35^{sc} (designated Gp34) purified from CJD-infected hamster brains exhibited strong WGA and RCA binding [968]. Following neuraminidase treatment, Gp34 stained with RCA faster but otherwise to an identical extent while WGA staining was decreased, indicating the presence of sialic acid residues attached to Gp34. Furthermore, beta-N-acetylglucosaminidase completely abolished WGA binding and markedly reduced RCA binding. In contrast to data obtained for scrapie PrP, Endo H reduced both ricin and WGA binding markedly while anti-PrP immunopositivity remained unchanged. Such results of sequential enzyme treatment suggest the following sequence of N-linked oligosacharides: terminal sialic acid-D-galactose-N-acetylglucosamine. This sequence is common for membrane glycoproteins.

PrPsc and PrPc poses potential disulfide bonds between Cys residues at positions 179 and 214 according to the gene sequence [52]. Both proteins bound a sulfhydryl-reactive derivative of biotin, MBB which is consistent with the presence of disulfide bonds [1023].

Both PrP^{sc} and PrP^{c} contain membrane anchors composed of a phospatidylinositol glycolipid (designated also GPI, glycoinositol phospholipid) [982]. Reverse-phase high performance liquid chromatography (HPLC) analysis of PrP 27–30 hydrolysates revealed a peak migrating with [3H] ethanolamine. This peak persisted after boiling the protein in SDS following extraction with chloroform-methanol or purification by SDS-PAGE. Gas chromatography/mass spectrometry (GC/MS) analysis showed the presence of ethanolamine, inositol, phosphate and stearic acid. Furthermore, a weak immunopositivity with PrP was found on Western blot when antibodies against a cross-reactive determinant (CRD) of the variant surface glycoproteins (VSGs) of Trypanosoma brucei were used. The CRD epitopes recognized by these antibodies are located on the phosphatidylinositol-containing glycolipid. Therefore, the CRD-immunoreactivity indirectly supports the presence of phosptatidylinositol glycolipid anchor on PrP^{sc}. A similar anchor was found on PrP^c in vitro by means of phospatidylinositol-specific phopsholipase C (PIPLC) which released nearly all of the PrP^c but not PrP^{sc} into the medium indicating that PrP^c is located almost exlusively on the cell surface while PrP^{sc} remains intracellularly [192, 982]. Thus both PrP^{sc} and PrP^c possess a phophatidylinositol glycolipid anchor. The glycolipid is covalently linked to the C-terminus by an amide bond to ethanolamine which, in turn, is linked by a phosphodiester bond to an oligosaccharide. The PrP C-terminus sequence is similar to that of C-terminal glycolipidation signals of other proteins [52, 820]. When endoproteinase Lys-C was used to determine the GPI attachment site, a peptide of the sequence – Glu-Ser-Gln-Ala-Tyr-Tyr-Asp-Gly-Arg-Arg-Ser^{231}-GPI was released. The sequence of the GPI attachment site was further verified by mass spectrometry after the GPI cleavage with HF. Noteworthy, the truncated peptide Glu-Ser-Gln-Ala-Tyr-Tyr-Asp-Glu devoid of the terminal 3 amino acid residues and the GPI was observed in approximately 15% of PrP27–30 molecules [982]. The origin of the truncated PrP variant is uncertain. Probably it is produced by yet to be discovered postranslational modifications, but RNA editing and alternative translation products from a single mRNA species acre also conceivable. Interestingly, the cleaved sequence Gly-Arg-Arg is a well known sequence recognized during a proteolysis participating in a maturation of several bioactive peptides. The relation of the presence of either of this PrP variants to scrapie pathogenesis is yet to be provided.

However, the protease resistance of PrP^{sc} is not caused by postranslational modification in a form of glycosylation [996]. *In vitro*, both ScN$_2$a and ScHaB cells are able to synthesize PrP of apparent molecular weight 26 to 35 kDa following scrapie virus inoculation. This protein, after proteinase K digestion, yielded proteinase K-resistant core peptides of apparent molecular weigths 19, 25, and 25–30 kDa. In the presence of tunicamycin, an asparagine glycosylation inhibitor, a single PrP variant of apparent molecular weight 26 kDa was observed which was transformed into a 19 kDa protease-resistant PrP variant after limited proteolysis with PK. Analogously, swainsonine, an inhibitor of alpha-mannosidase II, produced a 27–30 kDa protease resistant variant of PrP. The lack of asparagine-linked glycosylation was substantiated by PNGase F, which did not affect this PrP species. Furthermore, the 19 kDa PrP variant is detergent-insoluble.

The relation of asparagine-linked glycosylation to protease-resistance was further evaluated by means of the mutagenized recombinant PrP MHM2$_{\text{Ala182/198}}$ gene containing the entire mouse PrP open reading frame in which Leu^{108} and Val^{111} were replaced with methionines while two asparagine-linked glycosylation sites were abolished [996]. ScN$_2$a cells transfected with the MHM2$_{\text{Ala182/189}}$ gene produced unglycosylated PrP of apparent molecular weight 19 kDa, identical to protease-resistant core PrP synthesized in the presence of tunicamycin.

Thus, the inhibition of asparagine-linked glycosylation did not prevent the appearance of resistant variants of PrPs.

A putative interaction of PrP^{sc} with other proteins has been recently demonstrated by studies of Oesch et al. [819] who identified the protein binding the PrP^{sc}. Hamster brain proteins were separated by SDS-PAGE, transferred to nitrocellulose filters and exposed to radioiodinated PrP 27–30. Two PrP binding proteins of apparent molecular weight 45 kDa (PrP ligand, Pli 45) and 110 kDa (Pli 110), respectively, were found. Unexpectedly, the amino acid sequence of Pli 45 matched that of glial fibrillary acidic protein (GFAP), the protein of intermediate filaments of astrocytes. It is well known that proliferation and hypertrophy of astrocytes are one of the major features in SSVE neuropathology.

2.2.4. Functional studies of the PrP proteins

The function of PrP in normal cells is unknown but it is present in neurons and other brain cells [130], and it has a highly conserved amino acid sequence suggesting an important neurophysiological role. Cashman et al. [186] from Wiśniewski's group in IBR, New York, suggested that PrP^c may participate in lymphocyte activation. Human lymphocytes contain 2.7 kb PrP mRNA and PrP33–35^c. Capping of PrP^c on lymphocytes was detected by immunofluorescence with monoclonal antibodies MAb 3F4. A uniform distribution of PrP^c on lymphocytes was observed by cytofluorometry. PrP immunoreactivity increased when lymphocytes were activated concanavalin A (Con A), phytohemaglutynin, and anti-CD3 monoclonal antibodies. Furthermore, antibodies against PrP inhibited Con A-induced lymphocyte proliferation.

Recently, an avian analogue of PrP, chicken prion-like protein (ch-PrLP), was discovered (see chapter 2.3.6) [482]. An analysis of cDNA predicts a protein of 267 amino acids. Analogous to other PrP proteins (see paragraph: 2.2.2.2.) ch-PrLP contains a series of eight hexapeptide repeats (Arg^{42} to Pro^{89}) near the N-terminus, an uninterrupted stretch of 20 nonpolar amino acids (Val^{119} to Gly^{138}) flanked by charged residues near the middle and a hydrophobic region (Trp^{252} to Met^{266}) near the C-terminus. Three Asp-linked glycosylation sites ($Asp^{188,203,212}$) were also predicted. The ch-PrLP is highly homologous (33% and, if the conservative substitutions are taken into account, 43%) to mammalian PrP proteins. Furthermore, analogously to PrP^c, ch-PrLP is anchored through a GPI anchor to surface membranes. PIPLC cleaved GPI and released immunoreactive material from transfected neuroblastoma N2a cells. It is noteworthy, that ch-PrLP was the major protein in preparations of acetylcholine receptor-inducing activity (ARIA) purified from chicken brains, but whether ch-PrLP and ARIA are the same proteins remains to be established. However, their co-purification and co-localization to the anterior horn motor neurons suggest that ch-PrLP and, analogously, PrPs may serve as trophic factors for certain (cholinergic?) populations of neurons and the absence of such trophic influence (by shifting from PrP^{sc} to PrP^c) may lead to the development of disease.

2.2.5. *The scrapie associated fibrils or prion rods*

Scrapie associated fibrils (SAF) were first discovered in 1981 by Merz, Somerville, Wisniewski and Iqbal [764]. In their original paper, SAF were found in synaptosomal preparations from mouse brains infected with the 139A, ME7, 87V and 22A strains of scrapie virus and from hamster brains infected with the 263K strain of scrapie virus. Two types of SAF were defined on the basis of their ultrastructural appearance in negative stain electron microscopy. Type I SAF consisted of two twisted filaments, measuring 2–4 nm in diameter, forming a fibril 12–16 nm in diameter with a 2–4 nm space between filaments. Every 40–60 nm the fibril narrowed to 4–6 nm. Fibrils measured 100–500 nm in length and they were usually straight but occasionally sharply bent. Type II SAF consisted of four filaments in a fibril 27–34 nm in diameter. The four filaments were arranged in a parallel fashion. Some fibrils consisting of four filaments had type II SAF at one end and two pairs of filaments forked at the other end. Type I and II SAF were also detected, albeit with greater difficulty, in plastic-embedded material by thin-section electron microscopy. Diringer et al. [296, 496] developed an efficient method of SAF purification [295] from hamster brains infected with the 263K of scrapie virus. In contrast to measurements of Merz et al. [763–765], the SAF found by the Berlin group were shorter (50–300 nm) but otherwise similar [295]. They were composed of two helically twisted filaments measuring 4–6 nm in diameter. In the sucrose gradient used for SAF purification, the length of the fibrils increased from top to bottom. Merz et al. [762] detected SAF in synaptosomal mitochondrial fractions of naturally occurring and experimentally transmitted CJD included one case of Gerstmann-Sträussler-Scheinker syndrome. Both type I and type II SAF were found in natural and experimental CJD but type I SAF predominated. Subsequently. Merz and co workers in a collaboration with investigators from Gajdusek's laboratory have established [761] SAF as an ultrastructural marker for the whole group of subacute spongiform virus encephalopathies. SAF were found in scrapie-infected hamsters, two scrapie-infected squirrel monkeys, one kuru-infected squirrel monkey and six human CJD brains [761]. They were consistently absent from all control specimens and from brains with an experimentally induced pathology produced by cuprizone and triethyltin intoxications similar to that which is caused by unconventional viruses [761]. Isolation and visualization of SAF has been used as an inde-

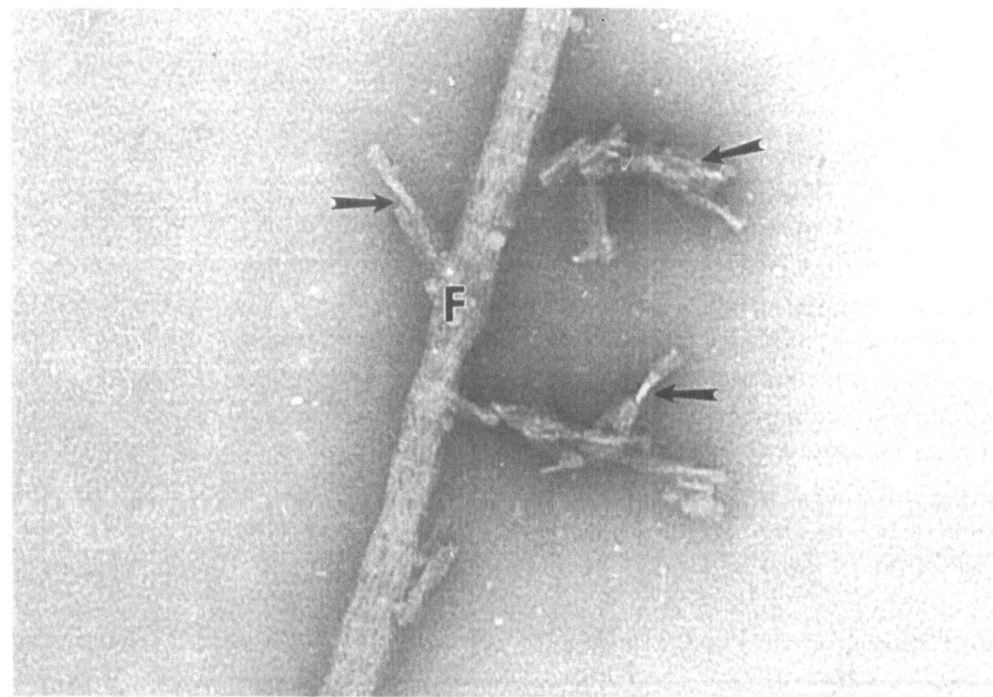

Fig. 9. Several *SAF* (arrows) attached to a collagen fiber (*F*). SAF (and in specimens illustrated in Figs. 9 through 13) were purified as previously described [146, 147] and negatively stained with phosphotungstic acid. Unpublished work of P.P. Liberski, P. Brown and D.C. Gajdusek, National Institutes of Health, Bethesda, USA. Original magnification, ×90 000

pendent confirmation of a CJD diagnosis in a recipient of human growth hormone obtained from pituaitary glands [421]. These SAF measured approximately 20 nm in diameter and formed periodically twisted arrays. In one of the original reports [761], SAF were not found in scrapie-affected sheep. However, they have been found in sheep scrapie by Gibson et al. [430], Scott et al. [943], Rubenstein et al. [921], and Skarpheoinsson et al. [966] although their detection can be more difficult than in several of scrapie in hamsters and mice [430]. Recently, SAF have been used to confirm that a new subacute spongiform virus encephalopathy – bovine spongiform encephalopathy (BSE) – was related to natural scrapie [505–506, 1041].

SAF from different scrapie sources exhibit subtle ultrastructural differences (Fig. 9–12). SAF isolated from scrapie-infected hamsters and mice, and purified by rate-zonal centrifugation, revealed different sedimentation rates; that of mouse SAF being lower than that of hamster SAF. Furthermore, SAF purified from mice infected with the ME7 strain of scrapie virus sedimented further into the gradient than those purified from mice infected with the 139A strain of scrapie virus. These findings

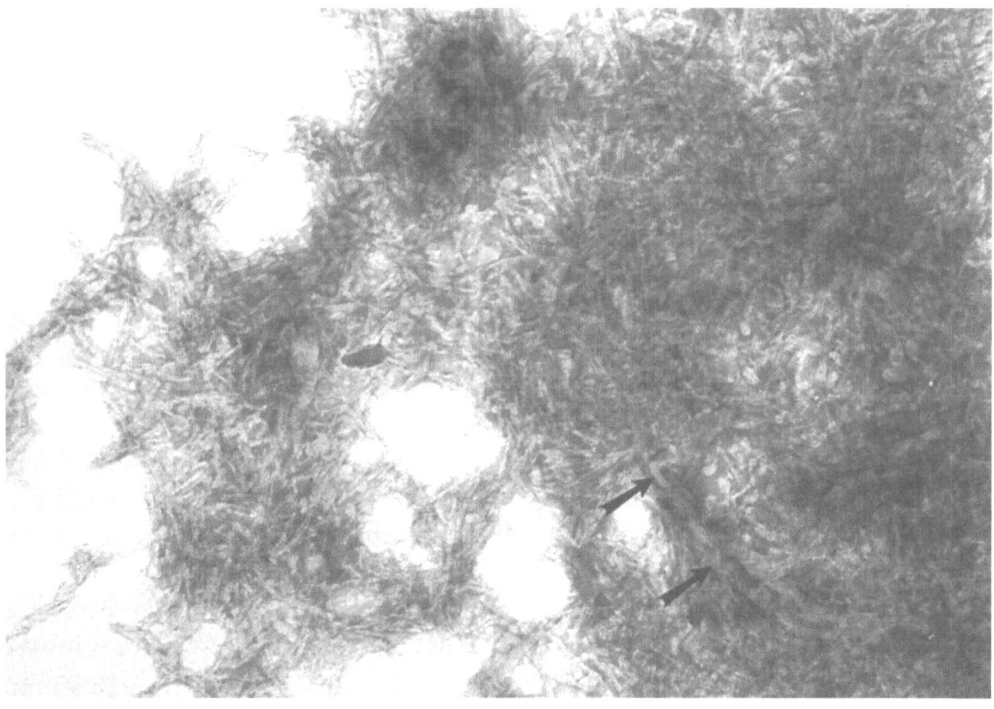

Fig. 10. A large cluster of SAF. Note protein-detergent complexes easily detectable because of their striations (arrows). Unpublished work of P.P. Liberski, P. Brown and D.C. Gajdusek, National Institutes of Health, Bethesda, USA. Original magnification, ×42 000

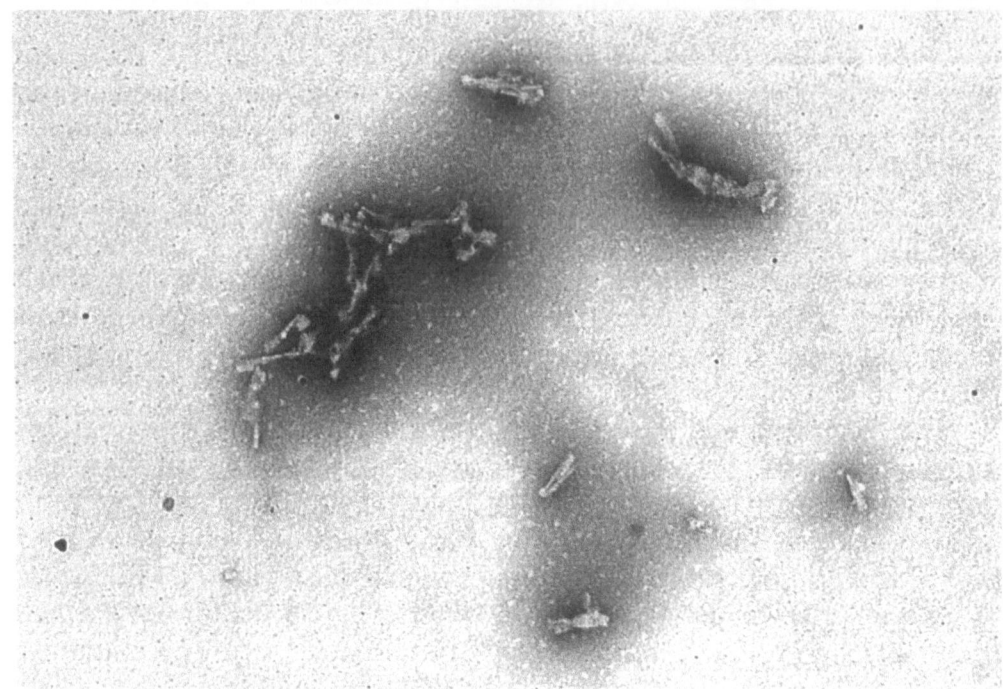

Fig. 11. Numerous pleomorphic SAF isolated from hamster brains infected with the 263K strain of scrapie virus. Unpublished work of P.P. Liberski, P. Brown and D.C. Gajdusek, National Institutes of Health, Bethesda, USA. Original magnification, ×42 000

suggest differences in length or degree of clustering [974–976]. SAF isolated from mice infected with the ME7 or 139A strains of scrapie virus were consistently longer (100 to 200 nm) and their yield was lower (100–150 SAF per grid square for ME7 and 50 to 100 SAF per grid square for 139A) than those isolated from hamsters infected with the 263K strain of scrapie virus (length of 50 to 100 nm; yield of 1000 to 2000 SAF per grid square). In another study, the substructure of individual fibrils was more easily visualized in mouse SAF than in hamsters SAF. This phenomenon was further characterized quantitatively using different models by Liberski, Brown and Gajdusek, unpublished observation (vide infra).

Prusiner et al. described "flattened rodlike structures" in preparations of scrapie-infected hamster brains [747, 750, 886]. By rotatory shadowing with tungsten, they measured approximately 25 nm in diameter and 100–200 nm in length. By negative staining with uranyl formate fibrils were thinner (10–20 nm in diameter) but of the same length. In contrast to SAF described by Merz et al. [761–764], the substructure of the fibrils which Prusiner's group designated "prion rods", was not easily

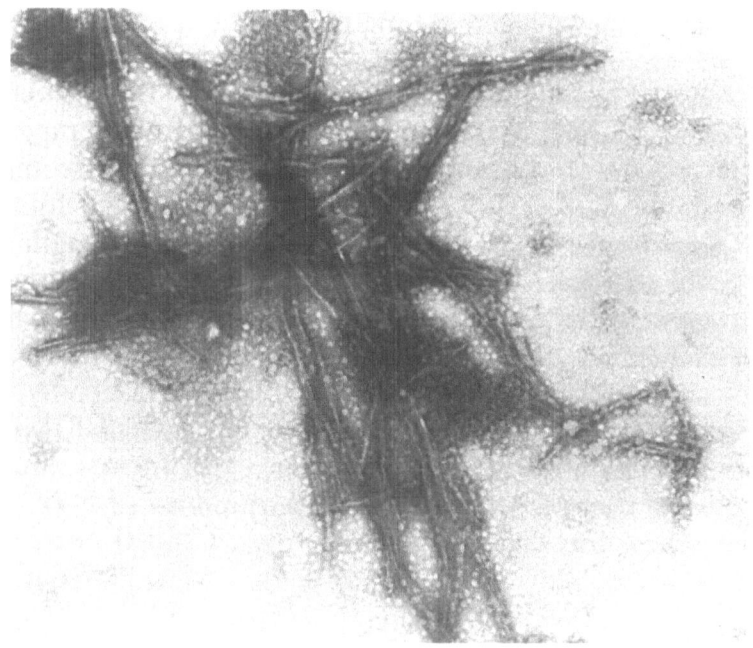

Fig. 12. Numerous SAF isolated from mouse brains infected with the ME7 strain of scrapie virus. Courtesy of Dr. Peter Gibson, MRC & AFRC, Neuropathogenesis Unit, Edinburgh, U.K. Original magnification, ×50 000

visualized. Interestingly, in the first report of "prion rods" [868] it was noted that similar structures were seen infrequently in control preparations. This claim has never been pursued further. Prion rods (SAF) purified from human CJD [97] cases measured 4 to 18 nm in diameter and 25–550 nm in length. Fibrils purified from mouse infected with the Fujisaki starin of CJD virus measured 3 to 17 nm in diameter and 31–214 nm in length.

The diversity of SAF ultrastructure (Fig. 9–11) was recently studied in detail by Liberski, Brown, Wolf and Gajdusek (unpublished observations). Most SAF purified from hamster brains infected with the 263K strain of scrapie virus formed large aggregates, easily detectable even at low magnification. A tendency of SAF to cluster around other particulate structures, mostly collagen fibers, was also noted. At higher magnification SAF appeared as rod-like fibrils, 33–207 nm in length (average 111 nm; SD, 13 nm) and 2.4–36 nm in width (average 13 nm; SD, 8 nm). Each fibril was composed of 2 to 4 protofilaments, of which 56% were twisted helically around each other, 33% formed parallel arrays; and 11% had an undefined substructure. A few fibrils were unwrapped at one end, forming flare-like structures. Occasionally, tiny smooth "spaghetti-like" filaments of wavy appearance and <2 nm in diameter were observed in close proximity to SAF or as aggregates observed against the background of the grid. SAF isolated from CJD virus-infected mouse brains were also mostly clustered and appeared somewhat longer (31–333 nm; average, 115; SD, 46) and thinner (2.4–23 nm; average, 10; SD, 2.4) than those isolated from scrapie-infected hamsters. When these measurements were statistically compared to those of hamster fibrils, the differences in width were found to be statistically significant ($P < 0.01$) but the differences in length were not ($P > 0.1$). In addition, the substructure of mouse fibrils was often difficult to classify: 60% of protofibrils were twisted helically, 11% formed parallel arrays, and 28% were undefined.

Precise ultrastructural criteria established for SAF by negative staining electron microscopy [761–764] originally revealed two types of SAF as mentioned above. In subsequent experiments, however, it became apparent that the ultrastructural morphology of SAF was more diverse than was at first supposed and that type I and II of SAF merged with intermediate forms. SAF purified from scrapie-infected hamsters and reported by Diringer et al. [295], Cho [210] and Liberski et al. [677] were shorter and thicker and those purified my Merz and her colleagues.

The morphometric data for SAF provided by Liberski et al. (unpublished observations) are in good agreement with those previously published by other investigators. SAF purified from hamster brains infected with the 263K strain of scrapie virus were significantly thicker than those purified from mice infected with the Fujisaki strain of CJD virus, and had better defined protofibril substructure. Because the NIH group used the same procedure to purify SAF from both hamsters and mice, these differences cannot be ascribed to technical artefact, as previously suggested [677]. Instead, they presumably reflect differences in the respective host amyloid protein subunits of the fibrils or a agent-produced modification of it, since a given host species respond to infection by either agent with the production of its own protein [694].

The higher yield and clearer structure of SAF from scrapie than from CJD specimens has also been noted by other investigators [718]. As the yield of SAF correlates with the infectivity titers [973], different yields can be ascribed to higher titers of infectivity in scrapie-infected hamsters than in CJD-infected mice.

The nature and biological significance of wavy filamentous structures occasionally observed in specimens of SAF isolated from hamsters infected with the 263K strain of scrapie virus but not from mice infected with the Fujisaki strain of CJD virus is unknown (Fig. 13). Similar structures have been previously observed in SAF prepared according to a protocol similar to the NIH group or that exposed to 10% SDS [749–750]. We have observed a dramatic increase in these structures in SAF preparations treated with formaldehyde and autoclaved [147], and

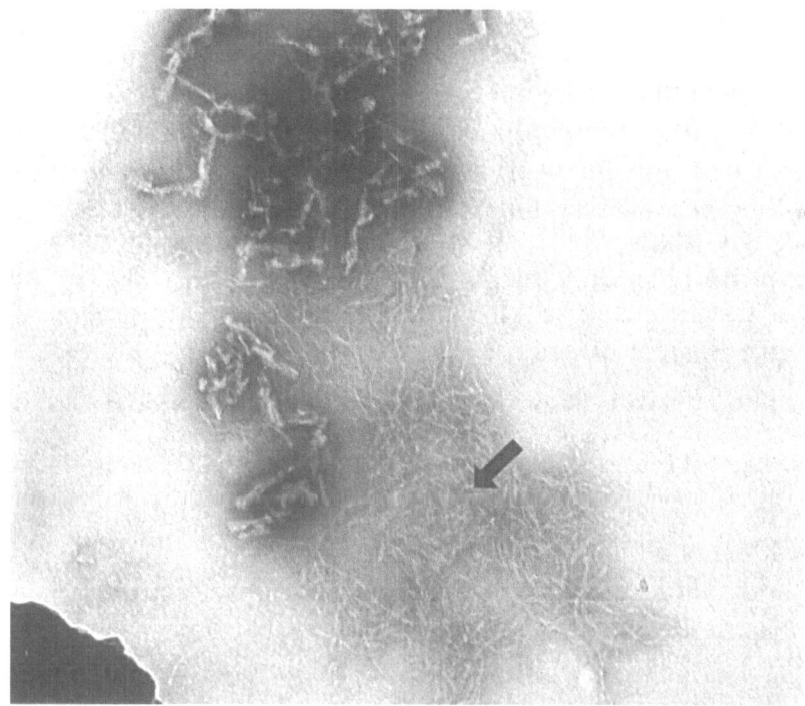

Fig. 13. "Spaghetti"-like filaments (arrow) in close contacts with SAF isolated from hamster brains infected with the 263K strain of scrapie virus. Unpublished work of P.P. Liberski, P. Brown and D.C. Gajdusek, National Institutes of Health, Bethesda, USA

it is therefore likely that they represent the morphologic consequence of partial degradation.

2.2.5.1. The chemical composition of SAF

SAF and prion rods SAF (prion rods) consist of PrP 27–30. Two independent reports by Diringer et al. [295] and Prusiner et al. [886] came to this conclusion in 1983. Silver stained SDS-PAGE of sucrose density gradient fractions purified from hamster brains infected with the 263K strain of scrapie virus revealed consistent bands that co-migrated with alpha-chymotrypsinogen indicating a molecular weight of approximately 26 kDa [295]. As the same fractions revealed numerous SAF (vide supra) it was suggested that the 26 kDa protein is a molecular component of SAF. Analogously, Prusiner et al. [886] showed that prion rods consisted of PrP 27–30. Radioiodinate labelling and silver staining showed that PrP 27–30 was the predominant protein in sucrose gradient fractions purified from hamster-infected scrapie brain. The same fractions contained numerous "clusters of rod-shaped particles" (hence, designated prion rods).

While the co-purification of SAF (prion rods) and PrP 27–30 strongly suggested the latter as a component of SAF, the immunoelectron microscopy provided unequivocal support for this hypothesis. Polyclonal rabbit antibodies against purified PrP 27–30 were first generated in Prusiner's laboratory [49, 50–51, 74]. Using immuno-gold electron microscopy Barry et al. [49] was the first to demonstrate the immunodecoration of prion rods (SAF) with anti-PrP27–30 antibodies. Furthermore, when clusters of prion rods were dispersed by of sonication, the spherical particles were generated and also immunolabeled with anti-PrP antisera.

Although SAF and prion rods are virtually identical, there is subtle ultrastructural variation beyond the rigid ultrastructural criteria established by Merz et al. [761–764] which are not always applicable (Liberski, Brown, and Gajdusek, unpublished data). Consequently, some investigators proposed that SAF and prion rods were different structures [288, 747, 750, 862, 864–865, 867]. The use of antibodies directed against PrP 27–30 or purified SAF made it possible to address this question directly. Merz et al. [760] used three different antisera to immunolabel SAF. Two of them were raised against SAF purified from mice infected with the ME7 strain of scrapie virus or hamsters infected with the 263K strain of scrapie virus. An additional antiserum was raised against purified PrP27–30 (hence, by definition against prion rods) in Prusiner's laboratory [74]. All three antisera immunodecorated SAF purified from mice infected with either ME7 or 139A strains of scrapie virus or hamsters infected with the 263K strain of scrapie virus, respectively. The antisera exhibited different reactivity and immunodecorated different proportions of SAF isolated from different scrapie models. The immunodecoration of SAF by antisera against SAF and PrP27–30 strongly indicates that PrP 27–30 is a component of SAF and, because it has been demonstrated that PrP 27–30 is a component of prion rods, SAF and prion rods are bona fide the same structures. The subtle ultrastructural differences between SAF and prion rods are probably due to the differences in the methods of purification and staining. The same conclusion was recently drawn by Hope et al. [505] after examining SAF purified from bovine spongiform encephalopathy. It is noteworthy that the drastic chemical treatments used during some purification protocols may yield PrP 27–30 but not SAF [Liberski, unpublished observations]. Furthermore, the synaptosomal-mitochondrial fractions from scrapie-infected hamsters revealed specific areas of membranes immunodecorated with anti-SAF antisera. The last finding is consistent with the association of scrapie virus and membranes.

The physicochemical properties of SAF (prion rods) have been studied in some detail, with the only agreement that PrP is a component of SAF. Whether any other molecule is a part of SAF has yet to be elucidated. It should be stressed however, that such studies are difficult to perform due to inherent tendency of SAF (prion rods) to aggregate. The morphology of prion rods (SAF), as revealed by negative stain electron microscopy, did not change following heating at <37°C in the presence of up to 2% of SDS [749–750]. When the concentration of SDS was increased to 5%, the surface of the fibrils appeared smoother. Increasing the concentration of SDS still further to 25% at temperatures between 25 and 100°C, produced thin filaments measuring <5 nm in diameter. These filaments are probably the same "sphagetti"-like filaments reported by Brown et al. [146–147], and Merz [759]. Exposure to ⩾2% of SDS at 100°C resulted in the complete disruption of fibrils with a concomitant drop of infectivity by a factor larger than 100.

The ultrastructure of fibrils was drastically changed at pH 10. While retaining their general rod-like pattern, the fibrils were transformed into arrays of globular structures. Such structures were even more pronounced when the pH was raised to 11. Moreover, when fibrils were exposed to guanidinum ions, they were transformed completely into clumps of amorphous material.

In contrast, the structure of fibrils exposed to proteinase K at 37°C was preserved while infectivity decreased by a factor of approximately 1000. This study was extended by Brown, Liberski, Wolf and Gajdusek in a series of reports [146–147]. First, several enzymes were used to examine their effect on the structure of SAF. These enzymes were: N-acetyl-beta-glucosaminidase A and B, carboxypeptidase A, B, and Y, chondroitinase ABC, collagenase, DNase I and II, hyaluronidase, leucine aminopeptidase, lipoprotein lipase, microccocal endonuclease, neuraminidase, phospolipase A_2, B and C, proteinase A and K, RNase A, H, and T_2, trypsin, endoglycosidase H and F, glycopeptidase F, and phospoinositol specific phospholipase C. None of the listed enzymes apart from proteinase K produced substantial reduction of infectivity as measured by the estimation of an incubation period. Moreover, the structure of SAF remained virtually the same under all experimental conditions. However, boiling SAF in the presence of SDS diminished the number of fibrils and produced a "fuzzy" appearance of those remaining which was similar to "smooth" surfaces of prion rods in the experiment mentioned previously [749–750]. In a final experiment, digestion with glycopeptidase F resulted in the appearance of multiple PrP variants of low molecular weight in SDS-PAGE without changing the infectivity as measured by the more accurate end point titration. The structure of SAF after several routine decontamination protocols (autoclaving and formaldehyde treatment) has also been studied [146–147]. The score of the number of SAF estimated from electron micrographs in each experiment paralleled the infectivity titers. The number of SAF in control specimens (4+) did not change following formaldehyde treatment with or without subsequent autoclaving. However, autoclaving or autoclaving followed by formaldehyde treatment reduced the number of SAF to 1–2+. With the latter treatment, most SAF, while retaining the rod-like pattern were transformed into structureless blebs and globules that were reminiscent of SAF exposed to pH 11, as reported by McKinley et al. [749].

Recently, an unexpected and still controversial association of SAF and "tubulovesicular structures" has been observed by Narang et al. [805–806] using a "touch or impression" technique. In this method a formvar coated electron microscopic grid is placed briefly on the freshly cut surface of the brain and then examined by negative staining electron microscopy. The impression grids prepared from brains of hamsters infected with the 263K strain of scrapie virus contained numerous tubules measuring 50 nm in diameter and more than 1000 nm in length. Narang et al. [805–806] claimed that these "tubulofilamentous structures" were the same tubulovesicular structures (approximately 35 nm in diameter) observed by thin section electron microscopy in all subacute spongiform virus encephalopathies [671, 680, 685]. Regardless of their identity, Narang et al. [805–806] claimed that exposure of these tubulofilamentous structures when exposed to SDS caused fragmentation and the release of an "inner core" which was morphologically similar to SAF. Furthermore, it was claimed that these inner core filaments were apparently immunodecorated with antisera against SAF. However, in a published electron micrograph (Fig. 5b) gold particles appeared to immunodecorate the "fluffy" material attached to the filaments, rather than the filaments themselves. One explanation of the observed phenomenon is that when the impression is made, PrP33–35[sc] taken from the cut surface of the brain undergoes conformational changes leading to SAF formation. This explanation is in agreement with data suggesting that SAF are formed *in vitro* during the extraction procedure. However, the story is further complicated by the fact that exposure of the tubulofilaments to proteolytic enzymes (collagenase 1, proteinase K, trypsin) changed their diameter to approximately 30 nm. After subsequent exposure to DNase, these "thin" tubules disintegrated and SAF appeared. Narang et al. [805–806] speculated that the tubulofilaments are complex structures composed of an innermost core of SAF coated with intermediate layer of DNA and outermost layer of protein (PrP33–35sc). It must be emphasized that this hypothesis is highly speculative and needs independent confirmation.

The origin of SAF remains uncertain. PrP^{sc} and PrP^c have the same amino acid sequence but differ in their resistance to proteinase K and in their ability to assemble (polymeryse) into SAF (prion rods). Such differences may stem from different but yet to be dicovered post-translational modifications or conformational changes. Furthermore, whether or not SAF occur *in vivo* is not entirely clear. Meyer et al. [766] examined microsomal fractions before or after detergent treatment. Prion rods (SAF) were not observed before detergent treatment and appeared following extraction with sarcosyl. Thus, it was suggested that SAF are generated after solubilization of membranes or were hidden within membranes. The last hypothesis seemed implausible, as SAF have never been observed in situ despite one unsubstantiated report to the contrary [454]. Recently, Somerville et al. [974–976] provided some evidence that SAF may, however, assemble *in vivo*. SDS-PAGE analysis of PrP revealed four major peptides of molecular weights 30–34 kDa, 27–29 kDa, 24–25 kDa, and 18–20 kDa, respectively. On 2 dimensional (2D) gels these peptides resolved into many charge-size variants comprising 6 regions: 3 basic and 3 acidic. Following proteinase K digestion the 30–34 kDa PrP variant was removed while on 2D gels, three more acidic PrP variants were conspicuously absent. The same pattern of PrP bands on either SDS-PAGE or 2D gels was found irrespective whether detergent treatment was used after or before proteinase K treatment. Thus, partial resistance to the proteinase K was not the result of SAF assembly and it could be interpreted that SAF already existed *in vivo*. However, it is still difficult to reconcile this interpretation with the fact that SAF have never been found *in vivo* in any of the electron microscopic studies reported so far.

Recently, two sets of contradictory data on SAF morphogenesis were published. McKinley et al. [755] provided evidence that prion rods (SAF) are formed *in vitro* following detergent extraction and limited proteolysis of membrane-asssociated PrP^{sc}. Membrane fractions containing PrP^{sc} yielded only amorphous vesicles before detergent extraction or proteolytic treatment. When exogenous fibrils were added to membrane fractions they were easily visualized, thus the apparent absence of any fibrils in these fractions did not result from technical difficulties. Detergent extraction alone has no effect on the size of PrP^{sc} as evidenced by Western blotting, but detergent dispersed vesicles into "small pleomorphic fragments". When these fractions were extracted with detergent and digested with proteinase K, PrP^{sc} converted to PrP 27–30 and fibrils are formed. Furthermore, the exact ultrastructural morphology of SAF depends on the particular protease and detergent used in the procedure. Typical SAF were formed following a limited proteolysis with proteinase K in the presence of SDS or Sarkosyl, while longer fibrils were formed following a digestion with either pronase or trypsin in the presence of the same detergent. By the same token, the last treatment produces PrP species larger than PrP 27–30. In contrast, no fibrils were formed following extraction with Nonidet P-40 or after using collagenase which does not digest PrP^{sc} or leucine aminopeptidase which removes only a few N-terminus amino acids. Taken together these studies strongly suggest the importance of both proteolytic treatment and detergente extraction in the SAF formation.

On the contrary, Isomura et al. [525] suggest that SAF exists *in vivo* as amyloid aggegates from which they are dissociated following detergent extraction.

In summary, PrP is an unequivocal constituent of SAF, and no other component has been identified.

2.3. The structure of the gene encoding PrP 33–35 (Prn-p) in different species

2.3.1. The hamster Prn-p gene

The cDNA encoding hamster PrP 27–30 has been cloned by Oesch et al. [820] using a set of 32 icosameric oligonucleotides, 5′-GG(T/C)TT(A/G)TTCCA(T/C)TG(A/G)TT(A/G)-TG based on reverse-translation of a seven amino-acid fragment of hamster PrP 27–30 (N-His-Asn-Gln-Trp-Asn-Lys-Pro-C) [878]. Screening of 150 000 colonies with a mixture of 5′-[^{32}P]-labeled probes revealed one positive clone of recombinant plasmid containg a 2 kb insert (HaPrPcDNA-1) of 1918 nucleotides preceded by 33 G residues and followed by 56 A residues. The N-terminal sequence of hamster PrP [878] corresponded to the sequence encoded from nucleotides 236 to 280 (numbering starts from the first nucleotide following the string of Gs). The hamster PrP open reading frame (ORF) extends from nucleotide 1 to 730 while the first ATG start condon was found at nucleotide 11 and the first TGA stop codon at nucleotide 731, respectively. A protein deduced from this sequence consists of 240 amino acids of molecular weight of 26 325. While PrP 27–30 consists of 165 or fewer amino acids it was already evident that the PrP gene ecoded a larger protein from which PrP 27–30 is generated by proteolytic cleavage with proteinase K during purification [820]. The amino acid sequence is characterized by a C-terminal stretch of hydrophobic amino acids and another stretch of hydrophobic amino acids near the N-terminus of PrP 27–30. Interestingly, two repeated 22 to 81 codon sequences of Gly-Gly-(Asn/Ser)-Arg-Tyr-Pro followed by 5 repeated sequences of Pro-(His/Gln)-Gly-Gly-Gly-(-/Thr)-Trp-Gly-Gln were found. These repeats are reminiscent of the eight repeats of tetrapeptide Gly-Gly-Gly-X (X stands for a hydrophobic amino acid) of 50 kDa keratin, a protein aggregating to yield intermediate filamens similar to SAF/prion rods composed of PrP.

Southern analysis using a radiolabeled cDNA insert of pHaPrP as a probe revealed a single 3.1 kb Eco RI fragment, 6.5 kb Bam HI fragment or 7.2 kb BgII fragment in both control and scrapie-infected hamster brains [820]. Furthermore, 2.3 kb and >15 kb Eco RI bands were found in murine and human DNA, respectively. These results were interpreted as being compatible with a single PrP gene presented in both normal and scrapie-affected animals.

Northern analysis of poly(A)$^+$ RNA probed with pHaPrP cDNA revealed a single 2.1 kb band expressed also in both control and scrapie-infected animals.

The cloning and sequencing of hamster PrP cDNA enabled the cloning of the chromosomal PrP gene [52]. First, a 2.1 kb HaPrPcDNA-s11 was cloned using a radiolabeled 790 bp MstII-TagI fragment of pHaPrPcDNA-1 as a probe. A 23-mer complementary to the 5'-end of PrP mRNA used to rescreen the first 40 positive clones. A complete nick-translated HaPrPcDNA-1 insert was used as a probe to identify 4 positive clones of 1.2×10^6 clones obtained from Eco RI-digested hamster DNA. The 3 kb insert of one of these (SPrP-8) was subcloned into pSP64 and completely sequenced. The sequence of the genomic clone SPrP-8 and that of HaPrPcDNA-1 were virtually identical suggesting the lack of introns in this segment. However, S1 nuclease mapping and a comparison of HaPrPcDNA-s11 and SPrP-8 sequences revealed a 3' splice site at position 473. Furthermore, the 5' terminal 72 bp segment of HaPrPcDNA-s11 suggested the presence of another exon upstream of SPrP-8. Using the Eco RI-Ncol fragment of SPrP-8 as a probe a 7.0 kb (SPrP-3) fragment overlapping the SPrP-8 over 2 kb was isolated. While the 7.0 kb Hind III fragment lacked the 5' SacII site present on cDNA and thus did not encompass the 5'-exon of PrP gene a 5 kb Hind III fragment (SPrP-7.1) containing the 5'-PrP exon encompassing 34 to 72 nucleotides of HaPrPcDNA-s11 was cloned using the 50-mer complementary to 5'-end of HaPrPcDNA-s11.

The entire ORF is contained within a single 3' exon separated from a 5' exon by 10 kb intron. The sequences spanning the 5'-splice site (CAG/GTAAGC) and 3' splice site (CCTTGTTCTTCATTTTGCAG/AT) are in good agreement with consensus sequences for the 5' and 3' splice sites. Both S1 nuclease mapping and primer elongation revealed multiple 5' termini encompassing a stretch of 25 nucleotides. Multiple initation sites map 55 to 80 nucleotides upstream of the 5' splice site. This region, upstream of multiple initiation sites is GC-rich and does not contain TATA boxes, a feature shared with "houskeeping" genes. Interestingly, two GS-rich PrP sequences are complementary with inverted sequences (GGGGCGG(G/A)GC and CCGCCC) recognized by the Sp1 protein, a promoter-specific factor binding to several viral early promoters.

The PrP mRNA is encoded by 56 to 80 nucleotides of the 5' exon (330–410) and 1960 nucleotides of the 3' exon. This mRNA measures 2000 to 2042 nucleotides to yield a 2.1 kb segment as revealed by Northern blot [820].

2.3.2. The murine Prn-p gene

The cloning of cDNA of the murine PrP (Prp-n) gene by Chesebro et al. [208] was reported at the same time that the hamster PrP cDNA was studied [820]. Using a mixture of synthetic oligonucleotides complementary to the published sequence of the N-terminus of PrP 27–30 [878], mRNA populations were screened for the presence of scrapie-specific RNA. A 2.4–2.5 kb RNA bands was detected in both controls and scrapie-infected mice. In the next step, the same oligonucleotide mixture was used to screen a cDNA library of scrapie-infected mouse brain. One clone (clone 9) was isolated and sequenced to yield a partial sequence of PrP 27–30. When this clone was used as a probe to screen mRNAs, a 2.4–2.5 KS band was detected in both scrapie-infected and control animals.

The full cDNA sequence of the murine PrP gene has been reported by Locht et al. [688]. The clone 9 contained a 1.7 kb fragment composed of 0.8 and 0.9 kb segments joined by an Eco RI site. The 0.9 kb fragment (subclone 9–13) contains the entire PrP ORF. A second clone (clone 7) was longer (1.7 kb) and contained the entire sequence of clone 9 preceded by an additional 350 bases of a leader sequence at the 5' terminus. Sequences of both clones differed at position 496 (A and G in clone 7 and 9, respectively), yielding an amino acid substitution (Met to Val). The cDNA contained a 5' leader sequence 99 nucleotides long , followed by 762 nucleotides encoding PrP, a 3' noncoding sequence of 1236 nucleotides containing 74 stop codons, and 54 deoxyadenosine residues at the 3' end. The 100–102 position ATG codon is most likely to be the initiation site for translation of the 254 amino acid protein (PrP) of 27 981 molecular weight.

Another partial cDNA of the murine Prn-p gene was characterized by Robakis and coworkers [906]. The poly(A)$^+$ mRNA was screened with a 44-mer complementary to the known sequence of the N-terminus of PrP 27–30 [878]. 6 positive clones were rescreened with HaPrPcDNA

[52] to yield one positive clone (clone XIV) of 2.2 kb. This 2.2 kb insert containing a sequence encoding PrP 27–30 was 79 bp longer than the original HaPrPcDNA [820]. An ORF extended from nucleotide 2 to nucleotide 808. An ATG start codon was found at position 47, 42 bp upstream from that previously reported [820].

Partial genomic PrP sequences have been also cloned by Westaway et al. [1042] using two restriction fragments of a murine cDNA clone as a probe [688]. The ORF was found to be uninterrupted, confirming other findings that the PrP ORF is located within a single exon [820]. The 5′ exon has not been sequenced, but, its presence has been suggested by the lack of hybridization of another probe, corresponding to the extreme 5′ cDNA terminus, to the EcoRi or BAmHI clones.

Southern blot analysis was used to investigate homologous sequences in chromosomal DNA of hamsters, mice, rabbits, sheep and humans. Eco RI-digested human and hamster DNA revealed 15 kb and 3.4 kb bands, respectively; mouse DNA yielded a 2.4 kb band, rabbit DNA a 3.0 kb band, while sheep DNA yielded 9.0 kb, 4.7 kb, and 2.4 kb bands. These results were supported by those of Westaway and Prusiner [1044] who reported conservation of PrP sequences. The PrP-related sequences were found not only in all mammalian species tested (hamster, rat, goat, sheep; Westaway and Prusiner [1044] also cited a PrP-related rabbit species as a personal communication by Kingsbury) but also in some lower organisms – Drosophila melanogaster, Caenorhabditis elegans and Saccharomyces cerevisiae. The hybridization of PrP-related sequences was weaker in these lower species than in mammals. In contrast, PrP-related sequences were not found in the DNA of Xenopus laevis, possibly because of the enormous size of the amphibian genome.

2.3.3. The rat Prn-p gene

The rat PrP cDNA has been cloned by Liao et al. [658]. A cDNA library from rat hypothalamus was screened with synthetic polynucleotide probes that revealed 87 positive clones. Several of them were subcloned using a M13 vector to yield a 1800 nucleotide long PrP cDNA which was sequenced. The ORF of rat PrP revealed a 93% homology with that of human PrP (*vide infra*), 96% with hamster PrP and 97% with murine PrP. The deduced PrP sequence consists of 225 amino acids, a series of completely conserved short peptide repeats of which those with the sequence Gly-Gly-Gly-X, (where X stands for a hydrophobic amino acid) may function in beta-pleated configuration. This conserved repetitive region is followed by a stretch of a 23 amino acids which may be a transmembrane domain and a cytoplasmic domain containing all nonconservative substitutions. Two potential n-glycosylation sites were also reported at positions 191 and 207.

2.3.4. The human Prn-p gene (PRNP)

Molecular cloning of the human PrP gene was reported by Kretzschmer et al. [624] (Fig. 14). A poly(A)$^+$mRNA, isolated from several neuroectodermal cell lines, was screened with hamster cDNA to yield a single 2.5 kb band on Northern blotting. A human retinal cDNA library was also screened with hamster cDNA. An Eco RI-digested 2.9 kb insert (HuPrPcDNA-1) was pUC8 subcloned and sequenced. Another 2.4 kb human clone (HuPrPcDNA-2) was also isolated. A consensus sequence of human cDNA 2432 nucleotides long, was established by comparison of the HuPrPcDNA-1 and HuPrPcDNA-2 sequences. Two ATG start codons were found at positions 50 and 71 of the 5′ terminus. The first ATG codon is surrounded by a consensus eucaryotic initiation site sequence: ATTATGG. As the sequence surrounding the second ATG codon is different from the consensus sequence, the translation probably begins at the former site. The deduced human PrP sequence consists of 253 amino acids to yield a protein of 27 665.67 molecular weight. The N-terminal Met-Leu-Val-Leu-Phe-Val sequence of a signal peptide was found. This sequence is cleaved at the putative signal sequence cleavage site. The 3′ non-coding stretch consists of 1624 nucleotides with numerous termination signals and 17 A residues at the 3′ terminus. This 3′ non-coding region of human PrP cDNA is 436 nucleotides longer than that of hamster PrP cDNA.

The ORF of human PrP differs from the hamster PrP ORF by one amino acid in length, and by 26 amino acids in sequence. The N-terminal portion of human PrP contains two short repeats (Gly-Gly-(Asn/Ser)-Arg-Tyr-Pro) followed by 5 long repeats (Pro-(His/Gln)-Gly-Gly-Gly-(-/Gly)-Trp-Gly-Gln). This region is highly conserved between human and hamster cDNA at all but position 56. Interestingly, although these conserved sequences must be functionally important, they are not required for transmissibility of the disease as they are cleaved by proteinase K to yield the core protein PrP 27–30.

2.3.5. The ovine and bovine Prn-p gene

A chromosomal ovine gene was cloned by Goldmann et al. [442] (Fig. 15). A genomic Suffolk sheep library was screened with a hamster cDNA to yield two (23/21 and 32/21) positive clones. The Eco RI and

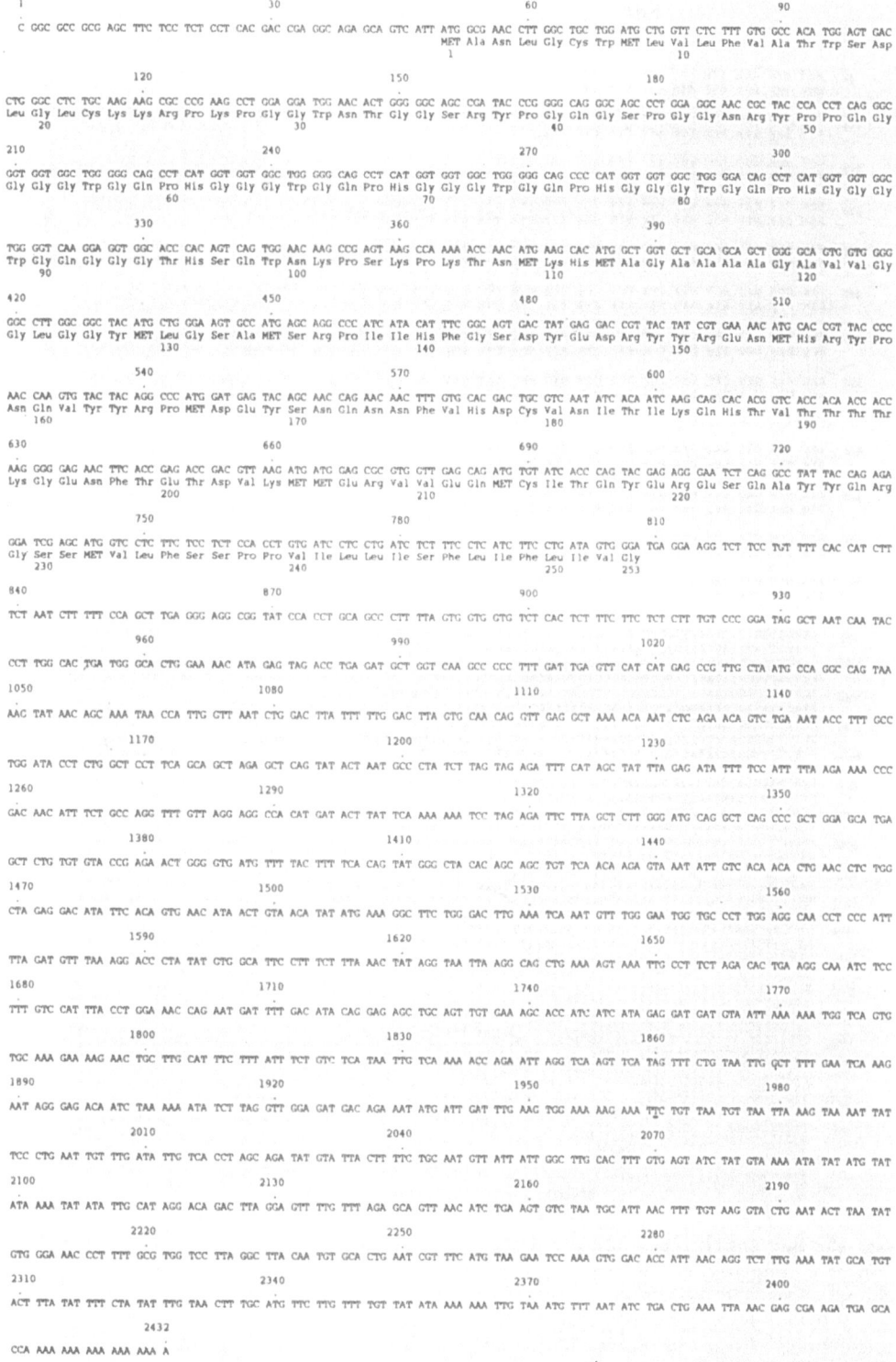

Fig. 14. Nucleotide sequence of the human PrP cDNA (consensus sequence) and the deduced amino acid sequence [623]. Courtesy of Prof. Stanley B. Prusiner, University of California, San Francisco and the editor of DNA

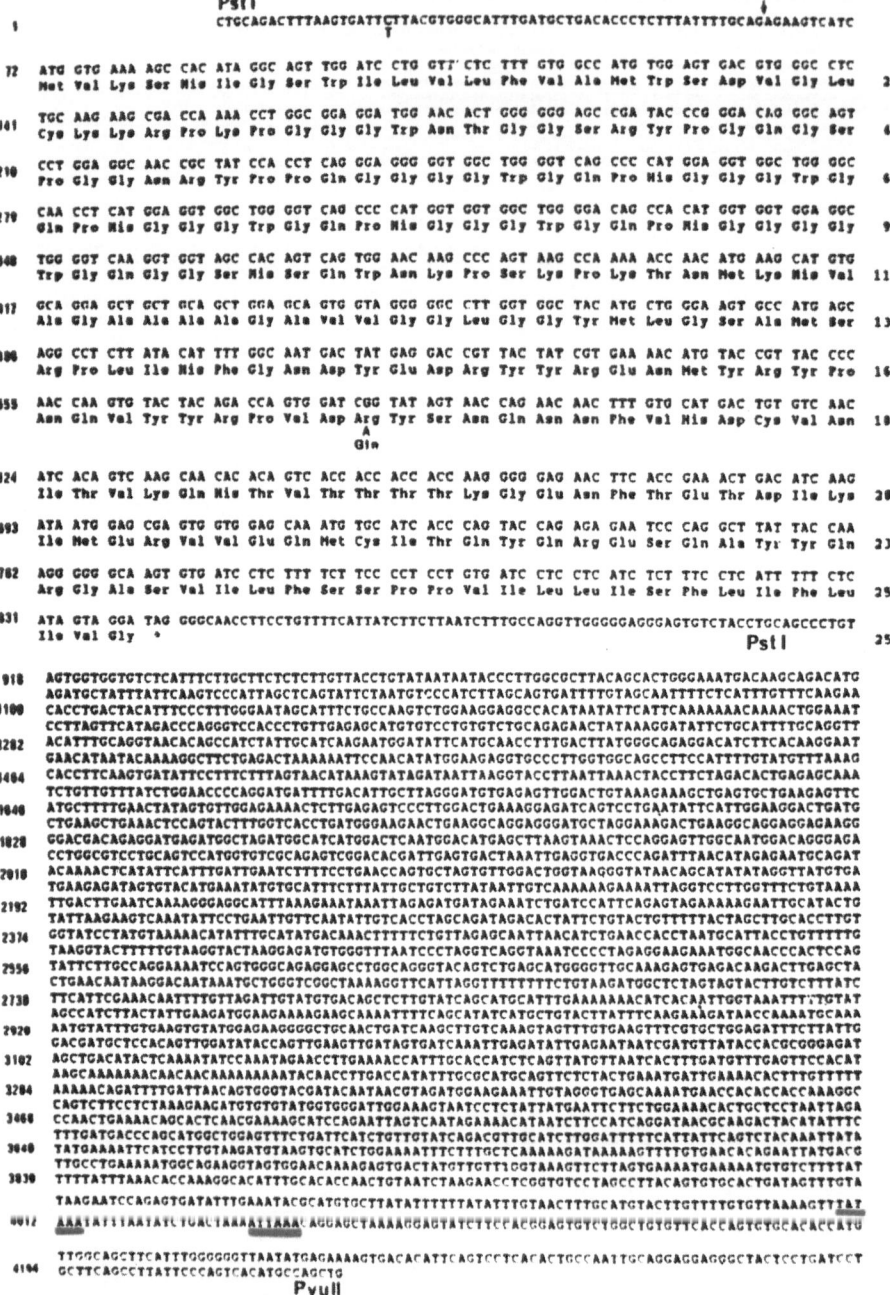

Fig. 15. Sheep PrP DNA and the deduced amino acid sequence [442]. The intron/exon border is indicated by an arrow, the putative polyadenylation signal and the region of the poly(A) tail attachment are underlined at positions 4000–4014 and 4034–4039, respectively. Courtesy of Dr. James Hope, MRC & AFRC Neuropathogenesis Unit, Edinburgh, Scotland and the editor of Proceedings of the National Academy of Sciences of the USA

Hind III restriction fragments of these clones corresponded to those found on Southern blotting [520–521, 1044]. The 23/21 clone was sequenced to yield an insert containing the ORF for PrP which had a 83% homology with hamster PrP cDNA. The sequence of another clone revealed two differences at position 20 and 583 (C and G in 23/21 and A and A in 32/21, respectively). The difference at position 583 brings about a substitution of Arg for Glu. At the 3′ terminus the sequence TATAAA (nucleotides 4009–4014) which resembles the consensus sequence AATAAA of the eucariotic polyadenylation site preccdes the poly(A) addition by 20–25 bp.

In agreement with the rodent Prn-p gene sequences, the ovine gene also contains an intron with the sequence TCT TTA TTT TGC AGA that was homologous to that of the consensus 3′ splice-site sequence. However, the 5′ exon has not been characterized.

```
(a)
  1  ATGGTGAAAAGCCACATAGGCAGTTGGATCCTGGTTCTCTTTGTGGCCAT
     M  V  K  S  H  I  G  S  W  I  L  V  L  F  V  A  M
     GTGGAGTGACGTGGGCCTCTGCAAGAAGCGACCAAAACCTGGAGGAGGAT
     W  S  D  V  G  L  C  K  K  R  P  K  P  G  G  G  W
101  GGAACACTGGGGGGAGCCGATACCCAGGACAGGGCAGTCCTGGAGGCAAC
     N  T  G  G  S  R  Y  P  G  Q  G  S  P  G  G  N
     CGTTATCCACCTCAGGGAGGGGGTGGCTGGGGTCAGCCCCATGGAGGTGG
     R  Y  P  P  Q  G  G  G  G  W  G  Q  P  H  G  G  G
201  CTGGGGCCAGCCTCATGGAGGTGGCTGGGGGCCAGCCTCATGGAGGTGGCT
     W  G  Q  P  H  G  G  G  W  G  Q  P  H  G  G  G  W
     GGGGTCAGCCCCATGGTGGTGGCTGGGGACAGCCACATGGTGGTGGAGGC
     G  Q  P  H  G  G  G  W  G  Q  P  H  G  G  G  G
301  TGGGGTCAAGGTGGTACCCACGGTCAATGGAACAAACCCAGTAAGCCAAA
     W  G  Q  G  G  T  H  G  Q  W  N  K  P  S  K  P  K
     AACCAACATGAAGCATGTGGCAGGAGCTGCTGCAGCTGGAGCAGTGGTAG
     T  N  M  K  H  V  A  G  A  A  A  A  A  G  A  V  V  G
401  GGGGCCTTGGTGGCTACATGCTGGGAAGTGCCATGAGCAGGCCTCTTATA
     G  L  G  G  Y  M  L  G  S  A  M  S  R  P  L  I
     CATTTTGGCAGTGACTATGAGGACCGTTACTATCGTGAAAACATGCACCG
     H  F  G  S  D  Y  E  D  R  Y  Y  R  E  N  M  H  R
501  TTACCCCAACCAAGTGTACTACAGGCCAGTGGATCAGTATAGTAACCAGA
     Y  P  N  Q  V  Y  Y  R  P  V  D  Q  Y  S  N  Q  N
     ACAACTTTGTGCATGACTGTGTCAACATCACAGTCAAGGAACACACAGTC
     N  F  V  H  D  C  V  N  I  T  V  K  E  H  T  V
601  ACCACCACCACCAAGGGGGAGAACTTCACCGAAACTGACATCAAGATGAT
     T  T  T  T  K  G  E  N  F  T  E  T  D  I  K  M  M
     GGAGCGAGTGGTGGAGCAAATGTGCATTACCCAGTACCAGAGAGAATCCC
     E  R  V  V  E  Q  M  C  I  T  Q  Y  Q  R  E  S  Q
701  AGGCTTATTACCAACGAGGGGCAAGTGTGATCCTCTTCTCTTCCCCTCCT
     A  Y  Y  Q  R  G  A  S  V  I  L  F  S  S  P  P
     GTGATCCTCCTCATCTCTTTCCTCATTTTTCTCATAGTAGGATAG
     V  I  L  L  I  S  F  L  I  F  L  I  V  G  *
```

```
(b)
160–186  [R1]  CCTCAGGGAGGGGGTGGCTGGGGTCAG
187–210  [R2]  CCCCATGGA---GGTGGCTGGGGCCAG
211–234  [R3]  CCTCATGGA---GGTGGCTGGGGCCAG
235–258  [R4]  CCTCATGGA---GGTGGCTGGGGTCAG
259–282  [R5]  CCCCATGGT---GGTGGCTGGGGACAG
283–309  [R6]  CCACATGGTGGTGGAGGCTGGGGTCAA
               Pro His Gly Gly Gly Gly Trp Gly Gln
```

Fig. 16. (a) Nucleotide and amino acid sequences of the protein-encoding region of the bovine PrP gene [443]; (b) nucleotide sequences of the octapeptide-encoding segments of the bovine PrP gene. The numbers on the left refer to the nucleotide positions of these segments in the sequence (a). Courtesy of Dr. James Hope, MRC & AFRC Neuropathogenesis Unit, Edinburgh, Scotland and the editor of Journal of General Virology

Northern analysis of ovine poly(A)$^+$mRNA revealed a 4.6 kb band in contrast to the 2.5 kb PrP mRNA of rodents. The ovine PrP sequence consists of a 24-amino acid signal peptide, a highly basic N-terminal region preceded by two keratin-like Gly-rich segments that are highly conserved, two possible N-glycosylation sites at positions 184 and 200, and a highly hydrophobic C-terminus.

The bovine PrP gene was recently characterized by Goldman et al. [443] (Fig. 16) using PCR and primers homologous to ovine PrP sequences [442]. In ethidium bromide-stained agarose gels two bands (allelic) of 495 bp or 471 bp were visualized. The differences between these alleles remains in the number of Gly-rich peptides encoded by either 24 (R2, R3, R4, R5) or 27 (R1, R6) nucleotides. One allele has six copies of these sequences (R1, R2, R3, R4, R5, R6) while another one has five (R1, R2, R4, R5, R6). R1, R2, R4, R5, and R6 are homologous to ovine sequences while R3 is not.

2.3.6. The avian analogue of the Prn-p gene

Recently, the avian analogue of the prion protein [1044], the chicken prion-like protein ch-PrLP), was purified and its cDNA characterized [482]. Nested PCR was used to amplify a 34-nucleotide sequence (from the 3′ nucleotide of the 7th amino acid codon to the 3′ nucleotide of 7th amino acid codon). This synthetic 34-mer was subsequently used to screen chicken brain cDNA. The 1.9 kb clone (analogous to three 2.2 kb clones) was sequenced and found to correspond to the protein of 267 amino acids. Northern blot analysis revealed a 2.9 kb mRNA most abundant in the brain and spinal cord. Larger, 3.5 mRNA was also detected. At the time of writing, the detailed structure of the cellular gene encoding ch-PrLP was unknown, but initial data indicate that the ORF lies within a single exon and that 5′ non coding sequences lie in a second exon; this structure is similar to those of PrP genes in different species.

2.3.7. The linkage of genes controlling the incubation period and susceptibility to scrapie and Creutzfeldt-Jakob disease to the gene encoding PrP 33–35 (Prn-p)

In every viral disorder, genetic make up of both the virus and the host play a role in many aspects of infection, and scrapie is not an exception [275]. However, most of the effects exerted by a few known genes are trivial in comparison with the role of *Sinc* gene in mice (and its analogue,

Sip in sheep) and Prn-p (there are strong evidence that *Sinc* and Prn-p is the same gene; vide infra). Furthermore, some genes act by the alteration of the effective virus titer, and thus influence the incubation period. Still other genes may affect the initial events of in scrapie infection, for example the reaction toward donor tissue, in terms of major histocompatibility complex genes [284]. The best example of this is the gene *PID* (prion incubation determinant) controlling the incubation period of both scrapie and CJD in mice [595–597]. Among many strains of mice tested, NZW mice were found to show the shortest incubation time when infected with the Chandler (139A) strain of a scrapie virus. Analogously, B10.Q and B10.AKM mice showed the shortest incubation time following inoculation with the Fujisaki (K.Fu.) strain of CJD virus. As these mice are identical only at the H-2 D subregion of the major histocompatibility complex located on chromosome 17, this suggested that this H-2 D subregion is the one involved. Furthermore, the B10.Br mice showed a prolonged incubation period under the same experimental conditions. B10.AKM and B10.Br mice differ only by one allele, q in the D subregion. Thus, the q allele in the D subregion resulted in the short, and the d allele resulted in a prolonged incubation period, respectively. An analogous genetic control was reported in Ile-de-France sheep where the association between OLA-A4, B6 loci and the gene for susceptibility for scrapie infection (*scr*[s], probably analogous to *Sip*) was found [769–771], but not in familial CJD, where no linkage with a particular haplotype could be found [620] but 7 out of 8 CJD cases shared the HLA antigens A28 and B8. However, as Dickinson and Outram pointed out [282], genes linked to the major histocompatibility complex may control the initial susceptibility "even not to the agent *per se* but to the H-2 type of the donor tissue in which the infective agent is likely to be sequestrated". The last question could not be answered on the basis of the original Kingsbury et al. [597] paper, because the proper controls using the Fujisaki (K.Fu) strain of CJD virus donors of different H-2 haplotypes were not explored.

2.3.7.1. Prn-i and *Sinc* in mice

The incubation period of murine scrapie is under control of a gene designated "*Sinc*" (see paragraph: 2.3.3.7) of which two alleles have been described [267–268, 270, 282–285, 558, 564, 566]. Dickinson and Mackay [275] identified the VM strain of mice (also designated 5M in earlier work) which, after inoculation with the ME7 strain of scrapie virus showed a prolonged incubation time of 40 weeks in contrast to other strains of mice showing incubation periods in a range 21 to 26 weeks. The gene, designated *Sinc*, has been identified by Dickinson, Meikle and Fraser [277–279] by classical genetical methods. Mice homozygous for the $Sinc^{s7}$ allele and inoculated with one of the ME7 class of strains of scrapie virus show a relatively short incubation period, while those homozygous for the $Sinc^{p7}$ allele show prolonged incubation periods under the same experimental conditions. In contrast, when $Sinc^{s7}$ homozygotes are inoculated with strains belonging to the 22A class of strains of scrapie virus the incubation periods are longer than those observed after inoculation the $Sinc^{p7}$ homozygotes. This classic paradigm for genetic control of scrapie incubation time was the basis of experiments showing that PrP gene and the *Sinc* gene in mice are probably the same [178]. It is noteworthy that Dickinson and Meikle [277] predicted "that there is only a short biochemical pathway between the *sinc* gene and scrapie replication"; a hypothesis proved later by the discovery of PrP 27–30 and its crucial role in scrapie virus infection [169–172].

Since VM mice, which are homozygous for the $Sinc^{p7}$ allele have been highly restricted by Dickinson's group, Prusiner's group identified another strain of mice: I/LnJ (both VM and I/LnJ originate from a common ancestor −[170–170]) which also appeared to be homozygous for the p7 allele of *Sinc* gene. When inoculated with the Chandler strain of scrapie virus, I/LnJ mice exhibited long incubation periods of approximately 254 days [170–171]. In contrast, NZW/LacJ mice exhibited a short incubation time (mean, 113 days) under the same experimental conditions. Furthermore, backcrosses of the heterozygotes (NZW/LacJ × I/LnJ)F_1 with the NZW/LacJ mice clearly produced a distribution of incubation times compatible with a single gene. This gene was designated *Prn-i* from *prion incubation* and the two alleles were designated $Prn-i^i$ (from I/LnJ mice with a long incubation period) and $Prn-i^n$ (from NZW/LacJ with a shorter incubation period).

Xba-1 restriction fragment length polymorphisms (RFLP) for the Prn-p gene distinguished between NZW/LacJ and I/lnJ (thus between $Prn-n^i$ and $Prn-i^n$ alleles) on Southern blots [170–171]. The I/LnJ fragment is 5.5 kb while NZW/LacJ fragment is 3.8 kb. Hence, the NZW/LacJ allele was designated $Prn-p^a$ and I/LnJ allele was designated $Prn-p^b$. In a separate experiment, Race et al. [889] reported that after the inoculation of the backrosses of F_1 (NZW/LacJ × I/LnJ) F_1 to the NZW/LacJ with Chandler strain of scrapie virus, mice with a short incubation period yielded a 3.8 kb band while those with a long incubation period (presumably heterozygous) yielded 3.5 and 5.5 kb bands, respectively. Interestingly, one mouse had a short incubation period and two bands on Southern blot, possible having a recombination between Prn-p and Prn-i

[889]. Such possible recombinational events between Prn-i and Prn-p had been reported earlier by Carlson et al. [170–172], and they constitute an important observation in determining whether Prn-i and Prn-p are the same or separate genes. Later it was shown by classical genetics that the allele in I/LnJ mice was the same as the allele of the *Sinc* gene [520–522].

This hypothethical relationship was investigated further by Southern blot experiments with the pEA974 probe from a cDNA clone of hamster PrP mRNA [520–521]. This probe hybridized to a 5.5 kb Xba-1 DNA fragment in VM (Sinc^{p7}) and a 3.8 kb DNA fragment in C57Bl (Sinc^{s7}) or DNA from VM (Sinc^{s7}) congenic mice. VM × VM (Sinc^{s7}) F$_1$ mice exhibited both 5.5 and 3.8 kb bands. However, the RIII and VL strains of mice (both Sinc^{s7}) showed 5.5 kb bands on Southern blots. Thus, the 5.5 kb Xba-1 fragment is not an accurate marker for Sinc$^{p/}$. Other PrP gene RFLPs have been detected. pEA974 hybridized to a 9.6 kb TaqI fragment from Sinc^{s7} mice (RIII, VM (Sinc^{s7}), C57Bl and VL) but to a 8.1 kb fragment in Sinc^{p7} mice (VM, IM, MB). Furthermore, HhaI, an enzyme that recognizes an unmethylated GCCC sequences, recognized a >30 kb fragment of VM mice (Sinc^{s7}) mice, a 20 kb fragment of Sinc^{p7} mice (VM, IM, MB) and 20 kb and 9.4 kb fragments with RIII and VL mice [520]. It should be stressed that the same restriction pattern was established regardless of whether or not the mice were inoculated with scrapie virus. In conclusion, strains of mice may be separated in three groups: a, Sinc^{p7} (Prn-p^b) with a 5.5 kb Xbal fragment, an 8.1 kb TaqI fragment and highly methylated HhaI sites (VM, IM and MB mice); b, Sinc^{s7} (Prn-p^a) with a 3.8 kb fragment, 9.6 kb TaqI fragment and highly methylated HhaI sites (VM (Sinc^{s7}), C57Bl mice) and c, Sinc^{s7} with a 5.5 kb Xbal I fragment, a 9.6 kb TaqI fragment and lower level of methylation [520].

The molecular cloning and sequencing of the Prn-p^a and Prn-p^b alleles revealed differences within coding sequences that opened up completely new approaches for slow neurodegenerative disorders [1042, 1045]. The ORF from NZW/LacJ mice extended from +1 to +762 nucleotides and differed from that of I/LnJ at two positions. At codon 108, the substitution of T for C resulted in amino acid change from Leu (NZW/LacJ) to Phe (I/LnJ), while at codon 189, an AC to GT substitution resulted in a change of Thr (NZW/LacJ) to Val (I/LnJ). The latter substitution removes a 5′-G GTNACC-3′ recognition sequence for the BstEII restriction enzyme. Thus, after BstII digestion and Southern blot analysis, the ORF of NZW/LacJ mice is cleaved to yield 5.7 and 3.5 kb fragments, while I/LnJ mice exhibit only one 15.5 kb fragment. This new RFLP, made it possible to probe mice associated with long and short incubation times by polymerase chain reaction and Southern blots.

Mice with short and intermediate incubation times (NZW, RIIIS/J, C57B1/6J, CAST/Ei, MOLF/Ei, Ma/MyJ) had codons 108 and 189 for Leu and Thr, respectively while those with longer incubation times (I/LnJ, P/J, IM, BDP/J[e]) had Phe and Val, respectively. From the comparison of both studies mentioned above, it is evident that Prn-p[a] corresponds to the Sinc[s7] and Prn-p[b] to the Sinc[p7], respectively.

Subsequent use of several restriction enzymes established further polymorphism within the PrP gene [169]. A restriction analysis using *Bam*HI, *Bst*EII, *Eco*RI, *Sac*I, *Taq*I and *Xba*I in 55 inbred strains of mice revealed six different restriction fragment length polymorphisms (RFLPs), which define haplotypes of the PrP gene [169]. These mice with b, c, d, e, and f haplotypes yield a 5.2 *Xba*I Prn-p fragment already used to define the b haplotype [171]. Analogously, *Bam*HI restriction site polymorphism detected by a probe against 5′ untranslated region yield Prn-p 10.9 kb fragment of haplotypes a, d, and e, or 12.2 Prn-p fragment of haplotypes b, c, and f. Furthermore, *Bst*II recognizes two polymorphic sites, one is located within an intron, another one identifies a substitution of Thr to Val at codon 189 [1042]. When different strains of mice where inoculated with the Chandler strain of scrapie virus, it was evident that all those with an incubation period longer then 200 days were of Prn-p[b] haplotype. Those of Prn-p[a] haplotype showed an incubation period in a range between 105 to 128 days, Prn-p[c] haplotype (RIIIS/J mice, previously classified as Prn-p[b] on a basis of *Xba*I RFLP, [171]) showed an incubation period of the same range and Prn-p[d] mice showed an incubation period of 170 days. However, the backcrossing experiments showed that, despite the polymorphism within the Prn-p gene, it has only two alleles.

Polymorphism of donor Prn-p gene influences the incubation period of recipient mice [172]. When the Chandler strain of scrapie virus isolated from Swiss (Prn-p[a]) mice and passaged through I/Ln (Prn-p[b]) was inoculated into I/Ln mice, the shortening of incubation time from approximately 300 days to 190 days was observed. Similar effects were observed when the Chandler strain was isolated from NZW (Prn-p[a]) or (NZW × I/Ln) (Prn-p[a]/Prn-p[b]) F$_1$ mice. In contrast, Prn-p[a] NZW mice inoculated with that isolate passaged through the Prn-p[a] mice did not show any changes of the incubation times while the inoculation with the isolate passaged through the Prn-p[b] I/Ln mice prolonged it. Furthermore, the shortening of incubation time following inoculation with the isolate passaged through the Prn-p syngeneic mice reduced drastically variation among individuals. If the original Chandler isolate is a strain mixture (and most investigators believe it is, analogously to the SSBP/1 [266]) than the observed phenomena may merely reflect host-permitted selection following crossing Prn-p-associated mouse strain barrier. The authors, however, prefered another interpretation suggesting that the observed variation is epigenetic and "reflects host-directed differences in amino acid sequences of PrP^{sc}" [172].

The physical map of the Prn-p gene has been also constructed using an association with the agouti (A), beta-2-microglobulin (B2m) and ITPase (*Itp*) loci. Using (NZW × I/LnJ)F$_1$ × NZW/LacJ backcross, the map distance of 14.6 cM was found between Prn-p and A locus. In (B6 × MA/MyJ)F$_1$ × B6 backcross (the genotype of B6 is B2m^{b}Prn-p^{a}a and that of MA/MyJ is B2m^{a}Prn-p^{d}A) the Prn-p was found 6.25 cM distal from B2m locus and 16.25 cM proximal from A locus. Furthermore, it was found that the map distance between Prn-p and *Itp* was 1.85 cM, between *Itp* and A 9.2 cM, and between Prn-p and A 1.11 cM. Thus the gene order on chromosome 2 was established as B2m-Prn-p-*Itp*-A. This region is homologous with a region on human chromosome 20 in which bot ITP an PRNP genes are located [169].

The direct proof that PrP expression controls scrapie incubation period came from studies using transgenic mice [947]. A hamster genomic cDNA library was made and screened to yield one clone (cosHaPrP) hybridizing to both exons I and II of the PrP gene. To generate transgenic mice, purified cosHaPrP inserts were microinjected into fertilized ova. This resulted in three founder animals – Tg 69, 71 and 81. Southern blot analysis of the genomic DNA from these animals yielded a 3.0 kb Eco RI fragment characteristic of hamster PrP gene. The genomes of Tg 69 and 71 harbored 4–8 copies of the transgene while Tg 81 harbored 30–50 copies. Northern blots showed the presence of 2.1 kb hamster PrP mRNA while immunoblots with monoclonals antibodies revealed HaPrP^c in mouse brains. After inoculation with the Sc237 strain of scrapie virus (analogous to the 263K strain of scrapie virus which is only pathogenic for mice after extremely long incubation period – [580, 582]) the incubation periods measured 70–75 days and were within the range of these in hamsters infected with the same (Sc237 or 263K) strain of scrapie virus. Non-transgenic control mice had not any signs of scrapie more than 200 days postinoculation. This study clearly indicates that species-specific PrP controls the incubation period of scrapie.

A similar approach has been attempted to clarify the relationship between Prn-p^a and Prn-p^b [1043]. The genomic cosmid clone (cos6.I/LnJ-4) derived from Prn-p^b mice and harboring approximately 6 kb of the 5′ and more than 15 kb of the 3′ flanking sequences of the Prn-p was microinjected into ferilized ova of Prn-p^a mice. Three transgenic lines (Tg93, 94 and 117) were successfully created. When inoculated with the "Chandler" (probably a mixture of strains) isolate of scrapie virus, instead of expected prolongation of incubation period, paradoxical shortening of incubation period (from approximately 125–137 days in non transgenic mice to 75–93 days in transgenic animals) was observed. This shortening correlates somehow with the amount of expressed PrP^c; Tg93L expressed only 2 to 4 fold more PrP^c than a non-transgenic control, and the incubation period in these mice is the longest (approximately 93 days). The paradoxical shortening of incubation period in Prn-p^b transgenic mice may explained by selection of a strain of scrapie with shortest incubation period in Prn-p^b mice. The absence of observable shortening of incubation period when this isolate was re-isolated in CD-1 (Prn-p^a) mice is of no explanatory power because the incubation period of any given strain is a relative value.

2.3.7.2. Prn-i and *Sip* gene in sheep

The linkage of the PrP gene and the incubation period gene in sheep (*Sip*) has been recently established [521] but the strong genetic background of natural and experimental scrapie has been known for many years and even used to support the hypothesis that scrapie is a genetic disease [837].

Only a few classic genetic experiments were performed in sheep to establish "susceptible" and "resistant" scrapie lines [265–266, 280, 282–285, 500, 816]. The experiment in Cheviot sheep produced, by selective breeding following subcutaneous inoculation with SSBP/1 (scrapie sheep brain pool/1, a line started by D.R. Wilson [1050] from a pool of 3 natural sheep scrapie subsequently passaged mostly through Cheviot sheep), two sheep lines which differed in scrapie incidence by 90% [266, 280]. Several back cross segregation experiments strongly suggested that a single gene, designated *Sip* controlls the incubation period in sheep. Similar experiments were performed in Swaledale [500], Herdwick [816], and Ile-de-France sheep [771]. The proof that the same gene (*Sip*) controls the susceptibility to both experimental and natural scrapie in different sheep breeds came from classic genetics experiments by Foster and Dickinson [340]. The incidence of scrapie in Suffolk × Cheviot sheep approached the predicted 50% following inoculation with either SSBP/1 or SUF81 (a homogenate of 5 brains of Suffolk sheep with natural scrapie). Analogous data were obtained for Cheviot × Herdwick sheep [340]. Thus, it is the *Sip* gene that controls susceptibility (by controlling the incubation period) for both natural and experimental scrapie. This conclusion is strongly supported by the fact that the outbreaks of natural scrapie were always confirmed to positive Cheviot and Herdwick sheep lines [265, 838, 839].

The *Sip* gene which is homologous to *Sinc* has two alleles: Sip^{psA} and Sip^{pA}. Sip^{sAsA} or Sip^{sApA} sheep exhibit short incubation times after

inoculation with SSBP/1 inoculum while those with the Sip^{pApA} genotype have extremely long incubation times under similar experimental
conditions and many sheep do not develop scrapie at all [266]. However,
50% of individuals from negative line may develop scrapie following
intracerebral inoculation [268]. Sheep from lines highly susceptible
(positive) to scrapie infection are homozygous for Sip^{sA} allele, or are
heterozygous while those from negative lines are *homozygous* for
Sip^{pApA} [266].

The *Sip* gene does not act uniformly, in terms of susceptibility to
infection and incubation periods, either in natural or experimental sheep
scrapie and that is similar to effects of the *Sinc* gene [282, 185]. Using
the strains of scrapie virus cloned from the SSBP/1, *Sip* acts with the full
dominance but the reverse is true concerning natural scrapie [266]. The
not transmissible to mice strain CH1641 is the best example of it [341].

The association between *Sip* hene and the Prn-i was studied by
molecular biology techniques [520]. When the pEA974 probe was used
in Southern blot experiments, sheep from scrapie negative Cheviot lines
had a 5.0 kb HindIII fragment and a 4.4 kb EcoRI fragment of the PrP
ORF (Fig. 17). Those from positive lines showed a 3.4 kb HindIII

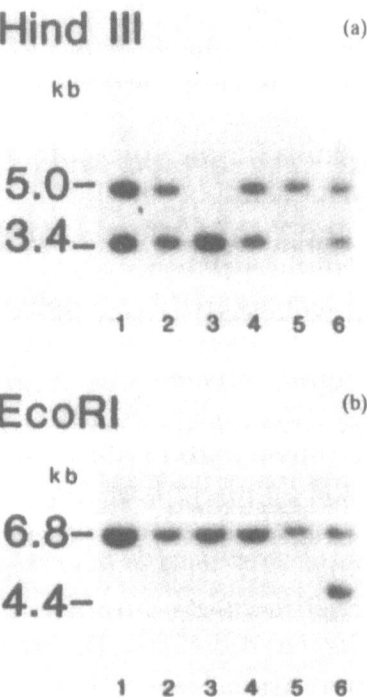

Fig. 17. Southern analysis of DNA restricted with (**a**) *Hind*III and (**b**) *Eco*RI. Lanes 1
to 4 were goats, lane 5 was cow, and lane 6 was a positive line of Cheviot sheep [521].
Courtesy of Dr. James Hope, MRC & AFRC Neuropathogenesis Unit, Edinburgh,
Scotland and the editor of Veterinary Record

fragment or both 5.0 and 3.4 kb HindIII fragments, and a 6.8 kb EcoRI fragment or both 6.8 kb and 3.4 kb EcoRI fragments. In contrast, Anglo-Nubian goats tested by Southern blots had either two HindIII fragments (5.0 kb and 3.4 kb) of PrP ORF and a 6.8 kb EcoRI fragment or a single 3.4 kb HindIII fragment and a 6.8 kb EcoRI fragment. These patterns are identical to those of sheep highly susceptible to scrapie and consistent with the observation that goats are uniformly susceptible to scrapie infection [266]. Furthermore, a similarly tested cow had a single 5.0 kb HindIII fragment and a 6.8 EcoRI fragment; these results, suggesting that cows might be equally susceptible to scrapie, is particularly interesting because of the recent epidemic of bovine spongiform encephalopathy.

2.3.8. The linkage of Gerstmann-Sträussler-Scheinker syndrome (GSSS) to the PrP (PRNP) gene

Gerstmann-Sträussler-Scheinker syndrome (GSSS) is a rare familial neurodegenerative disorder first described by Austrian investigators [407]. As stated in previous paragraphs, the nosological position of GSSD within the spectrum of CJD was established by successful transmission to nonhuman primates [729] and rodents [999, 1001–1002, 1004] with neuropathology indistinguishable from that of CJD. Furthermore, PrP-immunopositive plaques have been detected in GSSS [297, 605–609].

As GSSS is a transmissible disorder within the CJD spectrum but with a strong host genetic component, the PrP gene (PRNP) was the obvious gene in which to search for the any associated mutations [511–512]. First, both alleles of the PrP gene from a patient belonging to a US pedigree of GSSS were sequenced. The genomic library from peripheral blood leukocyte DNA was screened with human PrP cDNA [624]. Two PrP alleles were easily differentiated because of a Pvu II site present in only one of them [1061]. A codon 102 C to T mutation resulted in a Pro to Leu substitution. As this mutation also resulted in the creation of a Dde I restriction site, the Southern blots of the PCR-amplified the PrP ORF with subsequent restriction analysis was performed to screen DNA from other members of the affected families. The mutation of codon 102 was found in two affected members from the British pedigrees, but it was absent from 8 unaffected members. The Pro at position 102 is highly conserved among different species and is located within a surface loop of nine residues from the transmembrane alpha-helix [58]. The Pro to Leu substitution was subsequently found in two unrelated members of Japanese families [511–512], of large German family [140], in another

British family [223] and three members of the famous German Sch. family [441], one of the most studied GSSS family [126, 101–105, 934, 936, 941]. Recently, this mutation has been found in a living affected member of the family in which GSSS was first described [622]. The Pro to Leu substitution was found in 6 of 7 Japanese sporadic CJD cases with PrP-immunopositive kuru plaques [297]. The last finding is particularly important, as such over-representation of GSSS cases may explain, in part, the high proportion of Japanese CJD cases with PrP plaques [605]. In contrast, sporadic CJD cases without kuru plaques lacked this mutation [441]. Furthermore, the Pro to Leu substitution was detected in 4 unaffected members from GSSS families [297]. The interpretation of these results is unclear. As sporadic CJD cases with PrP-immunopositive plaques and Leu^{102} are regarded as GSSS cases, Leu^{102} may be a molecular marker for GSSS or indicating the susceptibility to the slow unconventional virus infection or even the cause of the disease. Interestingly, the Pro to Leu substitution of codon 102 would lead to profound changes in the secondary structure of PrP molecule [297]. It is conceivable that the differences in secondary and tertiary structure brought about by a Pro to Leu substitution are the postranslational modifications that converts PrP^{c} into PrP^{sc}. However, that hypothesis is not substantiated by the finding of unaffected GSSS family members bearing the Pro to Leu substitution. A close follow-up of these individuals may show whether Leu^{102} increases susceptibility to an infection with a CJD (GSS syndrome) virus.

Not all cases of GSS syndrome are linked to the mutation at codon 102. For instance, in a large Indiana kindred, none of the known mutation has been found [511, 992] (recently, however, a new mutation in the PrP gene: a substitution of codon 198 Phe with Ser [378] was reported in the Indiana GSSS family) despite the presence of multicentric plaques which are typically immunoreactive for PrP. Suprisingly, the GSSS cases from the Indiana family are also characterized by the presence of neurofibrillary tangles and neuropil threads [409] which, in turn, show immunoreactivity to the tau protein, a well known component of paired helical filaments of Alzheimer's disease [411]. Thus, the presence of Leu^{102} is not the only condition necessary for the development of GSS syndrome.

It was found recently, that the abnormal protein with Pro^{102} is indeed present within amyloid plaques of GSS syndrome [609]. A Kuru plaque core (KPC) fraction was purified from GSSS brain and resulted in two proteins, KPC-25 and KPC-30, which reacted with antiserum against synthetic peptide encompassing amino acids 138 to 152 of PrP. The N-terminal sequencing of KPC-25 and KPC-30 yielded several signals around the repetitive region of PrP (Gly-Gln-Pro-His-Gly-Gly-Gly-Trp).

Subsequent analysis with API, a protease specific for Lys-X, yielded 3 different peptides: P1(Thr-Asn-Met-Lys), P2(Leu-Ser-Lys) and P3(His-Met-Ala-Gly-Ala- -). The P2 peptide is homologous to that containing Pro[102].

The significance of the Pro to Leu substitution at codon 102 has been addressed by experiments using transgenic mice. Hsiao et al. [515] in Prusiner's laboratory constructed transgenic mice by microinjection of a mouse PrP gene with the Pro to Leu substitution at codon 101 (homologous to the codon 102 in humans). A region of approximately 100 nucleotides flanking the 101 codon was further modified to yield a sequence homologous to that of the hamster PrP gene, thus providing the means to screen the progeny DNA with a probe specific to that region. Three founders designated Tg(GSS MoPrP) were initially created, harbouring 64, 9 and 6 copies of the transgene, respectively. The Tg(GSS MoPrP)174 founder progeny remained healthy until 7 and 39 weeks of age when ataxia, rigidity and lethargy developed. At the time of writing the oldest animal developing signs of neurodegenerative disease was 272 days and the youngest was 57 days old. Neuropathologically, the brains of transgenic animals showed spongiform vacuoles accompanied by mild astrocytosis but not amyloid plaques (Fig. 18). Thus, the neuropathology of the transgenic animals was indistinguishable from that of naturally occurring and experimentally induced spongiform encephalopathies but not GSSS, because there were no typical plaques, particularly multicentric plaques. Interestingly, the amount of PrP detected by Western blots was negligible in comparison to than that encountered in scrapie-affected mice. The inoculation experiments have been performed to establish whether the PrP transgene induced not only spongiform changes but also infectivity; in other words whether PrP is the only component of the scrapie virus. It was reported recently from Prusiner laboratory [513] that the brain homogenate of founder Tg(GSS MoPrP) which developed signs of disease at 234 days of age, produced neurologic disease in 8 of 11 mice Tg(Prn-p^b) which express high level of mouse variant PrP, 234 days postinoculation and in 2 of 4 Tg(SHa PrP) which express high level of hamster PrP, 397 days postinoculation. Seven of 11 Golden syrian hamsters inoculated with 10% brain homogenate of Tg(GSS MoPrP) offspring showing signs of spontaneous neurological disease at 170 days of age, developed scrapie-like disease 221 days postinoculation. Furthermore, it was reported that Tg(GSS MoPrP) offspring transmitted disease to Tg(SHa PrP) after 190 days of incubation period. In contrast, Swiss CD-1 mice inoculated with either of homogenate remained disease-free at the time of this writing (approximately 400 days postinoculation). It is noteworthy, that hamsters and Tg(SHa PrP) inoculated with brain homogenate derived

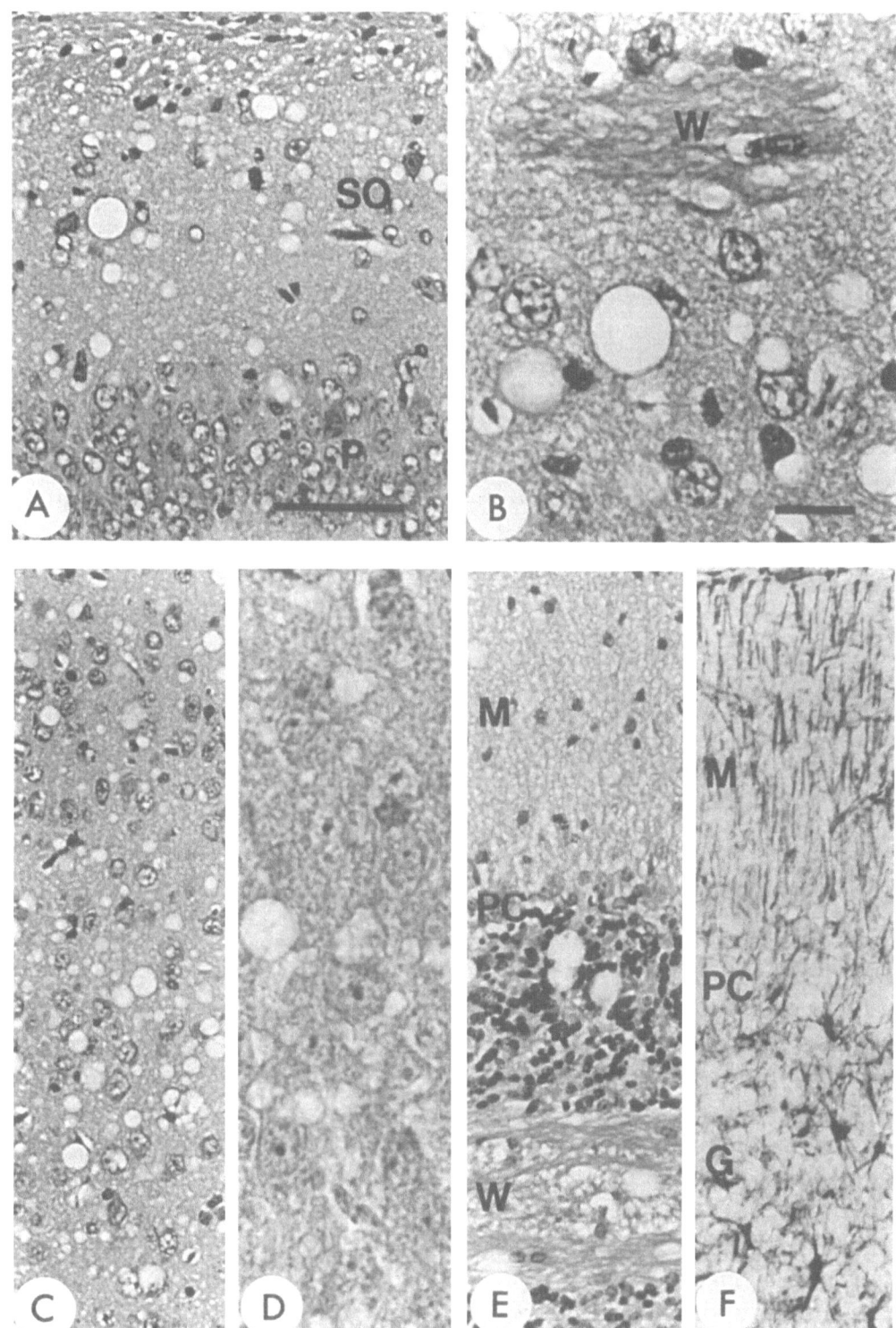

Fig. 18. Neuropathology of spontaneous neurodegeneration disease in a transgenic mouse with GSS syndrome mutation [515]. **A** The stratum oriens (*SO*) of the hippocampus. Note the pyramidal cells; **B** the gray and the white matter (*W*) of the putamen; **C** and **D** layers 3 and 4 of the dorsomedial frontal cortex; **E** the white matter (*W*), molecular (*M*) and Purkinje cell (*PC*) layer of the cerebellum. Note numerous spongiform vacuoles. Courtesy of Prof. Stanley B. Prusiner, University of California, San Francisco and the editor of Science

from Tg(GSS MoPrP) mice showed severe spongiform changes and presence of PrP^{sc}. In contrast, Tg(Prn-p^b) mice inoculated with the Tg(GSS MoPrP) founder showed subtle spongiform changes and no PrP^{sc}. The serial passage from hamsters inoculated with Tg(GSS MoPrP) offsprings and affected with the disease was established with an incubation period of approximately 75 days and a titer of infectivity of $>10^9$LD$_{50}$/g. In constrast, hamsters inoculated with Tg(SHaPrP), mice inoculated with brain homogenate of the Tg(GSS MoPrP) offspring, and Swiss CD-1 mice inoculated with brain homogenate from Tg(Prnp-p^b) that were inoculated with brain homogenate of the Tg(GSS MoPrP) founder remained disease-free. Thus it seems that PrP is necessary for the development of characteristic neuropathologic changes. However, the pathogenicity of the Tg(GSS MoPrP) isolate for hamsters but not for mice other than transgenic Tg(SHaPrP) mice is strikingly similar to analogous features of the 263K strain of scrapie virus used broadly in Prusiner's laboratory for more than a decade [580, 582]. As the contamination with this scrapie strain must be considered (such a possibility was mentioned in the original report [513], this experiment should be repeated in scrapie-free laboratory. If confirmed, PrP^{sc} is the only component of the infectious scrapie agent. This discovery would be of profound importance for molecular biology. If not, then another component responsible for the infectivity must exist and searched for.

2.3.9. Familial CJD

Before I shall discuss the recent advances in a molecular genetics of CJD, I will summarize data on familial CJD, forming a background, against which the molecular data should be analysed. As a few recent reviews and original papers are available [45, 136, 141–143, 410, 617, 730–731] only major points will be discussed here.

The familial form of CJD was recognized long before the era of slow virus disease [600] and the famous Backer family was the first to be discovered [757] . The proportion of familial CJD cases varies between published surveys. Masters et al. [730–732] found 218 familial of 1435 CJD cases refered to the Gajdusek's NIH laboratory; such a number constituted approximately 15% of a series [321]. However, data collected from series of consecutive (and thus unbiased) CJD cases from France [112, 132, 136–137, 141, 189], United Kingdom [1048], Finland [619, 478–479], Hungary [700] and Austria (Budka, personal communication) yielded a smaller percentage. Thus, Brown et al. [132, 136–137, 141] reported 19 (6%) of 329 consecutive CJD cases belonging to 6 affected families; furthermore, 19 additional CJD cases of affected families were not analysed in a cited study due to the time and space limits of its scope. Another 10 CJD cases were found in families in which sufficient data to prove CJD as a diagnosis of deceased family members were lacking. Analogously, Will et al. [1048] reported 6% of familial CJD cases in a recent survey in England and Wales, while Majteny [700] found 10% of familial CJD cases in Hungary (7 of 65 consecutive CJD cases). In contrast, a large number of 47% familial CJD cases was reported from Chile [182, 391, 393]. It is noteworthy, that a large proportion of familial CJD cases was observed among series of CJD cases of long duration [148]. Brown et al. [148] reported 3 to 4 times more familial CJD cases in such a series than among series of typical CJD cases (30% versus 6–7%, respectively). It should be remembered that these CJD cases of long duration are also characterized by other peculiarities, particularly a high proportion of cases with PrP-immuno-positive amyloid plaques [605].

Familial CJD seems to be slightly different from the sporadic type. In a series published by Masters et al. [730–731] the mean age of death (51 years) was significantly lower than that of spordic cases (58 years). In contrast, the duration of disease did not differ significantly between sporadic and familial CJD cases. Males and females were equally affected as well as cases from maternal and paternal lines [731]. The last finding is in striking contrast with that of natural scrapie where maternal transmission is a well known but yet to be explained phenomenon [266]. Anticipation, defined as a tendency for a decrease of the age of onset of disease among affected members of succeeding generations, a phenomenon well recognized in several hereditary disorders, was observed in the series of Masters et al. [730–731] and Kovanen and Haltia [619]. The minimal incubation period, calculated from data of spatial and temporal separation of familial CJD cases, ranged between one and four decades. When the putative incubation period was estimated on the basis of time lapsed between death of affected parent and the onset of CJD in the first affected progeny, it ranged between 4 and 40 years (mean 20 years). Furthermore, an analysis of affected siblings who tend to die at the same age but not at the same time suggested some form of vertical transmission. On the contrary, anticipation was found neither by Brown et al. [132, 136–137, 141], nor by Baron et al. [45] in a detailed study of 19 French CJD cases from 6 affected families. However, the same proportion of CJD cases of maternal and paternal lines, the younger age of onset and the incubation period in a range of 3 to 4 decades have been confirmed.

A detailed analysis of 19 familial French CJD cases has recently been reported [45]. It is noteworthy that 3 out of 6 CJD families were of

Sephardic Jewish origin, recently known to be associated with a high proportion of codon 200 mutation within the PrP gene (vide infra).

An analysis of age of death of affected siblings may help to differentiate between hypothetical modes of transmission: vertical, if affected siblings died at the same age but not at the same date or lateral, if they die at the same date but at different age. For example, Two brothers in a family Bel died at the age of 48 and 47 years in 1972 and 1977, respectively. Such a correlation may suggest some form of vertical transmission. Analogously, Baron et al. [45] reported case III-8 of a Sephardic Jewish family that was separated from the rest of the family for 42 years and died with CJD at "strikingly" the same age as her mother and brother. However, the reverse combination, suggesting a lateral transmission was also observed. Thus, as in natural scrapie in sheep [266], both modes of transmission may operate in familial CJD. CJD in persons married to members of the CJD families or conjugal CJD cases [537] may also substantiate the hypothesis of lateral transmission. For instance, case III-19 of a large Chilean CJD family [391, 393] was related by marriage and close contacts with other family members 13 years before the onset of CJD. A similar case was reported by Baron et al. [45]. Case III-2 of family P.A.D. was related by marriage to an unaffected family member but she has been raised among the affected family branch. She died with CJD at the same age as other affected family members. In summary, such an analysis does not provide a clear clue toward the elucidation of the mode of transmission in affected families.

2.3.10. The association of CJD cases of Eastern European origin and in Sephardic Jews with a mutation of the PrP gene at codon 200

The codon 102 Pro to Leu substitution linked to GSSS is not the only mutation encountered in naturally occurring spongiform encephalopathies. Goldfarb et al [435–436, 439–440] in Gajdusek's laboratory identified another mutation associated with some sporadic CJD cases of Eastern European origin. The codon 200 G to A mutation resulted in a Gly to Lys substitution (Fig. 19). The PrP gene region was amplified with PCR, subcloned in the pGEM5z plasmid and completely sequenced. Initially [435, 441], eleven patients were examined (3 familial CJD cases, 3 sporadic CJD cases, 3 GSSS, and 2 non-neurological controls). The codon 200 mutation creates a BsmA1 restriction site, and 6 additional familial cases, 30 sporadic CJD cases, 4 iatrogenic cases after human growth hormone therapy, 5 kuru cases, 3 GSSS cases and 121 controls were screened for this BsmA1 polymorphism. The codon 200 mutation was detected in 3 familial CJD cases (2 from family Ko. and 1 from

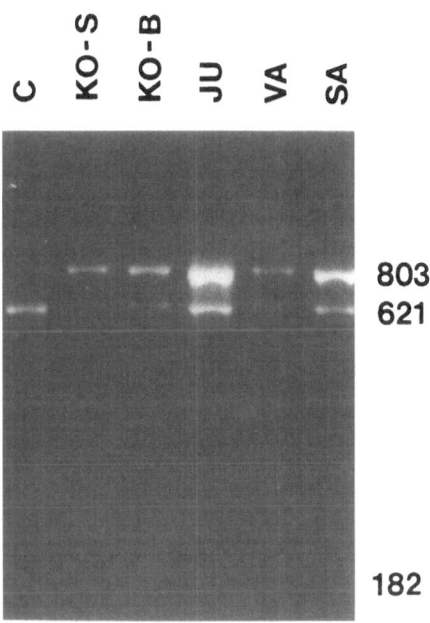

Fig. 19. PRNP coding region fragments after digestion with *Bsm*A1. C, control; KO-S, KO-B (sister and brother) and JU, CJD cases belonging to the KO and JU families, VA and SA, sporadic CJD cases. Note that control lane shows only 621 and 182 kb restriction fragments while CJD cases with the codon 200 mutation show 3 fragments (621 and 182 kb fragments from unaffected and 803 kb from affected alleles). Courtesy from Dr. L. Goldfarb, National Institutes of Health, Bethesda, USA and the editor of Lancet

family Ju.) and 2 sporadic CJD cases but not in the 3 other familial cases and the 28 sporadic CJD cases. Furthermore, this mutation was also detected in 11 CJD cases and 12 healthy first-degree relatives, and in 6 of 23 other relatives in a CJD clusters in Czechoslovakia (Orava and Lucenec regions) [436]. The Orava and Lucenec clusters ("Oravske kuru") are new phenomena, and consist of 22 CJD cases (3 transmitted to laboratory animals) from a population of 15 000. This occurrence of CJD exceeds by several hundred times the usual world wide incidence of roughly 1–2 cases per 1 mln per annum [132]. The ethnic origin of the CJD cases is of interest. The original siblings were of Polish origin, the third unrelated familial case was of Greek origin. One sporadic case was of Austrian origin and the remaining cases originated from Czechoslovakia.

Recently, 45 families characterized by the presence of the codon 200 mutation (Lys^{200}) were analysed [142, 436]. Lys^{200} was the most frequent mutation found in Gajdusek's laboratory (82% of 55 CJD-affected families). In Slovakia clusters ("Oravske kuru") 20 families positive for Lys^{200} (17/17 CJD cases, 23/68 first degree relatives and 1/80 healthy unrelated controls) were found. Furthermore, Lys^{200} was detected in 7

USA or Canadian families that emigrated to USA from Slovakia or other Eastern European countries and 6/6 CJD families (of 21 reported) from Chile. The calculated penetrance was 0.56 which may explain the presence of CJD-unaffected individuals with Lys^{200}.

Furthermore, the codon 200 mutation detected by Bsm A1 RFLP was found in 7 Sephardic Jews (2 sporadic and 2 familial CJD cases from CJD clusters among Libyan-born Sephardic Jews; 2 familial cases from Tunisia and Greece and 1 Greek sporadic CJD case) [440]. This original finding was recently extended by Golfarb et al. [436] who found the 200 codon mutation in 12 affected families and in 1/3 siblings and 13/21 children. In contrast, none of 23 unrelated Libyan Jews had the 200 codon mutation. Recently, Hsiao et al. [514] showed the 200 codon mutation in every one of 10 Libyan CJD cases but none of the CJD patients originated from Tunisia and Morocco. Furthermore, several healthy individuals from affected families at the risk for CJD, and all four children of one CJD patient from the family G were tested positive for the presence of the 200 codon mutation. The latter finding is of utmost interest further suggesting that the mutation itself is not sufficient for development of disease.

The phenomenon of CJD clustering among Libyan Jews have been known for several years [434, 808], but the clinical features of CJD cases of Libyan origin did not differ significantly from those of CJD cases of different origin. An extensive search for the putative iatrogenic factors responsible for the high frequency of CJD in this population has been published [434, 808]. Particularly, the high proportion of brain consumption (79 to 92% of CJD cases) was found but frequency of this phenomenon did not differ significantly from that encountered among control subjects [434]. It is noteworthy, that brains were consumed as stew (*m'chuma*), patty (*m'akod*) or slightly (5 minutes) grilled. Such a habbit differed from the analogous one observed among Tunisian Jews who cook or fry brains, however, the overall incidence of brain eating among the Tunisian Jews seemed to be even higher than that among Libyan Jews.

The familial clustering of CJD cases has been reported. Goldberg et al. [434] found two relatives among 14 Libyan CJD cases and 3 CJD were first cousins. In a subsequent study Neugut et al. [808] reported six related pairs among 20 CJD cases but in the remaining 13 CJD families CJD cases being the relatives of index cases were not found. Such a high proportion of familial cases clearly indicated the genetic susceptibility, and the finding of the mutation of codon 200 is a molecular confirmation of what has been suggested on a basis of classic genetics.

Thus, the 200 codon mutation appears to be a molecular marker for familial CJD cases of Eastern European and Sephardic Jews origin. But, as with the 102 GSSS mutation, the codon 200 mutation is not sufficient (approximately 50% penetrance) for the development of spongiform encephalopathy, because relatives with the same 200 codon mutation are clearly unaffected by CJD.

2.3.11. Other PrP gene mutations associated with sporadic and familial CJD cases

A codon 129 A to G polymorphism resulting in a Met to Val substitution was described by Owen et al. [827–828]. This mutation abolishes an NspI restriction site and creates a MaeII restriction site. A polymorphism has been detected in 18 out of 36 unaffected Caucasians [827–828]. The codon 129 mutation was found in both alleles of PrP gene, in 2 of 3 kuru cases, in 3 of 3 unrelated familial CJD cases, and in each of 2 iatrogenic CJD cases after growth hormone therapy. Furthermore, mutation at codon 129 in a single allele was detected in 3 of 15 sporadic CJD acses but also in 3 of 24 healthy controls. While the significance of codon 129 mutation was unknown at the time of its discovery, its seems that this polymorphism, while homozygous, predisposes to the development of CJD.

The first clue came from studies of iatrogenic CJD cases in recipients of human growth hormone (HGH) [133, 139]. Screening 7 British HGH-related CJD cases for all known mutations within PrP gene, Collinge et al. [225] found that 4 of 7 cases were homozygous for the codon 129 polymorphism. Additional 2 cases were heterozygous. In a subsequent study, Palmer et al. [835] reported that 21 of 21 sporadic (100% !) CJD cases and 19 of 23 suspected sporadic CJD cases were homozygous for the codon 129 polymorphism (16 were homozygous for Met^{129} while 5 were homozygous for Val^{129}). In contrast, more than 50% of the normal population were heterozygous for this polymorphism. If confirmed, the homozygocity in this codon may enhance susceptibility to infection with CJD virus, or may predispose to conformational changes of PrP^c to PrP^{Sc}, which is a crucial for scrapie pathogenesis.

Another polymorphism within the PrP gene was recently discovered in a population of North African immigrants to France [646]. This population is characterized by a high incidence of CJD. Using PCR amplification of the PrP gene, a one-allele deletion 20 bp long was detected in a 41-year-old Moroccan man and his two daughters. None of these individuals was affected with CJD at the time of this writing. This deletion was located between the NcoI site at nucleotide 277 and the PstI site at the nucleotide 397. The significance of this polymorhism, if any, is unknown.

2.3.12. A 0.15 kb insertion within the PrP gene observed with CJD and GSSS cases

An insertion of approximately 0.15 kb within the PrP gene was originally reported by Owen et al. [830–831] in affected members of the CJD family. Msp1-digested DNA was probed with a hamster PrP cDNA to reveal a single band of 0.75 or 0.9 kb associated with affected family members, while only the 0.75 kb band was detected in healthy unaffected individuals. The 0.9 kb band resulted from the 144 bp insertion in the ORF of the PrP gene. Furthermore, when this insertion was sought by PCR in 12 unrelated dementia cases it was detected in 2 cases of ill-defined dementia from the B. family. The B. family had not been diagnosed before as GSSS or CJD, but GSSS or atactic CJD is indicated by the presence of ataxia and dementia. The same large insertion was detected by PCR amplification and subsequent sequencing in 2 members of another family with undefined presenile dementia. Interestingly, a neuropathological examination was performed, but it was not compatible with GSSS or CJD. Thus, the mutation within the PrP gene may be associated with a clinically defined familial dementing illness without characteristic spongiform encephalopathy [224].

It was later found in 2 affected families that six long repeats in addition to 5 long repeats normally present within PRNP ORF [829–831] accounted for the 144 bp long insertion. PCR amplification of the entire ORF of the PRNP gene followed by size fractionation in 1% agarose gel resulted in a single 864 bp long DNA fragment in control samples and an additional 1000 bp long fragment in samples from CJD

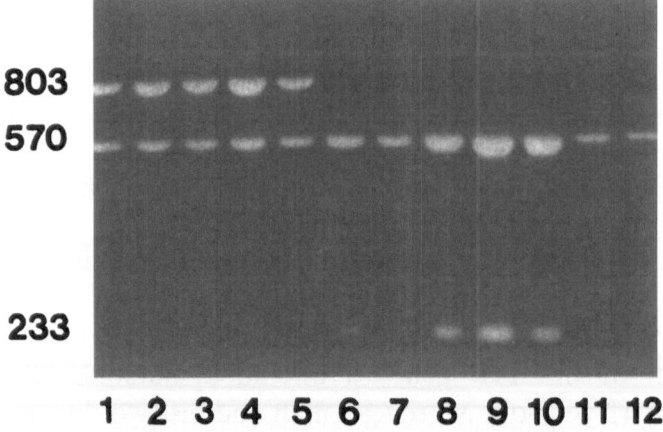

Fig. 20. PRNP coding region fragments after digestion with *Tth*111I. Lanes 1–5, CJD cases belonging to the family S; lanes 9–10 sporadic CJD cases; lanes 11–12, non-CJD controls. Note that control lanes show 570 and 233 bp lanes while familial lanes show in addition a 803 kb fragment from affected allele. Courtesy from Dr. L. Goldfarb, National Institutes of Health, Bethesda, USA and the editor of Lancet

cases. The normal fragment consisted of 5 repeats (R_1, R_2, R_2, R_3, R_4) between codons 51 and 91 while mutant fragment consisted of 11 repeats (R_1, R_2, R_2, R_2, R_3, R_2, R_3, R_2, R_2, R_3, R_4, R_3) in the same location. It is noteworthy, that these repeats are cleaved off during proteinase K digestion and that they are absent from truncated PrP 27–30. Thus, repeats are not directly related to infectivity. Furthermore, when 101 DNA samples of sporadic CJD, familial CJD, GSSS, Alzheimer's disease and other dementias were screened for mutations within the PRNP gene [829], 144 bp insertions was detected in 5 cases clinically diagnosed as CJD, Alzheimer's disease, Pick's disease, presenile dementia and Huntington's disease. Three additional insertions, similar to but not identical with 144 bp long insertion were recently reported in families of French and English origin [144]. It is noteworthy, that the mutant allele is amplified by PCR much less efficiently. When cloned it yields a large excess of "wild" allele. Thus, the proportion of CJD cases without insertions may be underestimated.

2.3.13. The codon 178 mutation in familial CJD

Goldfarb et al. [438, 812] found a mutation in codon 178 within the PrP gene in a large Finnish family with CJD [478–479] (Fig. 20). The pedigree includes 15 CJD-affected members over four generations. Following PCR amplification and sequencing, a G to A mutation at codon 178 was found resulting in an Asn to Asp substitution. The restriction enzyme Tth 111i normally cleaves the PCR product into two fragments of 570 and 233 bp, respectively. As the codon 178 mutation abolishes a cleavage site for Tth 111I, an 803 bp fragment was observed in affected individuals. A codon 178 mutation has been subsequently found by Nieto et al. [812] in familial transmissible CJD within American families of Dutch and Hungarian descent, and in a French family from Brittany. The 178 codon mutation has been identified in 1 CJD case from the Dutch-American family, 3 cases from the Hungarian-American family, 1 case from the French family, 2 out of 4 healthy first-degree relatives, but not in 51 sporadic CJD cases, 7 GSSS cases, 6 kuru cases, 6 Viliusk encephalitius cases, and 69 controls. It seems that 178 and 200 codon mutations are mutually exclusive as all CJD families so far tested exhibited either 178 or 200 codon mutations but never both.

2.4. The models of slow viruses

In previous paragraphs the molecular biology of slow unconventional virus infections was covered. In this chapter I will discuss current models of slow unconventional viruses against the background of existing data. It must be re-emphasized here, that scrapie, CJD, and related disorders should not be regarded simply as the interactions of biologically active macromolecules. Instead they are naturally occurring and experimentally induced diseases and thus, any model of the infectious virus must fit data on the virus-like behaviour of these slow unconventional agents and, even more important, it must explain the existence of different strains of scrapie virus.

2.4.1. Strains of scrapie virus

The most compelling evidence that the scrapie agent is a virus-like pathogen stems from the fact that different strains of scrapie virus exist. Strain of scrapie virus are defined in terms of their stable biological characteristics. When conditions of a given experiment are specified, the same strain can be isolated from different hosts, and the same host can be infected with different strains of scrapie virus. Furthermore, these characteristics may sometimes undergo changes to yield a new strain with new characteristics that are stable in subsequent passages. Such changes are consistent with mutations in a yet to be discovered disease-specific, nucleic-acid.

Approximately 20 strains of scrapie virus have been isolated from sheep and goats affected with clinical scrapie [267]. It is noteworthy that some isolates from sheep contain a mixture of strains (the best known example is the SSBP/1 source consisting of 22A, 22C and 22L) [266]. Not all strains of scrapie are transmissible to mice (the best example is the CH 1641 strain [341]) but all strains that are transmissible to mice can be divided into two groups: the ME7 group exhibits a short incubation period when passaged through $Sinc^{s7}$ mice (for example C57Bl) and a long incubation period when passaged through $Sinc^{p7}$ mice (for example VM mice); the 22A group exhibits exactly the opposite characteristics, a short incubation period in $Sinc^{p7}$ mice and a long incubation period in $Sinc^{s7}$ mice [267–268, 270, 281–285, 561–566, 572]. It seems that Sinc is the same gene as the Prn-i (p) gene. In other words PrP is the product of Sinc [178, 521, 572].

Passage through a species different from that used for the primary isolation (that is, crossing of the species barrier) is a useful method to isolate new strains. One of the best known examples of the isolation of a strain with completely different characteristics is the isolation of the 263K strain of scrapie virus [580, 582]. The passage history of this particular strain is very interesting and will be recapitulated here. In goats inoculated with a goat-passaged isolate of scrapie virus from sheep, two syndromes were observed [838] designated "nervous" and "itching" and later called "drowsy" and "scratching", respectively. Goats inoculated with inoculum prepared from the brains of "drowsy" goats developed the "drowsy" syndrome, and those inoculated with inocula prepared from brains of goats with "scratching" syndrome developed the "scratching" syndrome. Furthermore, the incubation period in mice inoculated with inocula from "drowsy" goats was shorter than that following inoculation with inocula "scratching" goats. Thus, the crude clinical studies of that time provided the first evidence of the existence of different strains of scrapie virus.

The "drowsy" goat strain isolated in mice was later passaged 5 times in rats. The clinical picture of mice inoculated with material from line passaged continuously in mice differed from that observed in mice inoculated with the re-isolate from rats. First, scratching was common in mice inoculated with the re-isolate, but was conspicuously absent from those inoculated with the material from continuous mouse passage. Second, the latter group of mice became obese and hypersensitive and had white matter vacuolation which was absent from those inoculated with rat isolate. The continuous rat passage line still retained its transmissibility to mice and after 12 passages in rats it was used to set up a passage line in hamsters. Between the 2nd and 4th passages in hamsters it was still possible to re-infect mice, but by the 6th passage the incubation period in hamsters became very rapid and transmissibility to mice was lost. These changes were due to the isolation of the mutant strain 263K, which was selected in hamsters because it had a shorter incubation period.

Other studies examined the properties of a single strain of scrapie virus that had been cloned in mice. For example, passage of the cloned 139A strain from mice to rats and back to mice did not change strain characteristics. In contrast, when the 139A strain was passaged in hamsters and then re-isolated in mice, the incubation period and the lesion profile were completely different from that of the original 139A strain [567]. This again suggests the emergence and selection of a mutant strain (139H/M) by passage in hamsters. A different mutant was obtained by passaging cloned 22C through hamsters and then back into mice [586]. But studies of two other strains of scrapie virus, that had been cloned in mice (ME7 and 22A), showed that passage through hamsters caused no permanent change in the properties of the re-isolates in mice which behave just like the original strains, ME7 and 22C. These latter studies are important in showing that scrapie virus has a genome which is independent of the host and can be preserved on passage between species. However, a mutant strain might be obtained if it has selective advantages (such as a short incubation period) in the new host.

Two sets of experiments suggest that scrapie virus must have an idependent genome, for which the orthodox candidate is obviously, a nucleic acid.

First, strains of scrapie virus undergo changes of certain characteristics, such as incubation period, lesion profile, presence and amount of amyloid deposits [155–160, 343, 345, 347, 348–351, 353–354, 358] which are compatible with mutations of more "conventional" pathogens. Three classes of strain stability have been established [154]. Class I stability strains (ME7, 22C) possess stable characteristics irrespective of the Sinc (s7 or p7) mouse genotype in which they are passaged. Class II strains (22A, 22F) possess stable characteristics if passaged through the Sinc genotype in which they were isolated, but change these characteristics gradually over several passages through a different Sinc mouse genotype. Class III strains (31A, 51C, 87A, 125A, 138A, 153A) exhibit sudden discontinuous changes of characteristics irrespective of the mouse genotype in which they are passaged. All six class III strains are characterized by similar incubation periods, large numbers of amyloid plaques, and high frequency of asymmetrical cerebral vacuolation. It is, thus, conceivable that all six class III isolates represent the same strain of scrapie virus.

"Class III breakdown" was defined as a "sudden shortening, in the course of single mouse passage [154], accompanied by a marked change in neuropathology". This occurred most frequently between the primary and the 7th passage and yield an isolate designated 7D. The 7D strain is characterized by a shorter incubation period, a more "generalized" lesion profile and an approximately 10-fold lower frequency of production of amyloid plaques, and all these characteristics are reminiscent of those of ME7. Thus it is highly probably that the 7D strain is the same as the ME7 strain of scrapie virus. In summary, these data show the selection of mutant strains of scrapie virus in the same host suggesting again that the genome of scrapie virus is host-independent.

It must be emphasized that the above-mentioned emergence of the new (ME7) strain of scrapie virus is independent of the host, but obviously the host genotype influences the selection of the new strain. It is thus erroneous to describe two strains as having "long" and "short" incubation periods, because these will depend on the Sinc genotype of the mice [1042]. Furthermore, the relative incubation periods of two given strains could be reversed on changing the mouse strain.

Second, different strains of scrapie can exhibit competition when inoculated at different times, either i.c. [273] or peripherally [272]. For example, when VM mice (Sinc^{p7}) are inoculated i.c. with the 22C (slow) strain a week before a second inoculation with the 22A (fast) strain, the mice were killed by the faster 22A strain, as shown by the short incubation period and the characteristic "lesion profile". In contrast, if the time lapse before the second inoculation is prolonged to 9 weeks, the incubation period of the 22A increases by 30 days because of the competition with the slow strain inoculated first. In another experiment, R III mice (Sinc^{s7}), inoculated i.p. with 22A (which now became a slow strain) followed by a second inoculation with the 22C (fast) strain 100 to 300 days later, do not develop disease caused by the 22C strain. The blocking effect of 22A was total such that the 22C strain failed to infect mice, and the mice died after the specified incubation period of 22A. The results are interpreted as two different strains competing for a limited number of multimeric "replication sites" – subunits of which are encoded by Sinc (or could even be PrP itself) [283–285].

In an independent experiment, Kimberlin and Walker [581] reported that, when Compton white mice were inoculated first with 22A (slow) at 10^{-2} dilution and then with 22C (fast) at 10^{-3} dilution, none of them were killed by the quicker 22C strain (in other words, 22A completely blocked 22C replication). However, when the 22A strain was inactivated by radiation, its blocking activity was reduced, and mice were killed by the 22C strain. Urea inactivation of the 22A strain of scrapie virus completely abolished the blocking activity. Thus, the "blocking" strain must be infectious to exert it effects.

The interactions between strains of scrapie virus and the *Sinc* gene are best exemplified by studies of *Sinc* congenic mice [162]. First, Sinc congenic mice were produced by serial backcrossing of Sinc^{s7p7} hetero-zygotes (C57Bl Sinc^{s7s7} × VM Sinc^{p7} mice) with VM Sinc^{p7} mice. At each of 18 backcrosses, Sinc^{p7} or Sinc^{s7p7} mice were mated with VM Sinc^{p7} mice and the progeny was inoculated with the ME7 strain of scrapie virus to assess the incubation period. The individual genotypes of the 20th generation were tested by additional matings with inbred Sinc^{s7} mice and inoculation of the progeny with the ME7 strain. When the two congenic VM mice strains were inoculated with the ME7, 22C, 22L, 79A, 22A, and 87V strains of scrapie virus, they developed clinical scrapie (except following inoculation with the 87V strain) after different incubation periods, and they were characterized by different lesion profiles. Furthermore, for each strain of scrapie virus the incubation period in the VM Sinc^{s7} mouse strain was similar to that in C57Bl (Sinc^{s7s7}). Moreover, the overall pattern of incubation periods in con-genic VM mice was the same as that observed in the non congenic counterparts. The incubation periods were shorter in Sinc^{s7} than in Sinc^{p7} mice following inoculation with the ME7, 22C, 22L, 79A, and 139A while the reverse was observed for 22A and 87V. This study provides an additional support for the theory that different strains of scrapie virus exist in hosts of the same *Sinc* genotype (strains wich "breed" true).

2.4.2. *The protein-only hypothesis*

The "prion" hypothesis formulated by S.B. Prusiner in 1982 [863] postulated that the scrapie agent is a *pro*teinaceous *in*fectious particle, mainly because infectivity was dependent on protein but resistant to methods known to inactivate nucleic acid. The idea, however, was not novel. A similar proposal was presented in 1967 by Griffith [457] who developed the suggestion of Alper et al. [17, 19] that scrapie virus may be devoid of disease-specific nucleic acid. Several investigators found that scrapie infectivity was sensitive to proteolytic digestion [209, 774]. As discussed in previous sections, scrapie amyloid protein, PrP 27–30, a proteolytic cleavage product of a precursor protein, PrP 33–35^{sc}, encoded by a host gene, is the only molecule discovered to date which co-purifies with infectivity. However, PrP 33–35^{sc} is not the only primary product of the cellular gene. PrP 33–35^c is a protein of identical amino acid sequence but different physico-chemical features. Particularly, PrP 33–35^c is completely degraded by limited proteolysis, while this is only partially the case with PrP 33–35^{sc}. Furthermore, following proteolytic treatment in the presence of detergents, PrP 33–33^{sc} yields a core protein PrP 27–30 which may be visualized by electron microscopy as scrapie-associated fibrils (SAF), also known as prion rods.

As discussed previously, point mutations in codons [117, 129, 178, 200] together with inserts [830–831] in the gene encoding PrP 33–35, are linked to sporadic [225, 827–828, 835] and familial [141–143, 224, 435–441, 513, 812] forms of CJD. Moreover, GSSS syndrome is associated with the codon 102 mutations [223, 297, 511–513, 515]. Furthermore, mice of short and long incubation periods following inoculation with the "Chandler" strain of scrapie virus exhibit polymorphism at two codons of the PrP gene [1042]. All these data together suggest that PrP and its gene play a pivotal role in scrapie infection, but the fact that susceptibility to scrapie infection is genetically controlled has been known for a long time [266, 837].

In terms of the protein only hypothesis, PrP^c must convert to PrP^{sc} to become infectious. Because both the primary amino acid sequences and all the known postranslational modifications of PrP^c and PrP^{sc} are the same, conformational changes have been invoked to explain the differences between these two isoforms but this conformational differences have not been found. There are other difficulties with this hypothesis.

Furthermore, following inoculation of a human isolate into mice, for instance, the protein purified from such experimentally infected animals is mouse PrP not the human PrP [98]. Thus, additional mechanisms must be found to account for the interaction of species-specific PrP^{sc} to yield a new species-specific PrP^{sc}. The possibility that the differences between PrP^c and PrP^{sc} are covalent has not been ruled out. Diringer et al. [294] reported that the different solublization behaviour of PrP^c and PrP^{sc} is not abolished by denaturation and these investigators concluded that the differences "can only be explained on the assumption of any type of covalent modification". However, the nature of these modifications is unknown.

Gabizon et al. [363, 365] reported that PrP and scrapie infectivity copartition into detergent-lipid-protein complexes and liposomes. Furthermore, scrapie infectivity and PrP^{sc} copurify following dispersion in liposomes and purification by affinity columns using monoclonal antibodies against PrP 27–30 [364]. While the latest finding is probably the strongest to substantiate the link between PrP and infectivity, the proper control to ascertain that no other molecule linked to PrP^{sc} is trapped inside these liposomes was not done. Such a control should use a "conventional" virus (for example a picornavirus) and it should examine the putative co-purification of it with PrP^{sc} following the formation of liposomes. Furthermore, the claim that anti-PrP antibodies neutralize scrapie infectivity is only too readily explained by aggregation.

Transgenic mice transfected with 4 to 50 copies of tandem-arrayed hamster PrP gene, become suceptible to infection with the 263K strain of hamster scrapie and produce hamster PrP not mouse PrP [947]. A hamster 2.1 kb mRNA was observed in these transgenic mice by Northern blot analysis, and modified hamster PrP was produced following inoculation with the Chandler strain of scrapie virus (uncloned 139A). However, mouse PrP was also produced by some transgenic mice. Furthermore, transgenic mice containing the PrP gene with the Pro to Leu GSS syndrome mutation at codon 101 developed spontaneous neurological disease indistinguishable from CJD (but not GSSS syndrome because of the absence of PrP multicentric plaques) [515]. These experiments, interpreted in the framework of the protein-only hypothesis, suggest that hamster PrP^{sc} readily converts hamster PrP^c but not heterologous mouse PrP^c. Thus, the so called species barrier may be dependent on this interaction between heterologous PrP species. However, transgenic mice with spontaneous neurological disorders apparently transmitted the disease only to Tg(Prn-p^b) and Tg(SHa PrP) and to syrian hamsters but not to non transgenic mice and they produce suprisingly negligible amount of PrP^{sc} [513].

While the protein only hypothesis "does not run counter to any physicochemical laws" [1038–1039] it does not account for existing data on the existence of scrapie strains. To explain the presence of at least 20 (perhaps more) different strains of scrapie virus, the same number of different PrP^{sc} conformational isoforms must exist. There must also be mechanisms for maintaining these different characteristics with a fidelity equivalent to transcription and translation. Such mechanism have yet to be discovered. Furthermore, the fact of strain competition and change of characteristics (class III breakdowns) must be explained by the interaction of PrP^{sc} isoforms and such interactions have not been detected. Thus, it seems that the protein only hypothesis still cannot explain all the existing data, but PrP doubtedly plays an important role in the pathogenesis of disease.

2.4.3. Virino and conventional virus hypothesis

While the "protein-only" hypothesis, however attractive, cannot explain the existence of different (at least not more than a few) strains of scrapie virus and their interactions with the host, both the virino and the conventional virus hypotheses offer simpler explanations for data which do not fit within the framework of the "prion" hypothesis. The virino hypothesis, first formulated by Dickinson and Outram in 1979 [283], was unfortunately linked to a finding, which was later shown to be irreproducible [703, 724]. This hypothesis suggests that the scrapie virus is a molecular chimera consisting of a host protein and disease-specific nucleic acid which is responsible for strain-specific characteristics. As modified PrP (PrP^{sc}) is host-encoded and there is no doubt that it is important for scrapie pathogenesis, this protein is a good candidate for the host shell protein of the virino hypothesis. The conventional virus hypothesis suggests that the scrapie virus is yet to be discovered and PrP^{sc} is an amyloid protein induced by the virus infection [122].

Manuelidis et al. [716] reported that 50% of PrP^{sc} (designated Gp34) is separated from the bulk of CJD infectivity by chromatography on a lectin (wheat germ agglutinin) column. The most important data, however, to support the notion that PrP may not be an integral part of scrapie virus stem from kinetics data. Braig and Diringer [122] reported peak incorporation of [^{3}H]leucine into SAF-enriched P$_E$ fraction. This [^{3}H]-labeled protein did not co-migrate with PrP on gels and preceeded the increasing infectivity titer and the PrP synthesis. One interpretation of such data is that scrapie virus infection initiates the conversion of PrP^c into PrP^{sc}; thus SSVE are "virus-induced amyloidoses". The kinetics studies of infectivity and SAF formation following peripheral and intra-cerebral inoculation revealed that bulk of SAF is formed after the increase in infectivity [245, 246]. These data indicate that SAF formation is a consequence of virus replication; in other words that scrapie is a virus-induced amyloidosis. Recently, Rubenstein et al. [922] reported similar relationship between SAF, PrP and infectivity in spleens from scrapie-infected hamsters and mice. As determined by negative-stain electron microscopy, SAF were first seen 7 weeks after i.c. inoculation, remained constant through the 12th week and increased rapidly with the onset of clinical disease, after which SAF plateaued. A similar curve was observed following i.p. inoculation. Regardless of the route of inoculation, the increase of SAF lagged behind that of infectivity. Furthermore, in i.c.-inoculated mice, spleen infectivity plateaued between 4 to 18 weeks postinoculation while at the same time SAF concentration increased 6 to 10 times. These data indicated that "a significant proportion of modified PrP generated during scrapie infection and is not associated with infectivity" because PrP^{sc} continues to increase after infectivity has reached a plateau.

Different kinetics of infectivity and modified PrP (PrP^{sc}) have been observed in experiments using amphotericin B, a drug which is effective against scrapie [857]. The incubation period of scrapie-infected hamsters that were treated with amphotericin B was significantly higher than that of untreated animals (78 and 57 days respectively). In these experiments, hamsters were injected with amphotericin B on the day of scrapie inoculation and then treatment continued for the next 50 days. In the untreated animals, both scrapie infectivity and the amount of PrP^{sc} increased exponentially at the same rate. By contrast, in amphotericin B-treated animals, scrapie infectivity increased in the same way as was found in untreated animals, but the increase in the amount of PrP^{sc} was delayed from 22 to 52 days postinoculation. These data suggest, that "PrP 27–30 is more likely a pathological bioproduct rather than a component of the infectious agent".

All these kinetics data indicate that, under several experimental conditions, PrP and infectivity may not be as tightly linked as the protein-only hypothesis may profess. The data obtained from the transgenic mice studies may be interpreted that, while PrP induces pathological changes in affected brains and is responsible for the species barrier effect, a small polynucleotide, probably viroid or subviroid in size [564] confers strain specificity but not necessarily infectivity [1039]. The conventional virus hypothesis suggests that scrapie virus is simply yet to be discovered, while the infection with this hypothetical virus is established, which initiates a conversion of scrapie precursor amyloid (PrP^c) into a final deposit (PrP^{sc}) [122, 911, 914]. The mutations discovered within the ORF of the PrP gene may predispose to such a conversion or they may be even essential for it. The conversion of amyloid precursor protein into deposited amyloid (beta-A4) protein, following mutation in the Dutch form of hereditary cerebral hemorrhage with amyloidosis, is a well-known precedence for such a process (see paragraph 5. – final conclusions).

In conclusion, after a decade since the initial discovery of PrP, the suggestion that PrP^{sc} is the only component of the scrapie virus is still debatable. The major obstacle to the universal acceptance of the protein-only (prion) hypothesis stems from the existence of strains of scrapie virus. Thus alternative hypotheses, the virino and the conventional virus hypothesis, were discussed. The final conclusion is, that despite the enormous progress in the molecular biology of scrapie, the complete structure of the scrapie virus and the pathogenesis of scrapie are still obscure.

3. The pathogenesis of slow virus infection

3.1. The general sequence of the pathogenetic events

In many respects, slow unconventional viruses behave as conventional viral pathogens. The spreading of these viruses to the central nervous system (CNS) and targeting within the CND are the most cited examples of such a "conventional" behaviour of the unconventional agents [559, 585, 596].

The role of the lymphoreticular system has been recognized almost from the beginning of scrapie research in experimentally infected rodents. Eklund, Kennedy and Hadlow, in one of the most cited experiments, reported the sequence of events following peripheral inoculation until the development of disease [305–308]. The experiment was performed in random bred Swiss mice inoculated subcutaneously (s.c.) with the $10^{5.7}$ intracerebral (i.c.) LD_{50} of the Chandler strain of scrapie virus (uncloned 139A strain; the original Chandler strain is probably a mixture of strains). A series of end-point titrations were performed at 1, 2, and 4 weeks postinoculation, and then at 4 weeks intervals up to 42 weeks. In the first weeks only the tissues of lymphoreticular system were infected. One week after inoculation, spleen contained $10^5 LD_{50}$ which was the majority of the $10^{5.7} LD_{50}$ inoculated i.c. into animal. That suggested that most of the original inoculum has been taken by cells in the spleen. Two weeks postinoculation none of the tissues contained any detectable titer of infectivity. However, by four weeks, scrapie infectivity was detected in the spleen and peripheral lymph nodes and maximum infectivity titers were at 8 weeks postinoculation; titers, $10^{7.2} LD_{50}$ and $10^{6.3} LD_{50}$, respectively. At the same time, the infectivity titer was first detected in the thymus and submaxillary salivary gland. Virus reached the central nervous system at week 12 postinoculation when an infectivity titer of $10^{1.4} LD_{50}$ was detected in the spinal cord. Thereafter the titer steadily increased to $10^{7.4} LD_{50}$ at the 42nd week. Infectivity was first detected in the brain at 16th week postinoculation and titer rose from $10^{4.4} LD_{50}$ to $10^{7.4} LD_{50}$. At this time, scrapie infectivity was detected also in the intestine (maximum titer, $10^{5.5}LD_{50}$), bone marrow (maximum titer, $10^{5.0}LD_{50}$), lungs (maximum titer, $10^{3.8}LD_{50}$) and kidney, liver and uterus (titer, $10^{1.0}LD_{50}$).

The results from this experiment was interpreted in the framework of virologic terms, that the virus spreads from the site of the original inoculum, probably via the blood, to eventually reach the final destination i.e. brain and spinal cord. The generalizations based on this single experiment in mice have been subsequently strongly criticized [822] by suggesting that different scrapie models (different combinations of the host and the strain of the agent) may produce different patterns of pathogenesis and even different a sequence of major events. However, data obtained for several other models confirmed that the overall pattern of scrapie pathogenesis was virtually the same for most of the scrapie models studied so far. A few exceptions from the general pattern will be discussed in following paragraphs.

Hadlow et al. [473–476] studied the pathogenesis of scrapie in goats infected with natural and experimental scrapie. The distribution of infectivity in different organs of goats with natural scrapie basically follows the pattern already known from experimental work in mice [305–306]. The highest titer ($10^{5.0–6.2}LD_{50}$) was observed in the midbrain, followed by diencephalon ($10^{4.4–5.7}LD_{50}$), medulla ($10^{4.6–5.8}LD_{50}$) and spinal cord ($10^{3.8–5.4}LD_{50}$). In the non-neural tissues, the titer in the spleen was $10^{2.8–3.2}LD_{50}$ while that in peripheral lymph nodes was $10^{2.2–4.0}LD_{50}$. A similar pattern of a distribution of infectivity was observed in Suffolk sheep with natural scrapie [475–476]. In sheep without overt clinical disease, the infectivity was first detected in two 3-month-old lambs in retropharyngeal, prescapular or mesenteric lymph nodes, tonsils and the spleen. The infectivity was widespread in the lymphoreticular tissues in lambs that were 10–14 months old. In one out of three 25-months-old sheep the infectivity was detected in the lymphoreticular system and, for the first time, in the central nervous system. Infectivity was not detected in the spinal cord, probably due to the sample error and the low titer, and the titer in the medulla and diencephalon was low; titers, $10^{2.3}LD_{50}$ and $10^{0.7}LD_{50}$, respectively. In sheep with clinical scrapie, the infectivity was widespread in the lymphoreticular system including the spleen, while the central nervous system, diencephalon, midbrain and medulla exhibited the highest and the cerebral cortex and the spinal cord presented the lowest titer.

The overall conclusion of these studies is that the spreading of the virus from the site of initial inoculation seem to be uniformly regular. The first replication takes place in the peripheral lymph nodes and the spleen from which the infectivity is transported into the central nervous system. The experimental studies clearly showed the modifications of this general pattern.

3.2. The role of the spleen

As described in paragraph 3.1 the scrapie virus replicates first in the lymphoreticular system, particularly in the spleen and the peripheral lymph nodes. The important role of the spleen was first elaborated by Fraser and Dickinson in 1970 [352]. The experiments were performed in C57Bl and BALB/c mice infected with the ME7 strain of scrapie virus and involved splenectomy performed before or after intracerebral or peripheral inoculation. Splenectomy performed before (7 to 58 days) or after (4 to 36 days) peripheral, but not intracerebral inoculation prolonged the incubation period by 12%. Such a prolongation pointed to the spleen as the site of the original replication of scrapie virus. However, the fact that splenectomy performed after inoculation prolonged the incubation period to the same extent as that performed before it suggested that only a proportion of the original inoculum was taken by the spleen and that other tissues, probably peripheral lymph nodes could take up the virus and support scrapie virus replication. The same effects of splenectomy on the incubation period, following peripheral (but not intracerebral) inoculation were also reported by Clarke and Haig [218].

Such a fundamental role of the lymphoreticular system in general and the spleen in particular has been confirmed by several subsequent experiments. Splenectomy regularly prolonged the incubation period following peripheral inoculations (except in neonatal mice) [355] with no evidence of replacement of splenic function in scrapie virus replication in VL and C57Bl mice. In contrast, BALB/c mice exhibited limited replacement after splenectomy was performed 21 days before inoculation. In an important series of studies, VL mice were splenectomized at birth or at 5, 10 or 15 days of age and inoculated intraperitoneally with the ME7 strain of scrapie virus from birth until 70 days of age. The results of this experiment showed that there was no replacement of splenic function, with regard to scrapie. However, mice splenectomized at birth showed smaller prolongation of incubation period following peripheral inoculation than those splenectomized at the later age (<10% and 10 to 25%, respectively).

The significant prolongation of incubation period was also produced when asplenic Dh/+ mice were inoculated peripherally with scrapie virus [269]. The increase of incubation period was compatible to that of wild-type mice (+/+) which were splenectomized surgically. The overall conclusion of these experiments was that splenectomy, either surgical or genetic, exerts a strong influence on the duration of incubation period following peripheral infection.

The identity of cell population(s) in the spleen which support the scrapie virus replication is unknown at the present time. Lavelle et al.

[651–652] separated lymphocyte subpopulations in albumin gradients and found that the highest specific infectivity (in a range of 2 to 6 LD_{50} per cell, [651]) was associated with fractions of lower densities. Furthermore, the infectivity was associated with lymphocytic but not macrophage fractions and the incorporation of tritiated thymidine was 10–20 fold increased in lymphocyte fractions between 6 and 9 weeks postinoculation [652]. In another study, CJD (the Fujisaki strain of CJD virus) infectivity was found associated with splenic macrophages and T and B lymphocytes [628]. The lower density lymphocytes (large lymphocytes and lymphoblasts) contained the highest titers and macrophages contained the lowest titer. Different spleen cell subpopulations exhibited also different susceptibilities for CJD virus infection *in vitro*. Macrophages and unstimulated T and B cells required 10^5 cells for 1 LD_{50} while stimulated T and B lymphocyte required less than 10^3 cells per 1 LD_{50}. The infection of lymphocytes with scrapie virus may be responsible of abnormal capping of spleen lymphocytes stained with FITC labelled anti-mouse Ig [1028–1029]. Interestingly, this abnormal capping coincided with a profound reduction of ATP in splenic lymphocytes and shortening of length of mitochondrial cristae as estimated by electron microscopy. The last finding is of potential significance in the light of recent data of an association of scrapie infectivity with mitochondria [9–10]. Analogously, the increased capping of Concanavalin A-labeled cells cultured from scrapie-infected brains has been reported [923].

Recently, Clarke and Kimberlin [220] directly addressed the problem of the distribution of scrapie infectivity between different cell populations in the spleen. The average titer of stromal cells (approximately $2 \times 10^4 LD_{50}$) was higher than that of pulp cells ($3 \times 10^3 LD_{50}$), and the majority of the infectivity associated with pulp was probably also released from stromal cells contaminating the pulp fractions. The actual stromal cell population which supports scrapie virus replication is yet to be discovered, but capillary endothelial cells, fixed macrophages or other elements of lymphoreticular system, for example follicular dendritic cells, may be involved. Clarke and Kimberlin [220] hypothesized that autonomic nerve terminals in the spleen may by those putative cellular elements which not only support scrapie virus replication but also provide a pathway for central nervous system invasion (neuroinvasion). That some stable (non proliferating) population of splenic cells maintains scrapie infection, is supported by the fact that total body irradiation did not produce any effect on scrapie infection [357].

There is evidence that susceptibility to infection with scrapie virus is developmentally regulated. When the ME7, 79A and 22A strains of scrapie virus were inoculated into neonatal C57Bl, VL or VM mice (depending on the particular virus/host combination), a proportion of cases had either very long or very short incubation periods [823]. Neonatal mice inoculated with a high titer of the ME7 inocula developed scrapie after shorter incubation time than that observed in weanling mice. These data were interpreted in terms of an absence of that putative cell population which is necessary to support scrapie virus replication. In contrast, when neonatal hamsters were inoculated with the 263K strain of scrapie virus, a significant reduction of incubation period was observed [751], and this effect was partially reversed by an administration of nerve growth factor (NGF) at birth. Furthermore, the neuropathology of the hamsters inoculated as neonates differed significantly from those inoculated as weanlings. In particular, both spongiform vacuolation and astrocytosis were more abundant, PrP amyloid plaques were not observed in hamsters infected as neonates, and an infiltration of gray matter with small mononuclear macrophages was found. The differences between mice and hamsters are difficult to explain. One possible explanation is that for the 263K strain of scrapie virus, which is highly neuroinvasive, replication in spleen is irrelevant for pathogenesis [583, 592], and experimental manipulation of these cells may be irrelevant to the pathogenesis too. In this respect, the hamster model may be similar to that of neonatal mice inoculated with a high titer scrapie inocula which develop scrapie after shorter incubation period [823].

3.3. The role of the spleen in neuroinvasion

As already reported in a classical paper by Eklund et al. [305–307], scrapie infectivity follows the same pattern of spreading: replication first takes place in the lymphoreticular system and the spleen and then spreads to the central nervous system (CNS). However, Eklund et al. [305–307] considered blood as the pathway of transportation of scrapie infectivity from the periphery to the CNS, and the latter claim has been subsequently challenged by a series of studies (*vide infra*).

Different routes of inoculation are characterized by different "efficiencies" measured by an estimation of operational titer by end point titration [585, 587]. The highest operational titer was obtained by the i.c. route, followed by intravenous (i.v.), i.p. and s.c. routes. The difference between i.c. and s.c. routes was approximately 25 000. Dose-response (dilution of the inoculum-incubation period) curves calculated for i.c. versus peripheral routes are very diverse with those for peripheral routes shifted to the right. However, when dose-response curves were adjusted to differences in route efficiencies (differences in effective LD_{50} in the inoculum), it appeared that peripheral routes virtually overlapped but differed from that i.c. route. The latter finding suggests that the overall pattern of scrapie pathogenesis is the same following peripheral inoculation, irrespective of the route used, but different from the i.c. route [585].

Following i.c. inoculation, much of the inoculum escapes and establishes an infection of the spleen [587]. However, infection of the spleen is irrelevant to pathogenesis which depends on the virus remaining in the brain. On the contrary, following peripheral inoculation with i.v., s.c., and i.p. routes, the infectivity was always found in the spleen before it was detected in the spinal cord and brain. The direct role of the spleen in neuroinvasion was studied by Kimberlin and Walker in CW mice infected with the 139A strain of scrapie virus (biologically cloned the "Chandler" strain of scrapie virus) [593]. Following i.p. inoculation, scrapie replication started in bone marrow, buffy coat cells, and peritoneal cells, followed by spleen (1st week), lymph nodes (3rd week), salivary glands (6th week), thoracic spinal cord (6th week), cervical cord (9th week) and lumbar cord (11th week). Thus, infectivity in the spleen always preceded that in the thoracic spinal cord and in the brain. While splenectomy had no effect on the efficiency of infection after i.p. and s.c. routes, which suggests that a proportion of the inoculum is taken up by other lymphoreticular cells, it prolonged incubation periods. In contrast, splenectomy reduced the efficient scrapie titer from $10^{6.5}LD_{50}$ to $10^{4.5}LD_{40}$, after i.v. inoculation. It means that the intravenous way which is the most efficient of the peripheral routes is highly dependent on the spleen. Furthermore, splenectomy delayed the onset of scrapie virus replication in the thoracic cord by 5 weeks and in cervical and lumbar cords by 2 and 4 weeks, respectively. However, when splenectomy was performed later than 5 to 6 weeks postinoculation, i.e. at the time when the infection in the thoracic spinal cord was already established, it had no effect on scrapie pathogenesis. Taken together, these data strongly suggest that there is a direct neuroinvasive pathway from the spleen to the thoracic spinal cord.

However, different scrapie models exhibit different roles of the spleen in neuroinvasion. For example, in hamsters infected with the 263K strain of scrapie virus, the infectivity spreads from the site of the inoculation to the spleen (1st week postinoculation), cervical lymph nodes (4 to 5th week), thoracic spinal cord (4th week) to reach the lumbar spinal cord and the brain at week 9 postinoculation. However, while this sequence of events is suggestive of the neuroinvasion from the spleen to thoracic spinal cord, the splenectomy had no effect on the incubation period. The last finding has recently been confirmed by showing that the plateau level of infectivity in the spleen did not preceed the increasing phase of infectivity in the brain [185]. Thus, despite being infected, the spleen is not an obligatory step in the pathogenesis of scrapie infection in this model. The possible explanation is that the infection of the thoracic spinal cord comes from sites of infection other than the spleen, such as autonomic nerves in the peritoneal cavity or prevertebral autonomic ganglia [585]. Analogously, splenectomy had no effect on the incubation period following intragastric inoculation [590]. In this model, the immediate uptake and replication of scrapie virus in the intestine lymphoid tissues was observed and it is conceivable that an autonomic innervation of Payer's patches provides a necessary neuro-invasive pathway from the periphery to the spinal cord.

Another well known exception is the 87V strain of scrapie virus passaged in the IM mice [152]. Following peripheral injection, no disease was observed 600 days postinoculation, in contrast to i.c. inoculation, which produced scrapie after an incubation period of 260 to 310 days [227]. Following i.c. infection, part of the inoculum was taken by the spleen and replicated there. A similar pattern was obtained following i.p. infection with a plateau in the spleen at the 16th week which sustained until the 80th week postinoculation. However, in contrast to i.c. inoculation, where virus replication was detected in brain between week 10 and 22 postinoculation, the peripheral route produced only negligible brain infection 50 to 80 weeks later. However, when a higher concentration of the inoculum was used, the infection eventually estab-lished in the brain [152]. This study indicated that the failure to establish a clinical disease following peripheral inoculation in this model is caused by a yet to be defined restriction of spread of scrapie virus from spleen to brain. Thus, the lymphoreticular system is capable to support replication of scrapie virus in the absence of overt clinical disease and such peri-pheral infection may contribute to the well known lateral spreading of scrapie infection. This phenomenon may be also important for spreading of natural CJD cases from already infected carriers who never develop clinical disease.

In contrast to the major effect of *Sinc* gene on the scrapie replication in CNS, the genetic control of scrapie replication in the spleen is very limited [583]. Replication curves (incubation period-time postinocu-lation) of the ME7 strain in the spleen of either IM ($Sinc^{p7}$, long incubation) or CW ($Sinc^{s7}$, short incubation) mice were virtually iden-tical despite marked differences in incubation periods. These curves showed a high titer already 1 week postinoculation (probably originating from primary inoculum) followed by a gradual decrease in incubation times (corresponding to the increase in infectivity titer) and then a plateau and a steady albeit small decrease in titers over the rest (10 to 25 weeks) of the incubation period. The last finding was interpreted as reflecting "gradual loss of infected cells from the non-replaceable populations in lymphoid tissue". Replication curves in cervical lymph nodes were also similar but the onset of replication in $Sinc^{p7}$ mice was delayed approximately 2 weeks in comparison to that in $Sinc^{p7}$ mice. In contrast, the replication in the brain was much delayed in $Sinc^{p7}$ mice as compared to $Sinc^{s7}$ mice (35 and 14 weeks postinoculation, respectively), and this difference reflected the ratio of lengths of the respective incubation periods. An analysis of additional scrapie models, including one uncharacterized isolate from sheep, revealed a reproducible pattern; a strong correlation between the onset of replication in the brain and the incubation period. It was also shown that a certain proportion of the incubation period must elapse (obligatory phase of replication in the lymphoreticular system) before the infectivit spreads to CNS. When the infection of the spleen or other lymphoreticular cells is already established, it initiates further spread of the virus to the CNS. The latter process is entirely neural and starts with infection of the peripheral nerve endings within the spleen or peripheral lymph nodes.

Phytohaemagglutinin (PHA), a cell mitogen known to increase the activity of lymphoreticular cells, enhances the susceptibility for scrapie infection when applied i.p. at the time of, or 3 hours before, scrapie inoculation [271]. However, the proliferative response of splenic lymphocytes to PHA or bacterial lipopolysaccharide is reduced 50 to 60% during scrapie infection [400]. Whether these two phenomena are related has yet to be established. A similar reduction of the incubation period was observed when the methanol extraction residue (MER) of BCG was applied 1–6 hours before i.p. inoculation with scrapie virus [569] and this effect was due to the enhancement of efficiency of scrapie infection. When a time lapse between MER and scrapie injections was prolonged to 1–5 days, the effect was not observed. Hamsters also exhibited a marked reduction of incubation period when MER was injected 2 hours before inoculation. It is noteworthy that the effect of MER on scrapie incubation is exactly the opposite to that observed following inoculation with conventional viruses. While MER enhances susceptibility for scrapie infection it protects against other virus infection [569]. Furthermore, the maximum effect of MER on conventional virus infection is seen when it is applied days or weeks before infection, while the effect on scrapie pathogenesis is seen when MER is applied 0.5–4 hours before. Sixteen other compounds have since been studied and 12 were found to reduce the incubation period when injected 2 hours before inoculation, but only when both the compound and scrapie were injected i.p. [594]. However, this pattern suggests only a non-specific enhancement of an unknown interaction of scrapie virus with peritoneal cells.

In contrast, a large dose of prednisone acetate [824] or arachis oil [825], well known immunosupressing agents, produce a reduction of susceptibility. The use of prednisone acetate several days before or after i.p. inoculation with scrapie markedly prolonged the incubation period and produced a proportion (approximately 20%) of survivors using a dose of the virus which would otherwise be completely lethal [824]. Arachis oil produced effects similar to those obtained with prednisone acetate. A similar effect has been noted for methisazone [578]. In conclusion, the unidentified cell populations of the lymphoreticular system seems to be involved in virus replication. However, the effects obtained with scrapie seem to be exactly opposite to those expected if conventional agents were involved.

Dextran sulphate 500 (DS 500) is another compound which increases the incubation period if injected 24, 6 or 2 hours before or 72 hours after i.p. or i.v. but not i.c. inoculation with scrapie virus [303–304]. DS 500 has a profound effect on the histology of the spleen and lymph nodes. Metachromatic material, consisting of DS 500, has been observed in the reticuolendothelial cells for 7 months following treatment with DS 500. Following i.p. inoculation of the 139A strain of scrapie virus into male STU mice, the infectivity titer in the spleen rose rapidly during the first 10 days to approximately $5 \times 10^5 LD_{50}$, then rose only tenfold for the next 20 days before stabilizing at the level of approximately $10^6 LD_{50}$ to $5 \times 10^6 LD_{50}$ [303–304]. In contrast, a single treatment with DS 500, reduced the titer of infectivity in the spleen by tenfold 20 to 40 days postinoculation, and the incubation period was increased by approximately 50 days. In the group receiving DS 500 three times a day, the titer of infectivity in the spleen was more than 3 log10 lower than that of untreated animals. DS 500 was most effective if administered before or on the day of the inoculation with scrapie virus. In contrast, when DS 500 was administered after inoculation, its effect rapidly declined. For example, the incubation period increased only 13 days following DS 500 treatment 10 days postinoculation. Conflicting data have been published by Farquhar and Dickinson [315]. While these investigators found DS 500 effective if injected before inoculation, they reported an effect also when given 2 weeks after scrapie inoculation with the 90% reduction of the infectivity titer. However, as these investigators did not measure the efficiency of infection, any extrapolations from the dose-response curves may be unjustified. The differences between both sets of experiments are not clear but the differences of different scrapie models are obvious.

The other compound, found to be effective in prolonging the incubation period following peripheral inoculation, is the heteropolyanion HPA-23. In a pilot study Kimberlin and Walker [578] reported a 20–38 days prolongation of the incubation period of mice inoculated i.p. with the 139A strain of scrapie virus by a i.p. injection of HPA-23. When scrapie virus was inoculated i.v., the estimated infectivity titer was reduced 250 fold [579]. The effect of HPA-23 was greatest when mice were inoculated at the time of drug administration, but an effect was still observed when HPA-23 was injected 2 to 4 weeks after inoculation [579]. When mice surviving the first inoculation of scrapie virus because of HPA-23 were re-inoculated i.v. or s.c. with the same scrapie inoculum, there was no longer any effect. Thus, the observed effect of HPA-23 was only temporary. These experiments were extended using different scrapie models in mice and in hamsters [594], thus suggesting that the effect of HPA-23 is an universal phenomenon in scrapie infection. The mechanisms of action of DS 500, HPA-23 and several other polyanions were summarized recently by Kimberlin and Walker [594]. HPA-23 given before or after scrapie inoculation reduced the effective scrapie titer of more than 2 log LD_{50}, but when HPA 23 was mixed with scrapie inoculum and injected s.c., only negligible effects were observed. The latter finding means that HPA-23 does not inactivate scrapie virus directly at the site of inoculation. Instead, when calculated dose-incubation period curves in control and HPA-23-treated mice were compared, it was evident that HPA-23 reduced the efficiency of infection, shifting dose-response curves to the right. A smilar effect of two other high-molecular weigth polyanions, carrageenan and DS 500, was also observed. A single dose of DS 500 administered six hours before scrapie inoculation reduced the effective scrapie titer by approximately 2 log LD_{50}. However, in contrast to HPA-23, when mice were injected i.v. with scrapie inoculum mixed with DS-500, the effective scrapie titer was reduced by approximately 3 log LD_{50}. Furthermore, the effects of DS 500 was still observed when this compound was administered 96 hours before or 8 hours after scrapie inoculation. The mode of action of different polyanions remains obscure. However, several lines of evidence suggest that they interact with yet to be identified phagocytic cells. DS 500 does not act as B-cell mitogen, as evidenced by the fact that DS 500 injected in the same time with PHA produced the same effect as DS 500 alone. It is possible, however, that DS 500 causes an aggregation of a proportion of initially circulating scrapie inoculum which is subsequently sequestrated or inactivated. The long lasting effect of DS 500 (up to 1 month [303–304, 315]) suggest that DS 500 is retained in the tissue, probably in phagocytic cells. Kimberlin and Walker [594] suggested that these cells may be splenic stromal cells already shown to have the highest titer of infectivity [220]. The latter conclusion was supported recently by another study of Diringer and Ehlers [294] who demonstrated a proportion of surviving animals when inoculated with scrapie after injection with another polysulphated polyanion, SP54 (M_r 3500–5000). The effective scrapie titer was reduced by 2 log LD_{50}, when SP54 was administered 2 months before scrapie inoculation. This effect of SP54 paralleled the absence of SAF in brains of animals which survived scrapie inoculation following treatment with SP54. Diringer and Ehlers [294] hypothesized that the effect of SP54 is mediated through yet to be defined interactions with phagocytic cells. However, in contrast to DS 500, SP54 was not demonstrated in splenic cells by staining with toluidine blue. Thus, if the latter finding is not caused by a technical effect, the first evidence of a host defence mechanism against slow viruses may be already detected.

3.4. The role of viremia

The putative role of viremia in spreading of infectivity from the site of peripheral inoculation (infection) to the central nervous system is still questionable and published reports provided conflicting data. In classical experiments of Eklund et al. [305–308] no infectivity could be found in the blood. In contrast, several investigators reported infectivity, however at low titers, either in the serum or in buffy coat cells, following peripheral or intracerebral inoculation. Clarke and Haig [219] reported infectivity in terminal stages (29–35 and 45–56 weeks) in sera of B.S.V.S. mice or Wistar rats infected intracerebrally with the Chandler strain of scrapie virus. Two additional infected rats killed 61 weeks postinoculation presented neuropathological changes of scrapie. Such a long incubation period suggested a low titer of the infectivity. In another experiment, intracerebrally inoculated mice were sacrificed 0.5, 1, 3, 6 and 18 hours postinoculation and the blood was collected from large blood vessels [327]. No infectivity was found half an hour after inoculation. When blood was taken one hour after inoculation, none of the inoculated animals developed scrapie, but neuropathological examination revealed changes consistent with it. A low titer of infectivity was found in serum 3 hours after inoculation as evidenced by the 30-weeks long incubation period. Although none of the animals inoculated with the blood taken 18 hours postinoculation developed scrapie, neuropathological examination revealed scrapie changes in the brain. This study clearly suggested that the peak of titer of infectivity in the blood 3 hours was reached postinoculation and decreased thereafter.

On the contrary viremia seems to be a sustaining phenomenon in guinea pigs experimentally inoculated with CJD virus [710]. Blood was taken from hearts of CJD virus-inoculated animals 1 to 28 weeks postinoculation at weekly intervals. In contrast to the experiments mentioned above, buffy coat instead of serum was used for inoculation. Infectivity was present in the blood at the 1st, 2nd, 3rd, 12th, 15th, 20th, 24th, 25th, and 26th week after inoculation. What had happened to the infectivity between the 13th and 15th week and between the 15th and 20th week postinoculation has yet to be discovered, and the problem of cross contamination must be seriously raised. It is interesting, that the animals inoculated with buffy coat cells showed more subtle clinical signs than those inoculated with brain homogenates. However, the end-point titration of brains infected with the buffy coat cells versus those infected with brain homogenates has not been done. Thus any conclusion concerning these differences in clinical characteristics cannot be made.

A more sensitive quantitative approach has been used by Diringer [292]. CLAC hamsters were infected i.p. with the 263K strain of scrapie virus, and blood was collected in the preclinical phase at 5, 10, 20, 30 and 40 days postinfection by cardiac puncture from the open thorax of these anaesthetized donor animals. 10–12 ml of blood were processed according to a method known to enrich infectivity [296, 496]. Probes of 50 μl representing about 2 ml of blood were assayed for infectivity levels by i.c.-infection of recipient hamsters. All recipients infected with blood samples obtained from donors 10–40 days post infection developed scrapie after incubation periods of 11–191 weeks. This corresponds to approximately 5–50 infectious units per ml of undilated donor blood. It was experimentally excluded that this constant low level of infectivity was derived from heart muscle tissue. These data on constant low levels of infectivity in blood were corroborated and extended into the clinical phase of disease by Casaccia et al. [185].

While all above-mentioned experiments suggest that viremia is sustained during the course of experimental scrapie and CJD, the most pertinent question is whether this viremic phase is significant for further spreading of the virus. Having used [99]technetium-labelled liposomes and compared their distribution after injection with that of scrapie infectivity, Millson et al. [775] reported that most of the peripherally inoculated liposomes were taken by the liver. Analogously, much of scrapie infectivity was found in the liver, but the titer here steadily decreased, suggesting that scrapie virus did not replicate there. Moreover, the infectivity in the blood was detected 5 to 30 minutes following i.c., i.v., and i.p., but not s.c. inoculation, suggesting that the infectivity from the blood is rapidly distributed to other extraneural tissues at every stage of scrapie infection. On the contrary, Diringer [292] and Casaccia et al. [185] showed steadily decreasing trend of the infectivity titer in the blood, and no peak before the infectivity started to mount in the spleen. Thus, it seems that while the infectivity may be detectable in the blood and establishes an early infection in the lymphoreticular system including the spleen, it is irrelevant for further pathogenetic steps. Kimberlin [585] hypothesized that infectivity in the blood may be responsible for the infection of yet to be defined cells (for example brain endothelial cells) which either do not support the replication further or do not infect neurons. In other words, infectivity in the blood is directed into "wrong" cells. In contrast, spreading via peripheral nerves (vide infra) directs infectivity into cells which support virus replication, namely neuronal cells.

3.5. The role of macrophages in scrapie infection

The distribution of liposomes labelled with ^{99}technetium clearly showed that, after i.p. and s.c. inoculation, a large amount of infectivity was found in the washings of the inoculation sites [775] suggesting uptake of the infectivity by macrophaes or other cells at the site of inoculation. However, limited data have been published so far on this subject. Carp and Callahan [174–176] developed a protocol to study interactions of peritoneal macrophages and scrapie virus. A homogenate of brain infected with the ME7 strain of scrapie virus was centrifuged and the supernatant was used to suspend peritoneal macrophages. Kidney cells served as controls. Samples were incubated for 2 hours, centrifuged, and both a supernatant and a pellet were inoculated into mice. The titer of infectivity was estimated from the length of incubation period. At the 10^{-2} and 10^{-3} dilution of brain homogenate the incubation period of mice inoculated with supernatants was significantly longer than that of mice inoculated with the pellet (washed macrophages). The findings suggest an interaction between peritoneal macrophages and scrapie virus probably via phagocytosis. In a subsequent experiment, a prolonged incubation of scrapie virus with peritoneal macrophages was used to determine whether a replication of virus within macrophages takes place [176]. The incubation period of mice inoculated with the mixture of scrapie virus and macrophages incubated for 4–5 days was always longer than that following incubation for 2 hours, suggesting that part of the infectivity was destroyed by macrophages. Furthermore, when mice were injected with thioglycolate (a compound known to increase the yield of peritoneal macrophages) 5 days before i.p. inoculation with scrapie virus, the incubation period of mice inoculated with a mixture of the virus and macrophages was further prolonged [175]. All above-mentioned data suggest that intraperitoneal macrophages actively inactivate the scrapie virus.

3.6. The neural spread of infectivity from the spleen to the central nervous system

Several experiments suggest that the pathway used by scrapie virus to enter the central nervous system from the spleen is neuronal, although the putative role of viraemia is still discussed. The dynamics of scrapie virus replication in different parts of the nervous system strongly support this hypothesis. In the first sets of experiments Compton white mice were inoculated with the Chandler (139A) strain of scrapie virus using i.c., i.p., or subcutaneous routes [588]. Animals were sacrificed at weekly intervals and different parts of the central nervous system were examined for scrapie infectivity using the incubation period assay and not the end point titration. Replication of the virus (measured by shortening of the incubation period in inoculated mice) started in the thoracic cord (removed between 4th or 5th and 11th thoracic vertebrae) 2 to 3 weeks earlier than in the brain or the lumbar cord. The infectivity titers in the thoracic cord were $<10^{2.5}LD_{50}$, $10^{2.9}LD_{50}$, $10^{3.6}LD_{50}$, $10^{3.7}LD_{50}$, while those in lumbar spinal cord were 10 to 100-fold lower – $<10^{1.5}LD_{50}$, $<10^{1.6}LD_{50}$, $<10^{1.9}LD_{50}$ and $10^{2.5}LD_{50}$ at 8, 9, 10, and 11 weeks, respectively. This consistently observed replication in the thoracic spinal cord which precedes that in the lumbar spinal cord and the brain may be explained by scrapie virus travelling along the sympathetic adrenergic nerves from the spleen to the CNS. In subsequent experiments, Kimberlin and Walker [589] studied the sequential pattern of the onset of replication of scrapie virus in different parts of the spinal cord and the brain in the same model. A clear pattern of virus spreading emerged. The virus started to replicate in the thoracic spinal cord between T9–T11. The onset of replication in other parts of the spinal cord and in the brain was significantly delayed and followed a distance from the site of inoculation to a given region. In the brain, the pattern of replication followed a posterior-anterior gradient, from the brain stem to the spinal cord. While such a distribution pattern of infectivity clearly suggested adrenergic sympathetic nerves innervating the spleen as a pathway used by scrapie virus to enter the central nervous system, attempts to delay or even block entrance by destruction of these nerves were completely negative.

The calculations of the speed with which scrapie infectivity spreads from the midthoracic spinal cord to the lumbar cord and to the different regions of the brain revealed uniform values in a range of 0.5–1 mm/day. These values are similar to those reported for slow axoplasmic transport and it is suggested that scrapie infectivity is transported by such a mechanism. Indeed, many neurotropic viruses such as herpes simples, rabies, pseudorabies, and borna viruses use axoplasmic transport to spread within the nervous system, but their speed of transport is much faster (50 to 250 mm/day) than that of scrapie virus [570].

It should be noted however, that scrapie infectivity is transported in both directions centripetally and centrifugally from the site of inoculation. The first, example, is the spreading of the infectivity from the midthoracic spinal cord, to the brain and down the lumbar spinal cord. To examine bi-directional spreading of scrapie infectivity, Kimberlin et al. [570] studied the pattern of sequential multiplication in brain, thoracic spinal cord (between T5 and T11 vertebrae), lumbar spinal cord (between L1 and L5 vertebrae) and the corresponding thoracic spinal nerves and thoracic and lumbar spinal dorsal root ganglia. As mentioned before, the replication in the thoracic spinal cord preceded that in the lumbar spinal cord. The onset of replication in the thoracic dorsal root ganglia was delayed by 13% in comparison with that in thoracic spinal cord and replication in the thoracic spinal nerves was even more delayed. These findings suggest centrifugal spread of infectivity. They also indicate that neuroinvasion from the spleen and visceral lymph nodes to midthoracic spinal cord does not involve the dorsal root ganglia or the spinal nerves. Instead, the postganglionic fibers to the coeliac ganglion and the preganglionic fibers to the sympathetic trunk were uggested as the likely pathways. Such a use of the peripheral autonomic pathway by unconventional viruses may explain the autonomic system involvement in cases of CJD [547].

To confirm that scrapie virus can travel along peripheral nerves to the CNS, an experiment were carried out using direct inoculation into the sciatic nerve [571]. Control animals were inoculated i.c. and also alongside the sciatic nerve. The incubation period for i.c. inoculated mice was approximately 150 days and all mice developed scrapie. In contrast, only 50–67% animals inoculated alongside the sciatic nerve developed scrapie, and the incubation period was longer (186–198 days). Mice inoculated into the sciatic nerve clearly separated into two populations. One group, exhibited incubation periods of 183 and 189 days (close to those inoculated alongside the sciatic nerves) and the second group, showed incubation periods of 161 to 177 days, similar to that of i.c. inoculated animals. Replication in the spinal cord after injection into the sciatic nerve occurred sooner than it would have done if neuroinvasion was from the spleen. This shows a direct pathway from the sciatic nerve to the spinal cord. It is noteworthy, that calculations of the speed of scrapie virus travelling along the sciatic nerve yielded a value of approximately 1.3 to 2 mm/day [570]. Thus, as the speed of infectivity travelling along peripheral and central neural pathways seems to be the same, the same mechanism of transport has been suggested for both neural compartments. Crushing of the sciatic nerve or the injection of lysoposphatidyl choline, a reagent known to produce paranodal myelin destruction, produced a proportion of early scrapie cases that exceeded by 50% the proportion following inoculation without damaging of the nerve. Thus it seems that part of the nerve, probably axolemma or connective tissue components, present a barrier for the virus. This barrier must be effectively bypassed to start the replication and spreading.

The low efficiency of infection of the sciatic nerve is shown by the fact that approximately 700 i.c. LD_{50} were inoculated into the nerve, but the effective dose was only $1-2$ LD_{50}. However, the incubation period of the $1-2$ LD_{50} injected into the nerve was actually shorter than the incubation period following a dose of $1-2$ LD_{50} inoculated i.c. ($190-200$ days). One possible interpretation is that the virus travelling along the sciatic nerve projects more directly to the so called "clinical target areas" than direct infection of the brain. This interpretation makes the prediction that injection of the spinal cord should give shorter incubation period than i.c. inoculation. This was confirmed by Kimberlin et al. [591], who found that the incubation period of the 139A strain of scrapie virus injected into the spinal cord was up to 30 days shorter than that following i.c. inoculation. Similar, shortenings of incubation periods were observed with intraspinal injection of CW mice with the ME7 strain of scrapie virus and with hamsters inoculated with 263K scrapie. Furthermore, the duration of replication of scrapie virus in the brain before clinical disease developed was shorter after intraspinal than after i.c. inoculation. These results confirmed that scrapie virus entering the brain from the spinal cord gains access more quickly to the "clinical target areas" than virus injected into the anterior brain. Attempts have been made to locate the "clinical target areas" by sterotactic inoculation into different brain regions [555–557]. Incubation period and neuropathology were assessed following inoculation into cerebral cortex, caudate nucleus, thalamus, substantia nigra and cerebellum. The shortest incubation period was obtained after inoculation into the cerebellar cortex and the longest one after inoculation into the cerebral cortex. However amyloid plaques were detected only at the injection site (cerebral cortex) which gave the longest incubation period. Amyloid plaques were absent in the cerebellar cortex probably because the short incubation period produced by injection at this site does not allow enough time for plaques to develop. It is interesting that vacuolation in this region was also numeral. This observation might suggest that the mechanism underlying the developing of the disease is more subtle. Alternatively, the authors suggest that the cerebellum itself is not an important target area but merely a relay station to other sites, where replication of the virus leads to clinical disease and death.

3.7. The neural spread of infectivity within the central nervous system

While spread of infectivity from the periphery to the central nervous system is still debatable, even if most data suggest sympathetic nerves as a pathway from the spleen to the midthoracic spinal cord, the spreading within the central nervous system seems to be entirely neural. In the pioneer experiment, Fraser directly addressed the question of spread of scrapie infectivity within the brain [342]. C3H mice were inoculated with the ME7 strain of scrapie virus into the eye and groups of mice were then sacrificed for neuropathology. In mice killed between 155 and 232 days postinoculation, asymetrical vacuolation was observed in the superior coliculus contralateral to the site of inoculation was observed. The proportion of cases with assymetrical vacuolation in the contralateral superior coliculus peaked at 161 days postinoculation and then decreased. The apparent absence of asymetrical vacuolation of the contralateral lateral geniculate bodies can be explained in terms of the lesion profile; the ME7 strain produces severe vacuolation in superior coliculi but not lateral geniculate body. This pilot observation was extended by Fraser and Dickinson [356]. In several experiments, intraocular inoculation of the ME7, 79A, 87A and 87V strains of scrapie virus into different strains of mice produced typical vacuolation in the superficial layers of contralateral superior coliculi. Injection into the right eye produces the first lesions in the left superior colliculus whereas inoculation of the left eye produces the lesions in the right superior colliculus. The severity of spongiform lesions in the contralateral superior colliculus reflected the lesion profile; it was the most prominent with the ME7 strain followed by the 79A and 87A strains. Furthermore, tectal lesions were delayed and appeared during the second half of incubation period with 79A while those of 87A were even more delayed, and survivors were oberved at the dose which would be otherwise completely lethal if injected i.c. It is noteworthy that the intraocular route was less efficient than the i.c. route and the incubation period was prolonged by 50 days using the ME7 strain. The effect of enucleation was studied by Scott and Fraser [944]. Eighty-eight percent of non-enucleated mice showed vacuolation in the contralateral superior colliculus 185 days postinoculation. In contrast, those enucleated 7 days post intraocular injection showed no lesions in the brain. Furthermore, only 20% of mice enucleated 14 days postinoculation showed vacuolar lesions in the contralateral superior colliculus, but this proportion rose to 80% when animals were enucleated 28 days postinoculation. In summary, enucleation performed 7 days postinoculation prevented a development of vacuolation in the superior colliculus and prolonged incubation period. These experimental results lend further support for the hypothesis of neuronal targeting of scrapie infection within the brain.

To examine whether the intraneural spreading is the universal feature of other slow virus infections, Liberski et al. [684] studied the Fujisaki strain of CJD inoculated intraocularly into NIH Swiss mice. The classical neuropathoogical techniques were suplemented by glial fibrillary acidic protein (GFAP)-immunohistochemistry. No neuropathological lesions were noted from 1 to 17 weeks postinoculation. At 18 weeks post-inoculation, typical vacuolation appeared asymetrically in the superior colliculus, particularly in the most superficial layers to which the retino-tectal fibers project and in the lateral geniculate body contralateral to the site of intraocular injection. Such lesions in the superior colliculus were also found in mice killed at 19, 22 and 27 weeks postinoculation. GFAP-positive, hypertrophic astrocytes were seen in the superior colliculus contralateral to the site of intraocular inoculation and accompanied the unilateral spongiform change.

Targeting of spongiform lesions and astrocytosis in the superior colliculus, particularly in the most superficial layers to which the retino-tectal fibers project, and in the lateral geniculate body, contralateral to the site of intraocular injection, strongly suggest that CJD virus, like scrapie virus, can spread first along the central visual pathways and then to other CNS regions. Whether this is accomplished via slow axonal transport, along the myelin sheath or via supporting glial cells is unknown.

3.8. Biochemistry and histochemistry of slow virus infections

The hypothesis that the clinical target areas in the brain are eventually responsible for the death of an affected host suggests yet to be discovered biochemical alterations responsible for different responses of different brain regions toward the infections with the scrapie virus. In the following chapters I will review data on such alterations and conclude that any disease-specific alteration has yet to be discovered.

3.8.1. Neurotransmitter alterations

3.8.1.1. Cholinergic systems

Studies on the cholinergic system in CJD and scrapie provided conflicting data. Early histochemical studies of frozen brain of a CJD patient showed a significant decrease of acetylcholinesterase (AchE) activity in cortical neurons and in the spinal cord, but not in the substantia nigra or in basal ganglia [360]. Furthermore, in C57Bl mice infected with the 139A strain of scrapie virus, AChE activity was reduced to 20% of that in control animals [524]. In contrast, choline acetyltransferase (CAT) activity in the cerebral cortex, hippocampus, basal ganglia, brainstem, thalamus, and crebellum did not show any measurable differences between scrapie-infected and sham-inoculated hamsters, while it was apparently reduced in several brain areas of scrapie-affected mice [744]. Analogously, no change in CAT activity was reported in C57Bl or CW mice infected with the 139A strain of scrapie virus [236, 360] or in the brain of patients with CJD [320]. However, the specific binding of [^{3}H]quinuclidynylbenzilate (QNB) to membranes extracted from hamsters infected with the 263K strain of scrapie virus was 25% lower than that of sham-inoculated animals. Both the "apparent" binding affinity (K_a) and the number of muscarinic receptors were affected. The unaltered level of CAT together with the alterations of muscarinic receptors sites suggest postsynaptic (dendritic) damage to the colinergic system. The other possibility is the increased synthesis or release of acetyl choline. Overall, these data indicate that damage to postsynaptic elements (dendrites) may be responsible for observed alterations in the cholinergic system.

Recently, several changes in the enzymes involved in the acetyl-choline metabolism were observed in PC12 neuroblastoma cells infected with scrapie virus, a system that proved to be a useful model for *in vitro* studies of scrapie [919]. CAT activity was decreased between the 2nd and 3rd week postinoculation until the 7th week postinoculation at which time it showed only 18% of the starting activity. AchE activity decreased even more rapidly; 60% of the enzyme activity were lost by the 2nd week postinoculation and additional 20% were lost between the 2nd and the 5th week. In contrast, enzymes involved in the mono-aminergic metabolism were unaltered. The biological significance of the reported alterations and the relationship to those encountered *in vivo* is unknown.

3.8.1.2. Monoaminergic systems

Several monoaminergic systems have been studied in different scrapie models. Norepinephrine level was reduced in rats infected with the mouse-passaged C506 but not with the rat-passaged 8745 (French sheep isolate) strain of scrapie virus [54–55]. Iy contrast, dopamine levels were reduced in both models of scrapie in the rat. Furthermore, levels of 3,4-dihydroxyphenylethyleneglycol (DHPG) and dihydroxyphenylacetic acid (DOPAC) were reduced in frontal and parietal cortex and in striatum, but not in the mesolimbic system of rats infected with the C506 strain of scrapie virus [55].

The serotoninergic system was studied in both CJD and scrapie. Nyberg et al. [817] found decreased levels of serotonin and increased levels of its metabolite, 5-hydroxyindoloacetic acid (5-HIAA), in the caudate, putamen, thalamus, mesencephalon, hypothalamus, brainstem and hippocampus of a CJD patient. Interestingly, such changes were not observed in the brain of patients with the so called "amyotrophic" type of CJD suggesting that the two syndromes are also biochemically different [927–928]. By contrast, the level of 5-hydroxyamine and 5-HIAA were unaltered in autopsy brain of another CJD patient [163]. In rats infected with the 8745 strain of scrapie virus, the serotonin level was decreased while that of 5-HIAA increased in cerebral cortex [54–55]. Furthermore, the concentration of serotonin was reduced in the frontal cortex, the hippocampus and in the mesolimbic system of rats infected with the C506 strain of scrapie virus [54–55]. It seems also, that the level was lowest in rats with the most advanced stages of scrapie.

A serotonin agonist, quipazine and its precursor, L-5-hydroxy-tryptophan (5-HTP) were used in attempts to ameliorate the signs of scrapie in hamsters infected with the 263K strain of scrapie virus [736]. 5-HTP, at a dose of 5 mg/kg and quipazine (1–2.5 mg/kg) produced small improvements in both ataxia and action jerks, which may be associated with disturbances of serotonin metabolism. Interestingly, higher doses of both drugs produced toxic effects of the activation of central serotoninergic neurons, the so called "serotonin behavioral syndrome". However, higher doses were necessary to elicit this syndrome in sham-inoculated than in scrapie-infected hamsters. The latter finding suggests that the hypersensitivity to serotonin is probably associated with extensive damage to brainstem structures in hamsters infected with the 263K strain of scrapie virus. The serotoninergic system was studied extensively by Pocchiari et al. [856]. While [^{3}H]imipramine binding (imipramine is a ligand which binds to presynaptic serotonin receptors) and [^{3}H]serotonin uptake and binding to class 1 serotonin receptors were unchanged in hamsters infected with the 263K strain of scrapie virus, the apparent affinity of binding to the serotonin receptor was lower in scrapie-infected than in sham-inoculated animals. In contrast, the concentration of serotonin class 2 receptors appeared to be decreased, while its affinity increased in scrapie infected animals. Thus, the two classes of serotonin receptors showed changes in opposite directions. Furthermore, the increase of 5-HIAA was reported in brain stem and spinal cord of CW mice infected with the 139A strain of scrapie virus [236]. By contrast, binding of [^{3}H]LSD to serotonin receptor was unchanged in the diencephalon while it was significantly reduced in the brain stem [236]. In summary, these data indicate an alteration in the postsynaptic elements rather than damage to presynaptic ones.

3.8.1.3. Histaminergic system

The histaminergic system was studied in hamsters inoculated with the 263K strain of scrapie virus by Nowak et al. [815]. The levels of histamine examined in cerebral cortex, hypothalamus, brainstem and "the rest of brain" (excluding cerebellum) showed no differences between scrapie-infected and sham-inoculated animals. By contrast, activity of L-histidine decarboxylase (HD), an enzyme located mainly in cytoplasm of nerve endings and responsible for the histamine synthesis, was decreased in the brains of scrapie-infected hamsters. The greatest decrease was observed in hypothalamus and "rest of the brain" while that seen in brainstem was not statistically significant. The observed changes in HD activity paralleled the topography of lesions in that particular model, except that the HD changes in brainstem were low while scrapie lesions in brainstem are prominent. However, histaminergic fibers originate mainly in the posterior hypothalamus, thus the absence of HD changes in brainstem may simply reflect the lack of histaminergic fibers there, while the remaining HD activity is associated with peri-vascular mast cells. Such dual localization of histamine in the brain may also explain the absence of any detectable changes of histamine level. Furthermore, the binding of a histamine class 1 receptor agonist, mepyramine, was unaltered in cerebellum of scrapie-infected mice [236]. In summary, it seems that the observed changes simply reflect wide-spread neuronal damage in scrapie-infected animals.

3.8.1.4. GABAergic system

Hamsters inoculated with the 263K strain of scrapie virus often exhibited aggressive behavior, thus the possible involvement of the GABA system was studied [855]. The injection of a single dose of 100 or 200 mg/kg of isonicotinic hydrazide (INH), a drug known to producing tonic-clonic seizures by an interaction with the GABA system, produced seizures at approximately 60 and 30 minutes, respectively, in control animals. In contrast, scrapie-infected hamsters developed seizures in half of the time. The onset of seizures following the injection of strychnine, a drug which interacts with the glycine receptor, was the same in scrapie-infected and sham-inoculated animals. However, the level of GABA in the cerebral cortex, cerebellar cortex and brainstem was unaltered in both scrapie-infected and sham-inoculated animals. Furthermore, no differences in the concentration of high affinity GABA receptors as examined by means of [^{3}H]binding, was found. In contrast, the level of cGMP but not cAMP was increased in the cerebellar cortex. Furthermore, the binding of [³H]-muscimol to GABA receptors was significantly increased in the cerebellum of scrapie-infected mice [236]. The altered reactivity to INH, together with an increased [³H]-muscimol binding, unaltered levels of GABA but increased level of cGAMP suggest that only a functional subpopulation of GABA (perhaps associated with axonal terminals) is affected.

3.8.2. Oxydative and lysosomal enzymes

Several mitochondrial and lysosomal enzymes were studied by the relatively crude biochemical and histochemical methods. In a pilot study, Millson [772] reported that the level of beta-glucuronidase, RNAse, and DNase were increased from 16 weeks postinoculation toward the terminal stage of disease, while the level of acid phosphatase was unchanged. In contrast, spleen enzymes were unaltered suggesting that these alterations were associated with changed brain metabolism (neuronal degenerations, astrocytic proliferation) rather than with virus replication. It was found subsequently that the level of acetyl-beta-D-glucosaminidase was increased, while that of beta-galactosidase was unaltered in scrapie-affected mouse brain. Furthermore, activation of lysosomal enzymes by mechanical distruction, repeated freezing and thawing, and heat shock released these enzymes more rapidly from extracts of scrapie-affected than from normal brains. A more complete biochemical study of lysosomal glycosidases showed that the levels of beta-glucuronidase, N-acetyl-beta-D-glucosaminidase and N-acetyl-beta-D-galactosaminidase were increased at the clinical stage of scrapie [773]. The levels of alpha-fucosidase and beta-xylosidase were increased at the terminal stages. Levels of alpha-mannosidase and beta-glucuronidase were increased only when assayed at pH 4.1 but not were assayed at more basic pH. The histochemical localization of N-acetyl-beta-D-glucosaminidase and beta-glucuronidase revealed that the increased activity of both enzymes paralleled the appearance of neuropathological lesions. The same set of enzymes was examined in a comparative study of scrapie and transmissible mink encephalopathy (TME) in a common host, namely Syrian hamster [574, 722]. Changes of alpha-fucosidase, N-acetyl-D-galactosidase, N-acetyl-D-glucosidase, beta-galactosidase and beta-xylosidase were small. In contrast, levels of beta-glucuronidase, and incorporation of [³H]fucose, [³H]N-acetyl-mannosamine and [¹⁴C]thymidine increased markedly during the course of both diseases. However, the marked differences between enzymes observed in scrapie- or TME-affected hamsters and scrapie-affected mice showed that they are a host-, not virus-specific. This conclusion is further supported by the finding that very similar changes in activities of several glycosidases, except for fucose incorporation, were reported in mice intoxicated with cuprizone [568, 575–576], a neurotoxin which produces brain lesions somewhat similar to those of scrapie [840]. Furthermore, different hosts show different patterns of normal activities of several glycosidases; those in mice and rats were similar to one another and different from those of Chinese and Syrian hamsters. When given hosts were infected with the Chandler strain of scrapie virus, the respective patterns of enzyme activities of mice and rats were similar, except for the absence of increased of activities of alpha-mannosidase and beta-xylosidase in rats. In contrast, the pattern of enzyme activities in scrapie-affected hamsters was characterized by lower levels except for beta-glucuronidase and beta-xylosidase which approached levels comparable to those encountered in scrapie-infected mice. Furthermore, when mice were infected with the hamster-passaged Chandler strain, the pattern of enzymes activities was similar to that found previously in mice. In conclusion, the pattern of glycosidase activities was a feature of the species of the host rather than the strain of the virus.

Only limited data on enzymatic alterations in CJD are available. Friede and DeJong reported decreased activity of DPN diaphorase and lactic dehydrogenase using histochemical methods on autopsy specimens of a familial case of CJD [360]. There was a correlation between the activity of both enzymes and a degree of tissue damage (particularly spongiform changes), and the neuropil was affected before the neuronal perikarya. Enzymatic alterations were greatest in the first, fifth, and sixth cortical layer. While spongiform vacuoles are mostly confined to deeper cortical layers, the involvement of the first cortical layer suggests that unspecific hypoxemic changes contributed to the reported enzymatic alterations [728]. Furthermore, the enzymatic alterations detected by histochemical staining preceeded the lesions demonstrated by classical neuropathology.

Different neuronal populations exhibited different decreases in enzymatic activities. Large neurons in the putamen, caudate nucleus, and deeper layers of the visual cortex were mostly unaltered. Those in frontal and parietal cortex showed variable degrees of enzyme reactions. In contrast, neurons in the insular and cingular cortices showed a uniform absence of enzymatic activity. In the spinal cord changes were widespread. The majority of anterior horn motor neurons and almost all neurons in the Clarke's column exhibited marked decreases of enzyme activities. In contrast to neurons, astrocytes showed markedly increased oxydative enzyme activity while oligodendroglia were unchanged. Robinson [910] studied NADH2 diaphorase, glucose 6-phosphatase, alpha-glycerophosphate – and lactate dehydrogenases, adenosinetriphospatase, acid- and alkaline phosphatases, and thiamine pyrophosphatase in frozen sections of CJD brain. In frontal cortex, particularly in the fifth cortical layer, the activities of the oxydative enzymes were markedly decreased. Acid phosphatase level was increased in several pyramidal neurons perikarya in the fifth cortical layer. Activities of ATPase, thiamine pyrophosphatase, and alkaline phosphatase were unaltered. A similar alteration of oxydative enzymes was noted in the substantia nigra, but only neurons showing shrinkage or displacement of nuclei had decreased activities of acid phosphatase and 5-nucleotidase. Federico and Annunziata [320] studied a panel of enzymes in brains of patients with typical CJD. Levels of N-acetyl-beta-glucosaminidase and N-acetyl-beta-galactosaminidase were increased in the white matter but were decreased in gray matter. The level of alpha-mannosidase was unaltered, and levels of beta-galactosidase and beta-glucuronidase were increased in both the white and gray matter. In subsequent studies, Annuziata and Federico [24] examined an extended panel of enzymes in the brains of 4 patients with CJD. In contrast to the previous study, levels of N-acetyl-beta-glucosaminidase and N-acetyl-beta-galactosaminidase were unaltered, while those of beta-galactosidase and beta-glucurolnidase were increased. Beta-glucosidase, alpha-fucosidase and alpha mannosidase were unaltered.

In summary, it seems that decreased oxydative enzyme activities are associated with widespread neuronal damage, and not with the replication of the virus. The increased level of some glycosidases is of potential interest, as it has been observed in both scrapie and CJD. It is possible that these enzymatic alterations are asociated with increased metabolism of several gangliosides or increased metabolism of astrocytes or microglial cells. It must be remembered, however, the PrP is a glycoprotein, and altered glycosidases activities may be associated with a postranslational modification of this protein.

3.8.2. Immunology of slow virus infections

One of the most characteristic features of the virus infections is the absence of any kind of immune response to the infectious virus [545, 745, 986]. However, there are a few immunologically mediated phenomena which are discussed in the following paragraphs.

3.8.2.1. Autoantibodies against neuroflament proteins

Using a previously described method to detect autoantibodies against cytoskeleton proteins of cultured neurons, Sotelo et al. [977–978] reported the presence of autoantibodies in sera of 13 out of 22 (59%) CJD patients, and 8 out of 28 (27%) kuru patients. Furthermore, antibodies were found in undiluted cerebrospinal fluid samples of the 13 CJD patients whose sera were positive for these autoantibodies. In a subsequent blind study, 45% of the CJD samples, 22% of the kuru samples, 13% of the samples from other chronic neurological disorders (Parkinson's disease with dementia, Alzheimer's disease, Guamanian parkinsonism-dementia, Pick's disease, subacute slerosing panencephalitis and brain lymphoma) and 10% of normal sera showed the presence of similar antibodies. The titer of antibodies was the highest in CJD samples (up to 1:1280), while the other positive sera had antibody titers in a range of 1:16 to 1:128. Subsequently, such autoantibodies were reported in the sera of 17 of 117 (14.5%) nonhuman primates infected with CJD virus and in sera of 9 of 71 (12.7%) primates infected with kuru virus. Control sera showed 4.6% immunopositivity [33–34]. It seems that autoantibodies appeared during the course of the disease. In CJD and kuru virus-infected chimpanzees, 7.5% showed autoantibodies prior to inoculation, while at the terminal stages, the proportion of positive sera increased to 29.3% and 16.7%, respectively. However, out of 12 chimpanzees with autoantibodies at terminal stages of disease, 6 already had these autoantibodies prior to inoculation, while 4 developed autoantibodies during the asymptomatic phase, and 2 during the clinical phase of disease. An analogous situation has been reported for kuru virus-infected chimpanzees. Interestingly, when the antibodies appeared, their titers remained at the same level during the subsequent course of disease. Furthermore, autoantibodies were detected in 3 of 15 CJD virus-infected Rhesus monkeys and 2 out of 15 CJD virus-infected African green monkeys, but none of kuru and scrapie virus-infected animals. This finding was independently confirmed by Tsukamoto et al. [1021]. In CJD virus-infected rodents, only 1 guinea pig showed autoantibodies. By contrast, 7 out of 20 (35%) sheep with natural scrapie exhibited autoantibodies.

The autoantibodies were directed against 10 nm neurofilaments as evidenced by an immunostaining pattern that was different from that obtained with antisera against actin, tubulin, and GFAP [977–978]. Furthermore, using an immunoblot technique, Bahmanyar et al. [33–34] showed that the reactivity of autoantibodies was mostly directed against the 200 kDa neurofilament protein. A weak immunoreactivity against 145 kDa and 70 kDa neurofilament proteins was also observed. Using immunohistochemistry on frozen material, it was shown that autoantibodies stained neuronal perikarya stronger than axons, which may suggest that they are directed against phosphorylated epitopes of neurofilament proteins. The biological significance of these autoantibodies is unknown at the present time. They are unspecific as evidenced by their presence, (albeit in lower proportions and much lower titers), in the sera of patients with other chronic neurological disorders. It is conceivable that once the central nervous system is damaged, neurofilament protein antigens are exposed and autoantibodies against neurofilament epitopes develop. Whatever the reason, autoantibodies are one of the very few immunological phenomena encountered in slow virus diseases. Whether they have any pathogenetic significance is unknown.

3.8.2.2. Increased concentration of IgG in serum of scrapie-affected sheep

Increased serum concentrations of IgG have been reported in Herdwick sheep affected with natural scrapie [229]. Thirty-three out of 41 sheep born in 1974 developed natural scrapie between 21 and 28 months of age. In these animals, the average IgG concentration was higher than in normal animals. A similar pattern was reported for sheep from 1975 and 1976 crops, but the concentration of IgG in any individual was highly variable, the patterns varied from a progressive increase in IgG concentration through the incubation period to a dramatic increase at the clinical phase of the disease. Overall, 65% of scrapie-affected Herdwick sheep showed anelevated IgG concentration. Subsequently, a higher concentration of IgG was reported in the sera of Herdwick sheep experimentally infected with the SSBP/1 (scrapie sheep brain pool, number 1) inoculum [226]. However, only a variable proportion of scrapie-affected animals developed elevated level of IgG; this proportion varied between 73 to 97%. Furthermore, it was demonstrated that although IgG1 is the predominant isotype in normal sheep sera, the IgG2 isotype predominates in scrapie-infected sheep and accounts for most of the increased concentration of IgG. An increased IgG concentration has also been found in IM mice infected with the 87V (plaque producing) strain of scrapie virus [228] but not in Suffolk sheep with natural scrapie [988].

The increase of IgG in IM mice was polyclonal, as evidenced by multiple bands revealed by means of isoelectric focusing. This change in IM mice infected with the 87V strain of scrapie virus is of particular interest because of the high number of cerebral PrP-immunopositive amyloid plaques in this model [228]. A detailed analysis showed that IgG_{2b} and IgG_3 were mostly responsible for the overall increase of IgG in this model. Furthermore, the rate of clearance of radiolabelled IgG from sera in IM mice was increased. However, while an elevated IgG level was observed toward the end of incubation period following i.c. inoculation, analogous changes were observed following i.p. inoculation. As the animals inoculated i.p. did not develop scrapie (but were infected), it appears that the elevated IgG level is rather not a cause nor a consequence of the clinical scrapie.

The biological significance of the increased IgG concentration is unknown, but it may be analogous to the increased presence of auto-antibodies in sera of patients with CJD and kuru and their experimental counterparts is nonhuman primates. As lymphocytic cuffs are negligible or absent in scrapie-affected brains [793], systemic B lymphocytes are a conceivable source of this IgG. Possibly, such a response is triggered by scrapie virus replicating in splenic lymphocytes, but there are no data concerning putative antigens against which increased IgG is directed, and no further work has been done on this subject.

4. Neuropathology of slow virus diseases

4.1. Natural scrapie

Scrapie in sheep and goats is a prototypic disease for the whole group of the subacute spongiform virus encephalopathies. Despite having been known for approximately 200 years [83–85, 86, 746, 844, 964], even before its pioneer transmission by Cuilee and Chelle [238–241], the literature of the neuropathology of the natural disease is suprisingly scanty [85, 263, 343–347, 840–841, 845, 1074–1078].

In contrast to experimental scrapie, the neuropathological picture of natural disease is heterogenous and thus, as Fraser [346] pointed out: "it may appear that basic differences exist between the relatively wide range of natural cases and the very narrow experimental spectrum of agents which have been used in sheep". Besnoit and Morrel were perhaps the first in 1898 to describe spongiform vacuoles as the neuropathological hallmark of scrapie [86].

Generally, neuropathology of natural scrapie is characterized by the same triad of "specific unspecific changes" (Gibbs, personal communication) as kuru, CJD and GSS syndrome (*vide infra*). Neuronal vacuolation in several nuclei of brain stem is the most prominent finding. Facial, cuneate, gracile, ambiguus, dorsal motor vagus, superior olivary, abducens and red nuclei, the reticular formation, the nucleus of the spinal tract of the trigeminal nerve, and Purkinje cells are most conspicuolsy affected [346]. Spongiform changes in the neuropil [734] are much less pronounced than in natural and experimental CJD and experimental scrapie.

Astrocytosis, initially detected by metal impregnation techniques and recently by GFAP-immunohistochemistry may be particularly severe. In those cases of scrapie characterized by ataxia, proliferation of Bergmann glia is conspicuous [61–63]. Indeed, scrapie was diagnosed in the past on the basis of severe astrocytosis in the brain stem but not vacuolation [472].

Neuronal vacuolation and gliosis are accompanied by several types of unspecific neuronal degenerations which lead to neuronal loss. Amyloid plaques are rarely detected (4 of 34 scrapie cases in Beck and Daniel series [63]. Real demyelination is not encountered but Wallerian degeneration characterized by moderate myelin breakdown is frequently seen [62–63, 344, 834].

4.2. Kuru

Kuru, the first human slow virus disease, discovered in 1957 by Gajdusek and Zigas [389–390; also, 368–369, 374–382, 387] and transmitted to subhuman primates by Gajdusek, Gibbs and Alpers [384, 388] is restricted to natives of the Fore linquistic group at Papua New Guinea Eastern Highlands. Since the cessation of ritualistic cannibalism responsible for the spreading of kuru, the number of cases has declined and only 2 or 3 cases are observed yearly at the present time.

Clinically kuru manifests as invariably fatal cerebellar ataxia accompanied by tremor, choreiform and athetoid movements, fecal and urine incontinence and, in some contrast to the neuropathological picture (*vide infra*), is remarkable uniform in clinical signs, symptoms and the course. Dementia, so characteristic to CJD, is not observed.

Pathological changes are confined to brain and spinal cord [60–66, 68, 612–613]. In any particular case different brain regions are involved with different intensity and no pattern of "systemic" degeneration can be singled out. However, cerebellum with a predilection for paleocerebellar structures, pontine nuclei, thalamus, and basal ganglia were predominantly involved.

The neuropathology of the first 12 cases of kuru were elaborated by Klatzo, Gajdusek and Zigas [612–613]. Neuronal changes observed in anterior horn cells of the spinal cord, different brain stem nuclei, cerebellum and cerebral cortex were mostly nonspecific in nature. Numerous neurons were shrunken, hyperchromatic or showed lack of Nissl substance resulting in a pale appearance. The amount of lipofuscin was frequently increased. Neuronal nuclei were pyknotic or karyorrheic. Central chromatolysis was frequently seen as well as intracytoplasmic vacuoles reminiscent of those encountered in natural scrapie in sheep and goats [832–833]. In the striatum several large neurons were vacuolated to a degree that produced a "moth-eaten" appearance of these cells. A few binucleated neurons were observed and rosettes of microglial neuronophagia were noted. In the pigmented nuclei of substantia nigra melanin was observed extracellularly and phagocytozed by microglial cells. The Purkinje cells showed balooning of their axons (torpedos) and these, in turn, appeared to be vacuolated.

Despite the fact that the polioencephalopathic nature of kuru was recognized in the earliest reports, degeneration of the myelin sheaths was noted. Well-defined myelin pallor was seen in the anterior and lateral colums of the spinal cord and counterstaining with Sudan IV revealed the presence of lipid-filled macrophages. The degeneration of myelinated fibers was also evident in brachia pontis and cerebellar white matter.

Astroglial proliferation was widespread but more intense in gray then in white matter. Interestingly, the degree of astrocytosis surpassed that of neuronal damage in several brain regions particularly in the white matter. In one case the formation of cysts similar to those encountered in some astrocytomas was noted.

The microglial proliferation was extremely widespread and numerous different forms of microglial cells including microglial rosettes were noted.

Perivascular cuffs composed of lymphocytes, plasma cells and, occasionally, mast cells were observed, mostly in Virchow-Robin spaces and, similarly to scrapie [793], the significance of this findings is unknown.

The most characteristic for kuru is the presence of amyloid plaques (currently known as *kuru plaques*). In the first report 50% of cases presented kuru plaques [612–613] and the number increased to approximately 75% of cases in subsequent studies [61, 63]. Kuru plaques measured 20 to 60 nm in diameter and were found in cerebellum, basal ganglia, thalamus and cerebral cortex in this order of frequency. In cerebellum they are most frequent in the granular layer but they are also seen in the molecular layer and the white matter. Plaques consist of a darkly stained central core surrounded by a delicate fibrillar halo merging with surrounding neuropil. Coronas typical to the senile plaques of senile dementia of Alzheimer type are not seen and dystrophic neurites are only occasionally observed. Kuru plaques are weakly argentophilic, PAS positive, metachromatic, congophilic and birefringent under polarized light. It is known from recent immunohistochemical studies that kuru plaques are composed by prion protein [852; Liberski, unpublished observations]. Interestingly, the spongiform changes regarded as the most characteristic for the whole group of disorders was not observed in the early kuru reports. In a subsequent report of 4 cases the spongiform changes ("*small* spongy spaces") were described for the first time [809] along with changes reported by Klatzo et al. [612–613]. In a few subsequent studies [948, 958] spongiform changes were easily detected, particularly in the limbic and paralimbic areas. As in CJD and scrapie, the intensity of vacuolation varied from case to case and between different brain regions. In cerebral cortex, spongiform changes were pseudolaminar and involved predominantly deep cortical layers but extended also to the superficial layers. In this respect, vacuolation in kuru is reminiscent of that CJD.

4.3. Creutzfeldt-Jakob disease (CJD)

4.3.1. Introduction

CJD is a slow neurodegenerative disease occurring world-wide in a sporadic or a familial manner with an annual mortality averaging 1 per million [599–601, 709, 728–732, 738, 915, 1019–1020]. Its nosological position has evolved over seven decades since the first description of this condition by Jakob [529–530] and Creutzfeldt [233], recently available in English translation [234, 528], and since subsequent historical contributions [129, 250–251, 493, 527, 531, 540, 599, 756, 985, 1026]. It seems that reports of similar cases preceded original publications of Creutzfeldt and Jakob [302, 334]. In 1968 Gibbs Jr, Gajdusek and their coworkers transmitted CJD to chimpanzees [418], thus establishing its position within the *subacute spongiform virus encephalopathies* even if the virus itself has never been completely characterized. It is not my goal to recapitulate the history of the evolution of the current concept of CJD [728] but rather to overview the general neuropathology of the disease emphasizing phenomena which elaborated on later in this review.

4.3.2. Classifications

Several classifications of CJD have been proposed, based on symptomatology and the topography of lesions. Daniel [247] subdivided the CJD spectrum into: 1, Jakob type (cortico-striato-spinal type) characterized mostly by severe dementia with pyramidal signs and symptoms; 2, Heidenhain type [490] with predominant involvement of visual cortex resulting in cortical blindness [120, 507, 632, 934, 963], 3; diffuse type, encompassing cortical and subcortical gray matter in a widespread manner, 4; ataxic form in which the cerebellum is mostly affected. Siedler and Malamud [962] and Malamud [701] singled out cortical, cortico-striatal, cortico-striato-cerebellar, cortico-spinal and cortico-nigral forms. It is clear that in contrast to kuru, remarkably uniform in its symptomatology and neuropathology, CJD, like natural scrapie, encompasses a plethora of signs and symptoms resulting from diverse topography of lesions and produced more than 80 synonyms for CJD in the literature [422]. Suffice is to say that detailed classification of CJD is, in my mind at least, not justifiable. Rather, the neuropathologic phenomena in CJD should be listed, remembering that in any particular case any particular phenomenon may be encountered in different regions and with different intensity.

4.3.3. Classical CJD

The classical triad of CJD neuropathology consists of vacuolation (spongiosis), neuronal degeneration resulting in neuronal drop out, and astrocytosis as reported in numerous studies [8, 11, 23, 73, 114, 179, 188, 232, 291, 312, 317–319, 360, 395–397, 621, 695, 965]. The changes are bilaterally symmetrical but may be focal and, occasionally, even unilateral [1068]. A correlation between neuropathological changes and computer assisted tomography [81, 933], as well as magnetic resonance [502] was reported.

Most characteristic and even pathognomic for CJD is the presence of spongiform changes [734] which remain well preserved even in exhumed cases [1012]. Spongiform changes consist of small round or oval vacuoles within the neuropil (ground substance of earlier investigators). Characteristically, vacuoles are often confluent and clustered to form "morula-like" (Budka, personal communication) structures. Deeper cortical layeres are involved to higher degree than superficial ones [728]; the vacuolation of superficial layers was regarded mostly, if not exclusively, as artefactual. Indeed, in transmission experiments no case characterized by the presence of vacuolation in only superficial cortical layers has transmitted the disease [728]. It was reported that spongiform changes were most prominent in cases of shorther duration. In contrast, in cases of longer duration, spongiform changes may be masked by disappearance of neurons and by proliferating astrocytes. In such cases examination of the most severely damaged areas may make the diagnosis of CJD difficult [728]. The presence or absence of spongiform changes produced considerable confusion before the transmission era, particularly when a spurious distinction was made between subacute spongiform encephalopathy [810–811] and CJD. The eight cases of Nevin et al. [810–811] were all characterized by the presence of spongiform changes not easily noted in earlier reports based on classical Nissl stainings. However, when the original cases of Jakob were counterstained with eosin by Masters and Gajdusek [728], spongiform changes were easily detected in 2 out of 5 Jakob's cases and few others subsequently published from Jakob's laboratory (vide infra). The same conclusions were drawn by Siedler and Malamud [962] and Malamud [701] based on the examination of 34 cases, all but one of which were characterized by the presence of spongiform changes. Recently, the author reviewed a series of 34 CJD cases collected at the Neurological Institute of Vienna University, Vienna, Austria [402, 535, 537; Budka and Liberski, manuscript in preparation] and in all of these cases spongiform changes were seen, although at different degree. Thus it seems that spongiform changes are the most disease-specific neuropathological changes but it must be remembered that such changes might be entirely absent similar to natural and experimental scrapie and transmissible mink encephalopathy [726].

The other type of vacuolation is the spongiform status [728] consisting of larger cavities of irregular shape within the neuropil between dense meshwork of proliferating astrocytes. It has been reported that status spongiosus is more characteristic for cases of longer duration and, in contrast to spongiform changes, is not CJD-specific [734].

Numerous changes in neurons have been reported in CJD. As all these changes are *bona fide* non specific, they are present with different intensity in almost all cases of CJD resulting in neuronal depopulation. Neuronal changes are most evident in deeper cortical layers but may be also present in the superficial ones and, obviously, in any affected structure. Noteworthy, the nucleus basalis of Meynert supplying the bulk of cholinergic fibers to cerebral cortex and involved in the pathogenesis of Alzheimer's disease is also haevily damaged in CJD [27, 183] and in GSSS syndrome [410]. In contrast, hippocampus is only rarely involved; when present – stratum moleculare-lacunosum is mostly affected [777–778]. Neurons exhibit swelling, with pale granular nuclei without discernible nucleoli and irregular or broken nuclear membranes and disintegrated cell body [810–811]. Other neurons, particularly small cells of the outer cortical layers are shrunken and were darkly stained. "Ghost-like forms of degenerating cells" were reported [810–811]. Neurons of the pigmented nuclei of the brain stem are mostly well preserved except that they contain abundant lipofuscin. In addition, neurons show chromatolytic changes and the presence of the *Pick's bodies* as evidenced by silver staining [61, 402]. The dendrites of Purkinje cells show several abnormalities revealed by metal impregnations [1011], including oblique or horizontal malorientation and extensive ramification of thick secondary branches (*antler* or *staghorn* patterns). Recently, the accumulation of phosphorylated epitopes of 68 kDa and 160 kDa neurofilament proteins in swollen neurons in CJD has been reported [796].

Astrocytosis is the third part of the classic neuropathological triad pathognomonic for CJD. Hypertrophic astrocytes detected by means of metal techniques (Holzer or Cajal's) or more recently by immunostaining against glial fibrillary acid protein (GFAP) are seen in all affected areas. In cerebral cortex they are particularly prominent in deeper cortical layers. Gemistocytic forms are frequently observed. When destruction is so severe to lead to the collapse of vacuolated neuropil, proliferating astrocytes may virtually replace all other cellular elements. In such a situation the spongiform changes may no longer be recognizable. In cerebellum the proliferation of Bergman glia is frequently observed.

Lymphocytic cuffings similar to kuru were exceptionally reported [758]. The significance of this finding, if any, is unknown.

As the above mentioned description fits well all CJD subtypes, several of them need additional comments.

4.3.4. Ataxic form of CJD

Neuropathologically, CJD is similar to kuru as pointed out in the original report on kuru pathology by Klatzo et al. [612–613]. An anectodal report diagnosed CJD in a visitor to Papua New Guinea [451]. The ataxic form of CJD was delineated by Brownell and Oppenheimer [151] and further documented in a few subsequent reports [118, 443a, 535, 918, 959]. Clinically, this form is characterized by cerebellar signs and syptoms dominated over dementia and other symptoms typical for CJD. In contrast to kuru, however, not only the paleocerebellar structures are involved but the destructive process encompasses the whole cerebellum. Granule cells are mostly affected, and in all regions smaller, shrunken, degenerating neurons are intermingled with normal looking granule cells. Purkinje cells are rather well preserved but swellings and torpedos are occasionally reported. The astroglial proliferation is conspicuous; in several regions (hippocampus) a dense fibrillary gliosis out of proportion to neuronal loss is seen. Abundant microglial cells have been reported. However, the presence of cerebellar pathology is not restricted to the ataxic (cerebellar) form of CJD [535]. In a series of 17 consecutive CJD cases involvement of cerebellum was reported in 12 cases. It consisted of a loss of granule cells with preservation of Purkinje cells, astrocytosis and spongiform changes, mostly in molecular layer and in the white matter. Torpedos and "empty baskets" were seen. It is thus evident that cerebellar pathology is much more common than previously recognized and only when it dominates the picture, clinically and neuropathologically, the diagnosis of ataxic form of CJD is justifiable.

The real bridge between kuru and CJD, however, is the presence of "kuru plaques". Kuru plaques, if present, are a hallmark of all subacute spongiform encephalopathies which were recently designated as the *transmissible cerebral amyloidoses* [385] because of this hallmark. They have been seen in approximately 75% of kuru cases but in a much lower percentage of CJD cases. The only exception are Japanese CJD cases up to 50% of which are characterized by widely distributed and numerous kuru plaques [605–608, 779–780, 1004, 1006, 1064–1065]. In CJD, kuru plaques are virtually indistinguishable from those found in kuru [5, 213, 488, 498–499, 510, 626]. As in kuru they are encountered mostly in cerebellum but have been also detected in cerebral cortex, subcortical nuclei and cerebellar white matter [848]. Kuru plaques are weakly argeentophilic, PAS positive, congophilic and show immunoreactivity for PrP [605–608]. In addition, exceedingly rare cases of CJD show congophilic angiopathy [546] but the molecular characteristic of the amyloid is yet to be established.

PrP positive kuru plaques are not the only type of plaques found in CJD. The other type, only rarely seen, are amyloid beta protein (A4) immunoreactive senile plaques, otherwise typical for aged brains and senile dementia of Alzheimer type. They have been reported in a very small number of CJD cases [463, 498, 674–676]. Typically, these plaques consist of a dense amyloid core surrounded by a corona, dystrophic neurites and glial cells; however, many plaques devoid of definite cores are also observed [498]. Similar plaques are detected in an otherwise typical case of CJD by Liberski et al. [674–676] and these plaques were also A4 positive [463]. Furthermore, Budka and Liberski (unpublished data) found several cases of definite CJD (suprisingly one of this was "unilateral" CJD [1068]) with A4 positive senile plaques in a series of 34 consecutive CJD cases collected from the files of the Neurological Institute of Vienna University. A similar observation has recently been made by Barcikowska and Liberski (unpublished observation). It seems that CJD cases with A4 positive senile plaques form a subset of the spectrum of the disease, a "gray" area between CJD and senile dementia of Alzheimer type or a coexistence of CJD and Alzheimer disease in the same patient [131, 134, 145].

4.3.5. Gerstmann-Sträussler-Scheinker (GSS) syndrome

A special type of a PrP-immunoreactive plaque, viz. the *multicentric plaque* is a hallmark of a separate familial variant of CJD, the Gerstmann-Sträussler-Scheinker (GSS) syndrome [729] first reported in 1936 by Austrian investigators [407]. Interestingly, the similarity of the neuropathological picture of GSSS to kuru has been recognized even before the results of transmission studies became available [949]. The nosological position of GSSS has been discussed for many years, particularly its relation to CJD and Alzheimer's disease [31, 258, 314]. Recently, due to the immunohistochemical localization of PrP within amyloid plaques of GSS syndrome [609, 1001–1002] and the discovery of a linkage between the occurrence of GSS syndrome cases and a mutation within codon 102 of the PrP gene [140, 297, 435, 441, 511–512] the nosological position of GSS syndrome has been established within slow virus disorders, possibly as a particular variant of CJD. Furthermore, the distinction between Gerstmann-Sträussler-Scheinker *syndrome* which exhibits spongiform changes and is transmissible to subhuman primates, and Gerstmann-Sträussler-Scheinker *disease* which is non transmissible and without spongiform change was proposed [729]. However, as spongiform changes in GSS syndrome are a variable finding, this distinction needs to be corroborated by others data.

The multicentric plaque of GSS syndrome (for detailed discussion see paragraph on PrP plaques) consists of several cores of different size surrounded by a common halo containing few dystrophic neurites and glial cells [100−104, 516, 630, 934, 936, 1030]. Thus, the multicentric plaque resides in a niche between the kuru plaque and the senile (neuritic) plaque of Alzheimer's disease. Noteworthy, both kuru plaques and senile plaques are encountered in GSS syndrome but only the presence of multicentric plaques is pathognomonic. Other typical features of GSS syndrome include degeneration of spinocerebellar and corticospinal tracts and posterior columns of spinal cord [630], cortical spongiform changes, mostly confined to molecular and outer granular layers [516], and limited astrogliosis. The presence of numerous tau-immunoreactive neurofibrillary tangles was also reported in a peculiar Indiana GSS family [409, 411].

4.3.6. Panencephalopathic form of CJD

Classically CJD is regarded as a neurodegenerative disease of gray matter – viz. *polioencephalopathy*. However, a small proportion of CJD cases, particularly those reported from Japan, are characterized by widespread white matter damage [80−81, 237, 611, 836, 1024, 1067]. White matter damage is much more extensive than to be explained as a sequence to severe cortical loss encountered in these cases. Mizutani et al. [779, 781−782] were perhaps this first to describe the *panencephalopathic* type of CJD. White matter except the U-fibers was conspicuously spongy and a concomitant decrease of axons was reported. Gemistocytic astrocytes were numerous but fibrous gliosis was not seen. Interestingly, spongiform changes as defined by Masters and Richardson [734] were only rarely seen but large clefts presumably resulting from collapse of damaged neuropil was observed. Furthermore, kuru plaques were seen in Japanese cases and it seems that the Japanese panencephalopathic type is linked to that characterized by multiple plaque formation. However, in contrast to the long duration of the latter cases, those belonging to the panencephalopathic type were rapidly progressive.

4.3.7. Amyotrophic form of CJD

The last subtype to be discussed is the amyotrophic form of CJD [14, 927–928]. One reason for confusion arising from the original Jakob description [529–530] is inclusion of the "amyotrophic form of CJD" within the CJD spectrum. Typical for such a misunderstanding is a case of Rafalowska and Strugalska [891] in which absence of any diagnostic features of CJD in an otherwise typical amyotrophic lateral sclerosis case did not preclude the classification of this case as "amyotrophic form of CJD". The "amyotrophic form of CJD" presents as dementia associated with lower motor neuron symptomatology and is characterized by neuropathology different from that previously described. Particularly the vacuolation limited to superficial cortical layers differed distinctly from that of typical CJD. Furthermore, the "amyotrophic form of CJD" proved to be non transmissible to subhuman primates. Thus, "the amyotrophic form of CJD" should be regarded as amyotrophic lateral sclerosis associated with dementia akin to the Parkinsonism-dementia-ALS syndrome of Guam; the name "amyotrophic lateral sclerosis" should be discarded.

4.3.8. CJD and other neurological disorders

CJD has been described to coexist in the same patient with other pathological conditions. The association of CJD and meningioma [489], Kohlmeier-Degos disease [940], normal pressure hydrocephalus [392] and retinitis pigmentosa in the same family in which CJD cases [776] were found is probably only fortuitous. CJD associated with Wernicke encephalopathy (WE) was described [404] in 3 cases in which neuropathological typical for CJD changes coexisted with petechiae within mamillary bodies, periaqueductal gray matter and the tegmental regions. Microhaemorrhages and proliferation of capillaries were observed.

Coexistence of CJD with other neurodegenerative disorders was also reported. When parkinsonian signs and symptoms were observed in CJD patients, true features of Parkinson's disease with the presence of Lewy bodies against the background of depopulated neurons of substantia nigra and locus ceruleus is a rare phenomenon [313]. Furthermore, degeneration of thalamus and inferior olives reminiscent of that in primary thalamus degeneration was seen in an otherwise typical case of CJD [778]. Features of progressive supranuclear palsy were also observed in a few members of familial CJD [82]. Atrophy of optic nerve and reduction in number of optic nerve fibers and ganglion cells in both retinas in CJD [602, 655] reflect a more generalized retinopathy reminiscent of that encountered in experimental conditions.

Involvement of the peripheral nervous system in CJD is exceedingly rare. Khurana et al. [547] described 2 CJD cases characterized by autonomic dysfunction. Neurons of middle and inferior cervical and thoracic sympathetic ganglia showed vacuolation, homogenization of cytoplasm and pale staining of nucleus. Axons accumulated an abnormal amount of neurofilaments as shown by electron microscopy. Similar changes were noted in axons of vagus nerve. Pathology of trigeminal ganglia was also reported [465], exhibiting proliferation of satellite cells, vacuolation of ganglion cells, numerous proximal axonal swellings with neurofilament accumulations as showed by immunohistochemistry for 200 kDa neurofilament protein, and abnormal expression of prion protein (PrP) [465]. Peripheral neuropathy featuring slow nerve conduction velocity and segmental demyelination seen in sural nerve biopsy was recently reported but is most difficult to distinguish from sequels of secondary metabolic as multifactorial damage occurring in patients who are severely ill with CJD [924]. In another CJD case features of demyelination with the formation of "onion bulbs" was reported [1024].

4.4. Elements of neuropathology of slow virus disorders

4.4.1. Spongiform vacuoles

4.4.1.1. Introduction

The spongiform vacuole, currently accepted as the neuropathological hallmark of the subacute *spongiform* virus encephalopathies (SSVE) has not always been so regarded or not even always observed, as in aged mink of the Chediak-Higashi genotype infected with transmissible mink encephalopathy (TME) [726]. Some early investigators have speculated that astrocytic hypetrophy "could be the initial response to the injurious agent rather than merely a reflection of primary damage to the nerve cell" [472] but such a view evolved into a current one, suggesting that vacuolation is a primary pathological change in a whole group of spongiform encephalopathies.

In this paragraph we shall discuss the spongiform vacuoles at the light and electron microscopy levels.

4.4.1.2. Distribution of vacuoles within grey and white matter: "lesion profile"

The differences in the topography of neuropathological lesions caused by different types of viruses is well known [821]. Perhaps the best known examples are the involvement of the medial temporal lobes in herpes encephalitis or the anterior spinal motor neuron in acute poliomyelitis. However, the existence of well defined immunological method of virus/ strain typing has never forced "conventional" neurovirologists to elaborate sophisticated histopathological methods to obtain results comparable to those in the field of "unconventional" viruses.

The differences in topography and intensity of vacuolation have been already noted in early scrapie research. Palmer [832–833] reported such differences in brain sections caudal to the level of superior colliculi of Suffolk sheep with naturally occurring and experimentally induced scrapie. He has noted that the intensity of vacuolation is higher in natural than in experimental scrapie and that particular nuclei were involved differently in the same animal. Zlotnik [1076], in an extensive study of 24 out of 1200 cases of natural and experimental scrapie in sheep (Suffolks and Cheviots) and experimental scrapie in goats reported severe differences in intensity and topography of vacuolation among different nuclei of medulla, pons, mesencephalon, and dien- cephalon. For example, the cuneate nucleus was always affected in experimentally infected Suffolk sheep but never in Cheviot sheep and goats. On the contrary, this nucleus was involved in 50% of both Suffolk and Cheviot sheep with natural scrapie. In Creutzfeldt-Jakob disease the differences in topography and intensity of lesions were known even before the etiology of the disease was established by the seminal work of Gajdusek and Gibbs [600–601]. The diversity of biological properties of different strains (isolates) of CJD virus has also been recognized [417] but formal strain typing on the basis of neuropathological studies has never been performed and was obviously impossible using nonhuman primates as rare and expensive as chimpanzees. Furthermore, the lack of detailed knowledge of host genetics and the virus strain (notorious in natural infections) precluded the use of the above-mentioned differences as tools for identifying of strain differences. This goal has been achieved by Fraser and Dickinson [350–351, 353–354] who developed the "lesion profile" method for virus strain typing.

The "lesion profile" (LP) reflects the intensity of vacuolation scored on a scale ranging from 0 (no vacuolation) to 5 (confluent vacuoles) within nine positions in grey and three in white matter [267–268, 270, 343–347, 350–351, 353–354]. Subsequently, other investigators used the LP method, modified by means of different regions at which the intensity of vacuolation was scored [667, 735]. LP proved to be a useful means for typing strains of scrapie virus; actually the diversity of virus strains as discovered on the basis of the length of the incubation priod and LP is regarded by some [154, 282–285] but not all investigators as proof that the scrapie virus possesses an independent genome.

In contrast to the incubation period in mice, the LP is only slightly influenced by the gene *Sinc* (most probably identical to the recently discovered gene *Prn-i*; *vide supra*). All four strains of scrapie virus belonging to the ME7 group (ME7, 79A, 22C and 87A) produced LP in C57BL mice (Sinc^{s7}) similar to that obtained in VM mice (Sinc^{p7}) [354]. However, subtle differences of LP in different Sinc genotypes have also been noted. For example, the 79A strain of scrapie virus produced LP of approximately the same density in both Sinc^{s7} and Sinc^{p7} homozygotes within all but cerebellum region [354]. Furthermore, the F$_1$ (C57BL X VM) LP for the ME7 strain of scrapie virus is of intermediate density between both parental LP except for the medulla. F$_1$ LP for the 22A and 22C strains of scrapie virus are more similar to that produced in C57BL than in VM mice [354]. In other regions, both the 22A and 22C show full dominance. Thus, the genetic control of the distribution of vacuolation reflected by the LP is more complex than simple control of the incubation period [282–285].

LP is stable over a wide range (6 orders of magnitude) of a virus titer contained within an inoculum [354]. It thus seems, that vacuolation is closely associated with virus replication, indeed it follows scrapie virus replication [39] and thus it serves us useful tool to study virus spread and targeting. Fraser [342, 356, 944] pioneered this type of experiment, using vacuolation as a marker for scrapie virus spread. Following intraocular inoculation, the vacuolation initially occurred in the contralateral superior colliculus and then the lateral geniculate body. Liberski et al. [684] used this approach to study neuronal targeting in experimental Creutzfeldt-Jakob disease. Taken together, the data suggest that scrapie and CJD virus use axons as means of transportation within central nervous system.

In summary, the topography of spongiform lesions is closely associated with virus replication and spread and may serve as a useful tool for virus strain typing and for studying the pathogenesis of this group of disorders.

4.4.1.3. Ultrastructure of vacuoles

4.4.1.3.1. Creutzfeldt-Jakob disease (CJD)
Neuronal elements (Fig. 21–22) are currently regarded as the structures bearing spongiform vacuoles in CJD. This current level of understanding has been achieved by painstaking refinements in electron microscopic methodologies with elimination of fixation artifacts.

Two distinct types of vacuoles are found in CJD (Fig. 23–24). The first type, developing within myelinated axons and thus designated "myelinated vacuoles", will be discussed later. The second type of

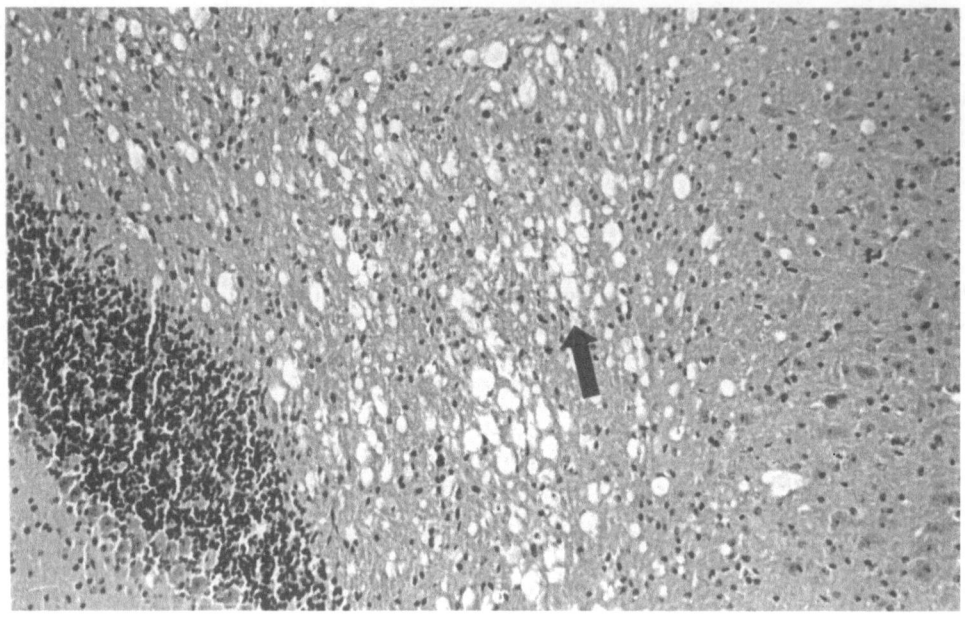

Fig. 21. Vacuolation of cerebellar white matter (arrow) in mouse brain infected with the Fujisaki strain of CJD virus. H & E. Original magnification, ×160

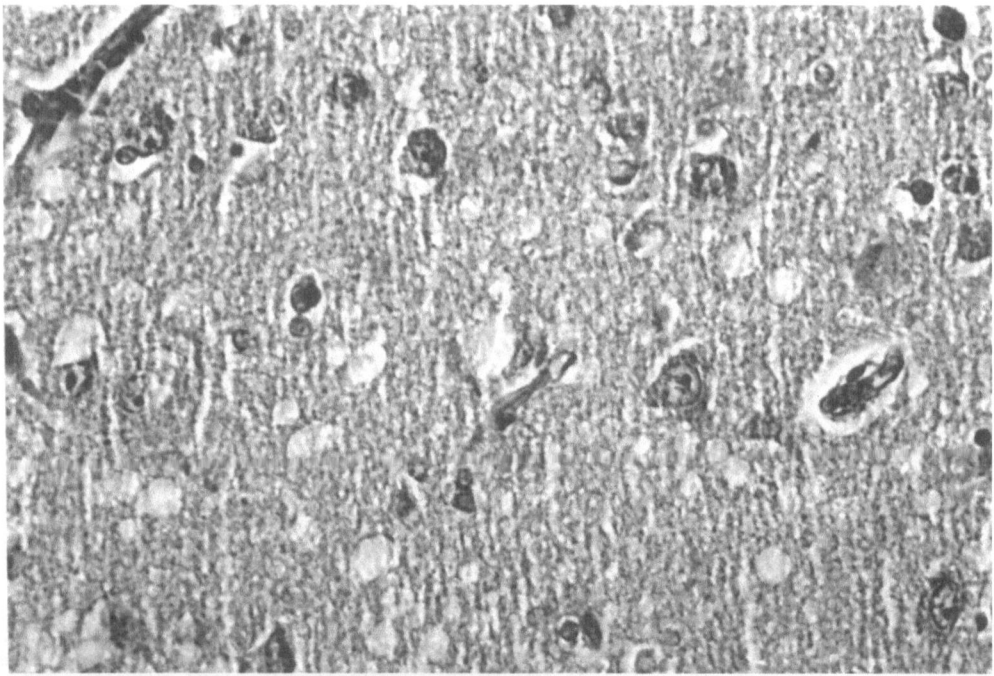

Fig. 22. Spongiform vacuoles in a case of GSS syndrome. Material courtesy of Dr. D.C. Gajdusek, the National Institutes of Health, Bethesda, USA

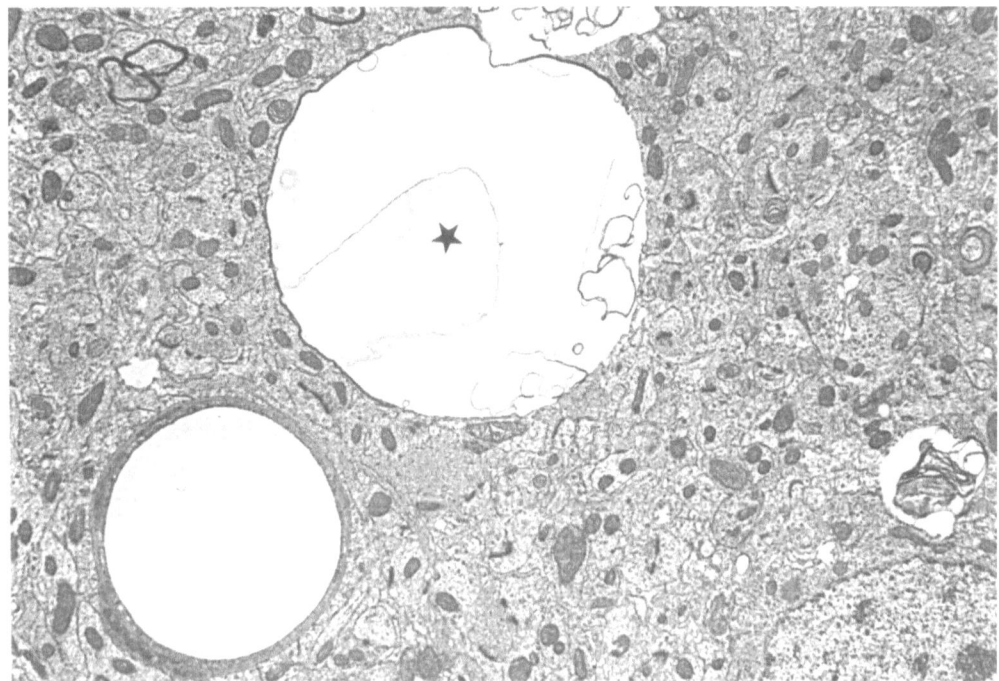

Fig. 23. Single vacuole (star) in mouse brain infected with the Fujisaki strain of CJD virus. Unpublished work of P.P. Liberski, R. Yanagihara, C.J. Gibbs, Jr, and D.C. Gajdusek, the National Institutes of Health, Bethesda, USA. Original magnification, ×4500

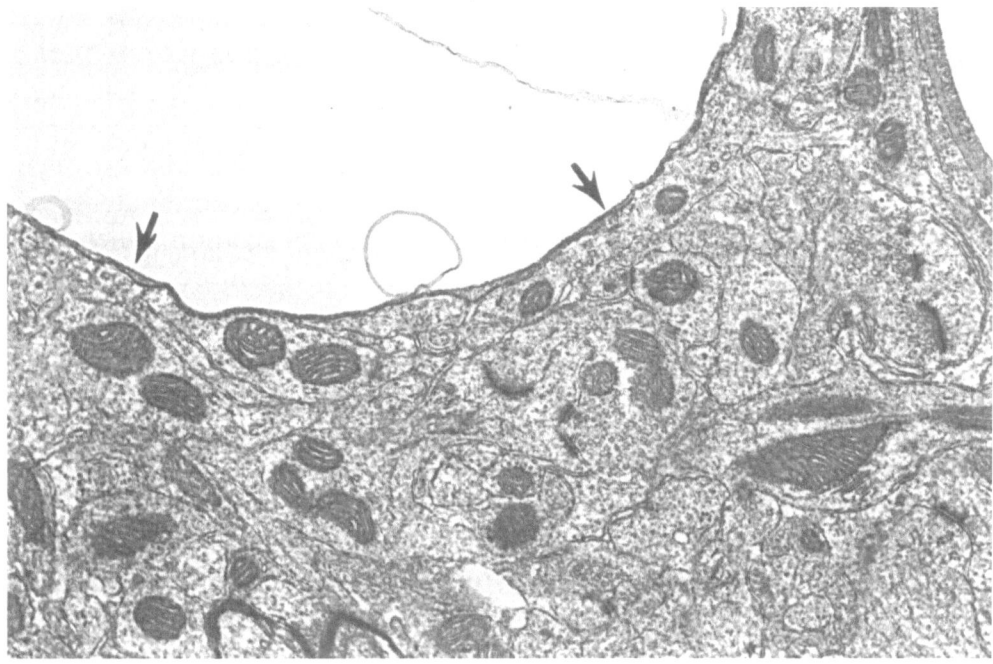

Fig. 24. A margin of vacuole (arrows) within a myelinated process of mouse brain infected with the Fujisaki strain of CJD virus. Unpublished work of P.P. Liberski, R. Yanagihara, C.J. Gibbs, Jr, and D.C. Gajdusek, the National Institutes of Health, Bethesda, USA. Original magnification, ×7000

abnormal vacuoles, found within neuronal processes not ensheathed in myelin and designated "unmyelinated vacuoles" [679, 686] are membrane-bound and originate from neuronal elements. Whether they originate in dendrites or in axonal terminals or preterminals was not always clear. Many secondary vacuoles (that is, vacuoles within other vacuoles), vesicles and curled membrane were found within these primary vacuoles. However, spongiform vacuoles, the most characteristic neuro-pathological change of all the subacute spongiform virus encephalopathies, are by no means disease-specific, and they contribute to the "specific-unspecific" neuropathology of this class of disorders (Gibbs, personal communication). Thus, the presence or absence of vacuoles must always be interpreted in the context of the entire neuropathological picture.

Marin and Vial [719] were the first to study two cases of CJD ultra-structurally and identified membrane-bound electron lucent "cavities" within swollen astrocytic processes. These vacuoles contained filamentous material and secondary membrane structures. Furthermore, it has been pointed out that the "cavities" differed from those enlargements of the extracellular space encountered in the edematous brain. In sub-sequent studies the intracellular vacuoles have been ascribed almost exclusively to dilated astrocytic processes [124, 448, 548, 970, 1013–1014]. Suprisingly, the astrocytic nature of spongiform vacuoles was recently reported by Miyakawa et al. [784]. However, published electron micrographs clearly documented brain tissues suffered from suboptimal fixation or perhaps even autolytic changes which is not unusual in human neuropathology. Ribadeau-Dumas and co-workers [903–904] were perhaps the first to consider neuronal processes as responsible, in part at least, for the formation of intracellular vacuoles but a few vacuoles were ascribed to astrocytic processes. In a subsequent paper, Bignami et al. [87, 89, 91] ascribed spongiform vacuoles to both neuronal and astrocytic processes and this dual view of vacuole histogenesis predominated.

All cited studies have suffered from suboptimal fixation which itself produced the swelling of astrocytic processes. Furthermore, it has been clearly shown by Gibson and Tomlinson [431] that postmortem vacuola tion within the human cortex resulting from astrocytic swelling increased significantly up to 30–35 hours after death and thereafter decreased vacuolation was observed. Such spurious astrocytic swelling largely con-tributed to the spongiosis as appeared in early ultrastructural studies particularly in human autopsy material.

The major breakthrough has been accomplished by Lampert et al. collaborating with Gajdusek's laboratory on properly fixed material from nonhuman primates inoculated with CJD and kuru viruses [640–644]. These investigators unequivocally showed that vacuoles corresponded to swollen cytoplasmic processes located within an otherwise intact neuropil. Not without some difficulties, most of these processes were identified as dendrites by virtue of attached presynaptic terminals, however, a minority were still classified as astrocytic processes. Vacuoles were always membrane-bound and the membranes were frequently interrupted and curled. Furthermore, the higher electron density (osmophilia) of membranes lining the vacuoles was reported. Swollen processes (membrane-bound vacuoles) frequently collapsed and underwent digestion by macrophages. It seemed that the local clearing of neuronal cytoplasm contributed to vacuole formation. The notion that neuronal processes constitute the major if not the exclusive cellular element responsible for the histogenesis of vacuoles has been confirmed and extended in subsequent studies. In an experimental model of CJD in guinea pigs and hamsters, Manuelidis et al. [550–553, 711, 712–713] ascribed the spongiform change mostly to neuronal processes. Neuronal perikarya undergoing the local clearings of the cytoplasm were also seen. However, some involvement of astrocytic processes was noted, but published electron micrographs of swollen astroctyes suggested that the swelling resulted from improper fixation or anoxia/ischaemia rather than from the pathological process. It is clearly evident that hypertrophic astrocytes in optimally fixed material do not swell.

Recently, the structure of spongiform vacuoles has been studied by means of scanning electron microscopy (SEM) and compared with that visualized by classical thin-section transmission electron microscopy (TEM) [214]. Two types of altered membranes within vacuoles, corresponding to those changes first seen by Lampert et al. [640–644] were reported. First, unit membrane splitting yielding two dense subunits was seen by means of TEM. Such an alteration corresponded to ulcerations or defective membrane areas visualized by means of SEM. The other change consisted of "puffy amorphous membranes" and that corresponded to "rough elevated areas" seen on SEM. Pathologically altered intravacuolar membranes were associated with small "blisters" of different sizes and shapes. Blisters corresponded to vesicles visualized by TEM. Altered membranes have also been examined by means of the freeze-fracture technique [299]. In the cerebellum of scrapie-infected mice, vacuoles were mostly confined to the granular cell layer and numerous blebs protruding into vacuoles were visualized in freeze-fracture replicas [299]. Furthermore, membranes of adjacent cells or those lining vacuoles also protruded into lumina of vacuoles. On freeze-fracture replicas such membranes were "smooth" (i.e. devoid of intra-membrane particles on both membrane halves). In contrast, such particles were clearly seen on other parts of the cellular membranes. Structural alterations of membranes documented by their ultrastructural appearance may reflect profound changes in their lipid composition.

Total ganglioside content was reduced by 26 to 79% while that of cerebrosides was elevated by 49% or reduced by 35% to 50% in CJD brains, and such a reduction in sialic ganglioside level correlated well with the extent of cortical degeneration [53, 994–995]. Ganglioside constituents were also markedly altered [994–995]. Thin-layer chromatography revealed a reduction of polysialogangliosides, mainly GM_1 and GM_{1a} and an increase in GM_3, GM_2 and GD_3 gangliosides. While GD_{1a}-GalNac and GT_{1a} are minor constituents in normal brain they were markedly increased in CJD and thus easily separated by thin-layer chromatography in CJD brains [994–995].

Furthermore, the dramatic decrease to 4% of control values of C_{20}-sphingosine was found. Analogously, the level of polyunsaturated fatty acids, particularly 22:4 and 22:6 was reduced. Similar changes were reported in guinea pigs with experimental CJD [1072]. Total ganglioside concentrations were reduced to 18%–23% of control values with a concomitant increase of GM_3, GD_3 and GD_2 and a decrease of GD_{1a}, GD_{1b} and GT_{1b}. Alterations of the chemical composition of cell membranes may lead to deep changes of their physicochemical properties as detected by the spin-label method [1031]. All these facts together point to an alteration of lipid membrane composition, which may lead to structural alterations visualized by thin-section or scanning electron microscopy. Furthermore, a testable hypothesis has been put forward by Gibson and Liberski (unpublished) and Liberski et al. [679, 686] that chemical changes in membrane composition may form "*loci minoris resistentiae*" reacting to the tissue processing by vacuole formation. The lack of any intermediate structure responsible for vacuole formation (in other words, from the very beginning only completely formed vacuoles are detected [679]) may lend support for such an hypothesis, and thus high-voltage electron microscopy on unfixed specimens is clearly warranted.

Another interesting hypothesis concerning the histogensis of vacuoles has been proposed by Machado-Salas [697]. "Empty cytoplasm" of several dystrophic neurites (generic term for both abnormal dendrites and axons) visualized by means of Golgi techniques was attributed to spongiform vacuoles cut in a transverse plane of electron micrograph. Still other neurites, mostly axons, exhibited "apparently distended external membrane surrounding a totally empty space". The author suggested that dystrophic neurites and spongiform vacuoles represented stages in the formation of bullous swellings as documented by Golgi techniques.

Intranuclear vacuoles, a different type of vacuoles encountered in CJD, have been described by Manuelidis et al. [550–553, 712–713] in experimental CJD and subsequently in human CJD cases [193, 549].

These vacuoles were regarded as abnormal subcellular organelles and even a feature associated with particular CJD virus strain(s) in a sense as particular lesion profile is strain specific. However, as these vacuoles have been found by Liberski et al. [679] in both CJD virus-infected and sham-inoculated mice and because their numbers did not increase dramatically toward the terminal phase of disease, they should be regarded as artefactual of no pathological significance.

4.4.1.3.2. Scrapie, bovine spongiform encephalopathy (BSE) and chronic wasting disease

In contrast to a relative abundance of data on scrapie vacuolation at the light microscopy level, little information is available at the ultrastructural level, particularly in natural scrapie. Bignami and Parry [91] reported membrane-bound vacuoles within neuronal perikarya, dendrites and axonal terminals (but not in axons or astrocytes) in natural scrapie. Some vacuoles were extremely large and only a part of it could be traced in one low-power field. Thus, vacuoles in natural scrapie did not differ significantly from those of CJD but "membrane-bound cytoplasmic papilliform processes and vesicles, 100–200 nm in diameter" were frequently observed within vacuoles and such structures have never been reported in natural diseases of man. The histogenesis of vacuoles could not be established in the cited paper. Field and Raine [331] were the first to study experimental scrapie in mice ultrastructurally and, suprisingly, were not able to find any structure correlated with the "globular vacuolation" of the neuropil. However, they were also the first to describe vacuoles within the myelin sheath. The first description of spongiform vacuoles came from Gajdusek's laboratory [644]. The spongiform change of light microscopy corresponded to two different classes of ultrastructural phenomena. First, focal "clearings" of the neuronal cytoplasm were seen. Swollen neurons contained no subcellular organelles but "finely granulo-filamentous material". The second type of vacuoles were comparable to distended neuronal processes mostly dendrites which fused together forming "larger" vacuoles. The extracellular space, particularly between subcortical myelinated axons, was also enlarged. Furthermore, ruptured, curled membrane fragments were also reported. The ultrastructure of spongiform vacuoles was elaborated in subsequent papers [40–41, 195, 201–202, 660, 669]. Vacuoles were detected primarily in postsynaptic neuronal processes. Vacuoles were membrane-bound and exhibited secondary vacuoles (smaller vacuoles within larger vacuoles). The membranes lining the vacuole were frequently interrupted, producing curled membrane fragments. When cut obliquely membranes presented a "fluffy" appearance.

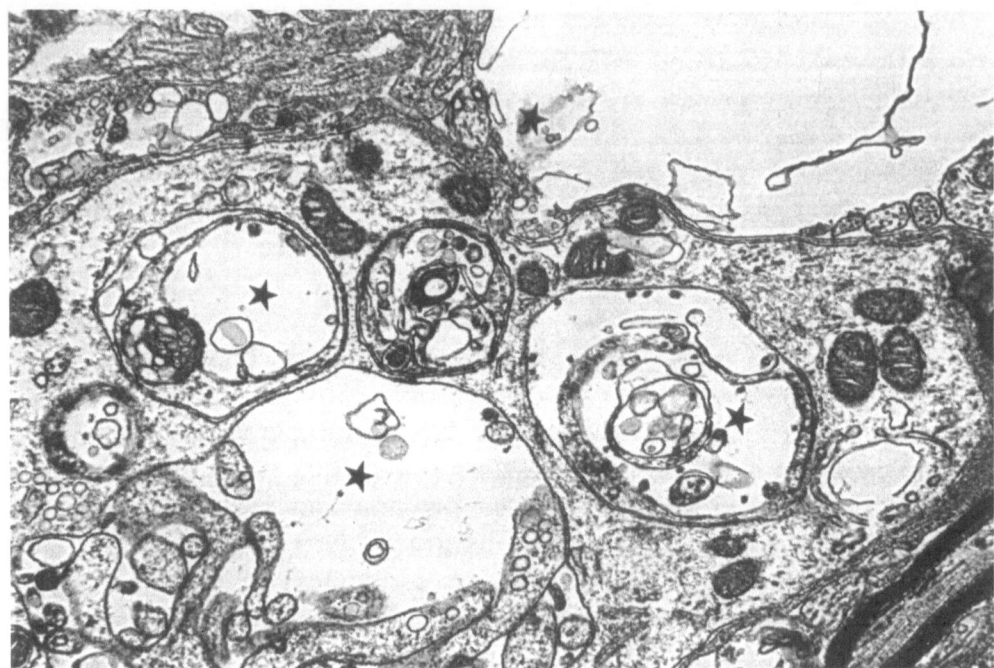

Fig. 25. Numerous vacuoles (stars) within neuronal processes of a BSE-affected cow. Material courtesy of Dr. Gerald A.H. Wells, Central Veterinary Laboratory, Ministry of Agriculture, Fisheries and Food, Weybridge, U.K. Original magnification, ×12 000

Only extremely limited ultrastructural data have been published so far on bovine spongiform encephalopathy and chronic wasting disease in mule deer and elk. Liberski et al. (unpublished observations) documented (Fig. 25) numerous membrane-bound vacuoles, predominantly in dendrites and rarely in myelinated axons. As in natural scrapie, some vacuoles, particularly those in the brain stem, were very large, and several neuronal processes contributed to their formation. Vacuoles contained abundant curled membranes, vesicles, secondary vacuoles and "fluffy" material. Generally, the ultrastructural pathology of BSE is the same as that of the other subacute spongiform virus encephalopathies.

4.4.1.3.3. Ultrastructure of intramyelin vacuoles in the
panencephalopathic type of Creutzfeldt-Jakob disease (CJD) and other
types of subacute spongiform virus encephalopathies
Classical CJD is regarded as a neurodegenerative disease of grey matter – viz. *polioencephalopathy*. However, a small proportion of CJD cases, particularly but not exclusively those reported from Japan, are characterized by widespread white matter damage (*vide supra*). White matter damage in these cases is much more extensive from that expected to be merely from severe cortical loss.

The panencephalopathic type of CJD is characterized by numerous vacuoles within myelinated axons. The ultrastructural appearance of intramyelin vacuoles has been recently reported by Liberski et al. [686] in mice with experimental CJD and by Manuelidis et al. [711]. Vacuoles, greatly distending the myelin sheaths, were observed within myelinated fibers. Shrunken axons with normal-appearing axoplasm were adherent to the innermost layer of the myelin sheath, occasionally only by a thin "neck". Some axons were covered with a few layers of myelin, and the myelin sheath lining the vacuole was clearly a part of that covering the axon itself. Two or more axons within one large vacuole, possible representing vacuolation of a common myelin sheath wrapped around a cluster of axon, were observed infrequently. The myelin sheath lining the vacuoles seemed normal with an intact periodicity, but splitting either at the major dense or intraperiod lines clearly contributed to vacuole formation. Vesicular myelin degeneration and intraaxonal vacuoles were also occasionally seen.

Only a limited number of papers deal with the ultrastructure of intramyelin vacuoles in CJD and related disorders. Field and Raine [331] documented markedly distended "axis cylinders" in experimental scrapie. The myelin sheath lining vacuoles was thinner than that lining normal axons but the periodicity was retained. These authors were also the first to document numerous processes wrapped around a common myelin sheath. The same phenomenon has been described in CJD by Marin and Vial [719]. In a series of papers, however, Manuelidis et al. [548–553, 712–713] reported intramyelin vacuoles as a rare ultrastructural finding in experimental CJD in guinea pigs and hamsters. Noteworthy, these authors stated that "the myelin sheath did not appear to be involved with the primary development of vacuoles" [552] but by the same token, illustrated unequivocal intramyelin vacuole in Fig. 6B of their report. The reason for such inconsistency is unexplained. Furthermore, Kim and Manuelidis [550–553] stated that splitting at the major dense or intraperiod lines did not occur in CJD-affected guinea pigs. However, to detect splitting ultrastructurally, frequently serial sectioning of any intramyelin vacuole must be done, and thus a sampling problem is the most plausible explanation for the apparent absence of splitting at the major dense or intraperiod lines in these CJD models.

Splitting either at the major dense or intraperiod lines was clearly documented in several experimental models of CJD by Tateishi group [931, 999–1000, 1006] and subsequently in a few CJD models as "unexpected" finding by Manuelidis et al. [711]. However, such a mechanism of myelin sheath destruction is not restricted to CJD. The formation of intramyelin vacuoles (also designated myelin ballooning or myelin dilatations), usually from splitting at the intraperiod lines, has been previously observed in rhesus monkeys with vitamin-B 12 deficiency [7], in rats following widespread axonal damage [639], in mice infected with Semliki forest virus [954], in "quaking" mutant mice [1037], and in mice intoxicated with cuprizone [492] or sodium cyanate [1007]. Noteworthy, when cultures of mammalian peripheral nervous system were exposed to a lethal dose of sodium cyanate, myelin split first at the intraperiod lines and later at the major dense lines [737]. No predilection for any layer of the myelin sheath was observed in diphtheria toxin-induced experimental demyelination of cats and rabbits [1054]. Hence, these two mechanisms of myelin dilatation may not be mutually exclusive.

4.4.1.4. Morphogenesis of spongiform changes

The morphogenesis of spongiform vacuoles is yet to be clearly delineated and the only experimental data published to date are still controversial. Beck et al. [60, 67–68] reported the only detailed studies of spider monkeys inoculated with several isolates of kuru virus sacrificed at various times postinoculation (the earliest clinical signs of kuru were observed 94 weeks postinoculation). Histopathologically, none of the experimental animals exhibited any signs of spongiform change. However, the peculiar subcellular organelles designated *abnormal configurations of plasma membranes* (ACPMs) were reported in all kuru-infected but not sham-inoculated animals. At lower magnification, ACPMs appeared as thick, electron-dense plaques along the course of otherwise normal-appearing plasma membranes. Interestingly, ACPMs were seen in animals 4 to 40 weeks postinoculation, but not 2 weeks postinoculation or at the clinical stages. Their numbers varied between brain regions; limbic cortex, corpus striatum and paleocerebellum showed the highest number of ACPMs.

   ACPMs were most frequently observed at membranes of two adjacent dendrites and, less frequently, between two presynaptic terminals, presynaptic and postsynaptic terminals, neuronal perikarya or neuronal and astrocytic processes. In the paleocerebellum, necks of dendritic spines were the most frequent region of their occurrence. ACPMs consisted of several electron-dense layers measuring approximately 3.5 nm in width separated by electron-lucent bands measuring 1.8 nm in width, respectively. Frequently, electron-dense layers collapsed on each other forming a complicated multilayered structure. The most complex ribbon consisted of 13 layers yielding a total width of 41 nm. Interestingly, small blebs or vacuoles were seen attached to the outermost lamella of the ACPMs. From several published electron micrographs it seems that vacuoles were in the process of detaching from the ACPMs and several vacuoles were observed "flowing" freely within the cytoplasm. It should be stressed however, that the reconstruction of electron microscopic images from a single section may be difficult and a reconstruction from serial sectioning was not attempted.

Frequently, ACPMs were found in close apposition with numerous coated vesicles and coated pits. Coated pits in kuru-affected animals outnumbered those in sham-inoculated controls showing a decreasing gradient through the incubation period (394/1170 um^2 at 2 weeks postinoculation, 220 at 4 weeks postinoculation, as compared with 2 in sham-inoculated controls, respectively). Another unusual finding, was the abnormal number of somatic spines on Purkinje cells.

In the original publication, the authors suggested that ACPMs could represent abortive intracellular junctions "forming *de novo* due to reactivation of embryonic growth mechanism".

The very presence of ACPMs has remained controversial. Gray [453] suggested the artifactual nature of ACPMs, as he could almost reproduce these structures using on the grid staining with uranyl acetate and lead citrate and omitting block staining assuming, erroneously, that this step was also omitted from the original investigation. Gray [453] interpreted ACPMs as focal areas of suboptimal fixation. In response to Gray's criticism, Beck et al. [67] stressed the use of block staining with uranyl acetate, thus disputing that the appearance of ACPMs could not be ascribed solely to the omission of this step, as Gray suggested. Furthermore, if the formation of ACPMs was methodology-dependent artefact, it is difficult to reconcile its presence only in kuru-infected animals and not in sham-inoculated controls, despite processing the samples in an identical fashion.

The controversy, actually, has been never resolved. Kim and Manuelidis in a serial experiment could not detect ACPMs in experimental CJD in guinea pigs [552]. Recently, Liberski, Brown and Gajdusek (unpublished data) addressed this problem in an independent experiment on two squirrel monkeys and one spider monkey infected experimentally with kuru and killed approximately in the middle of the expected incubation period. While large numbers of coated vesicles and coated pits were easily confirmed in this experimental model only a very few multilamellated structures corresponding to ACPMs, as defined by Beck et al. [67–68] could be found. As the cited experiment did not address any of the methodological problems raised by Gray [453], even the finding of a few ACPMs-like structures does not provide unequivocal proof that these structures are not artifactual.

Recently, Beck offered an alternative hypothesis to explain the formation of ACPMs (and *eo ipso*, the morphogenesis of spongiform vacuoles) [59]. Assuming, that the ultrastructural pathology of early kuru stages indeed recapitulates early embryonic development, an experimental model was developed in which rats were inoculated with brain homogenates originating from 10-day-old rats (the peak of synaptogenesis in this species). The pathological changes were confined to the brain stem and, surprisingly, recapitulated those encountered in early kuru stages. First, several neurons exhibited small vacuoles budding from one end; noteworthy neuronal vacuolation is a hallmark of natural scrapie and bovine spongiform encephalopathy and are also encountered in human CJD. Astrocytosis was a prominent finding. Ultrastructurally, both coated vesicles and coated pits were easily found in abundance; and ACPMs were also observed. Some vacuoles were in a close contact with intracytoplasmic vacuoles. What is the biological significance of these new data? While the morphogenesis of vacuoles and the very presence of ACPMs clearly need further evaluation, the overall similarity of early kuru stages and those observed following inoculation of normal brain taken at the peak of synapthogenesis suggests "re-activation of embryonic growth mechanism at different stages of ontogenesis" [59]. Noteworthy, kuru has also been designated "galloping senescence of the juvenile" to stress its pathology recapitulating accelerated aging [377]. Certainly, such an entanglement of early embryonic processes and late aging phenomena may be a hallmark for these whole group of disorders.

4.4.2. Astrocytic reaction

4.4.2.1. Introduction

Astrocytosis or reactive gliosis is a prominent feature of the subacute spongiform virus encephalopathies. However, whether the apparent increase of astrocytic nuclei, as seen under the light microscope, results from hypertrophy or also a hyperplasia (proliferation) of astrocytes is unclear. Also the class of astrocytes participating in this reaction needs clarification. On the basis of their morphology astrocytes are divided into protoplasmic and fibrous types and this classification has recently been supported by immunohistochemical data. These two different classes are designated *type 1 astrocytes* lacking the A2B5 surface antigen and *type 2 astrocytes* expressing the A2B5 antigen. These two classes correspond to protoplasmic and fibrous astrocytes, respectively [497]. Type 2 astrocytes originate from bipotential glial precursor cells which differentiate into either these cells or oligodendrocytes, while type 1 astrocytes differentiate from progenitor cells of a separate linkage. At the moment, however, we do not known whether type 1 or type 2 astrocytes are responsible for the glial reaction encountered in slow virus-affected brains.

4.4.2.2. Kuru, Creutzfeldt-Jakob disease and Gerstmann-Straussler-Scheinker syndrome

4.4.2.2.1. Kuru

Astrocytic hypertrophy and proliferation have been stressed as a hall-mark of kuru [60–66, 612–613]. Astrocytic proliferation was widespread [612–613] and more abundant in the gray than in the white matter. Usually, astrocytosis paralleled neuronal destruction, but it has also been observed in regions with only minimal neuronal changes. In the pons, severe gliosis was observed in the tegmental and basal portions with a conspicuous sparing of the pyramidal tracts and medial lemnisci. Gliosis was severe in the midbrain, basal ganglia, thalamus, subcortical white matter and in the cerebellum where the vermis was mostly affected. Conspicuously, Fananas' cell proliferation has been noticed. In the cerebral cortex proliferation of astrocytes was in excess of other pathologic changes. Furthermore, astrocytosis was diffusely present in the anterior horns of the spinal cord. Some astrocytes showed clasmatodendrosis. In contrast, Neuman et al. [809] did not observe severe astrocytosis in three additional kuru patients and more recently Scrimgeour et al. [948] found only mild astrocytic changes with the presence of rare bi-nucleated forms in the cerebral cortex. Thus, like all other neuropathologic changes in subacute spongiform virus encephalopathies, astrocytosis is a variable phenomenon.

Experimental kuru in chimpanzees is characterized by widespread astrocytosis and both hypertrophy and proliferation of astrocytes have been observed [65–66, 640, 642] Gliosis seemed to parallel the severity of spongiform changes and neuronal loss, being most abundant in markedly affected sensory cortex and less in better preserved motor areas. Striatum, diencephalon, the white matter and cerebellum showed severe gliosis. Ultrastructurally [640–642], astrocytes showed focal, probably artefactual, clearings of the cytoplasm and accumulations of glycogen granules. Hypertrophic astrocytes containing glial filaments were easily noted. In a separate unpublished study of early changes in New World monkeys infected with kuru, Liberski, Brown and Gajdusek found, using GFAP-immunohistochemistry and electron microscopy, moderate astrocytosis (unpublished data). Astrocytes were of the hypertrophic type and contained myriads of glial filaments. Interestingly, astrocytes were observed adjacent to cerebellar granule cells undergoing faulty myelination. The biological significance of this phenomenon is unknown.

4.4.2.2.2. Creutzfeldt-Jakob disease (CJD)
and Gerstmann-Sträussler-Scheinker (GSS) syndrome

While an overview of the astrocytic reaction has been provided previously the details of this reaction, particularly at the ultrastructural level, will be discussed here (Fig. 26).

Astrocytosis is one of the very few options within the limited repertoire of the central nervous system to react toward any injury. Astrocytosis is observed among almost all neurodegenerative conditions and CJD is no exception.

Martin and Vial [719] were the first to report the ultrastructure of human CJD. They described hypertrophic astrocytes containing numerous bundles of glial filaments measuring approximately 10 nm in diameter. A proportion of astrocytes suffered from suboptimal fixation which produced areas of "watery" cytoplasm. Astrocytic cytoplasm contained numerous lipofuscin granules. A similar distinction between hypertrophic astrocytes containing glial fibrils and those of "watery" cytoplasm was made by Gonatas et al. [448] Torrack [1013–1014], Brion et al. [124], Bubis et al. [164] and Ribadeau-Dumas and Escourolle [903–904]. Astrocytic nuclei frequently contained inclusion bodies [846].

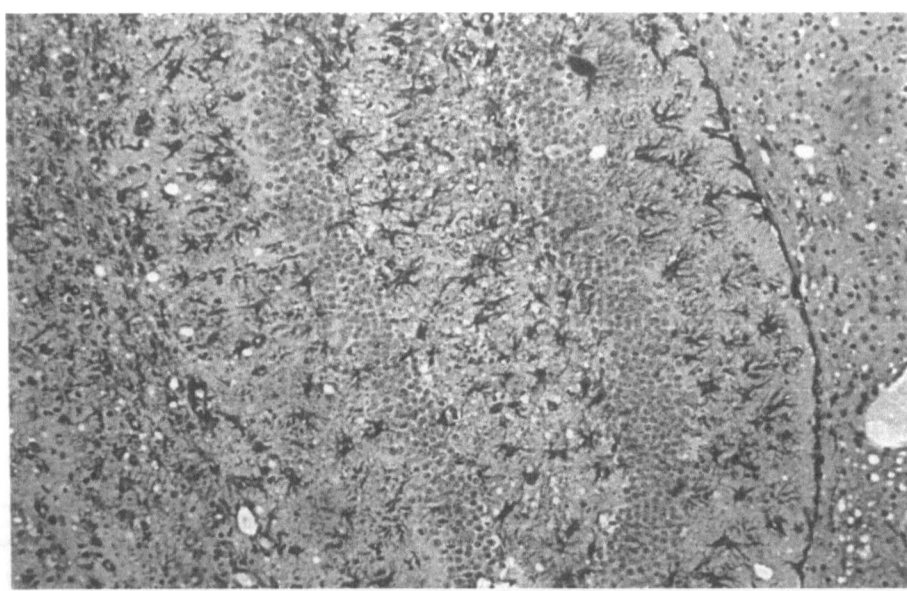

Fig. 26. Severe astrocytosis in the hippocampus formation of a mouse infected with the Fujisaki strain of CJD virus. GFAP immunohistochemistry. Original magnification, ×160

The first type consisted of granular and filamentous profiles frequently forming paracrystalline arrays. These inclusions most probably represented deformed chromatin. The other type corresponded to the IVth type of "nuclear bodies" according to the classification Bouteille et al. [121, 676]. As the first type of inclusion resulted most probably from suboptimal fixation, the second type or the type IV nuclear body needs a few comments. Such inclusions have been frequently found in infectious and neoplastic disorders and are regarded as a nonspecific reaction of a cell toward the noxious stimuli. Still another type of intranuclear inclusion reported by Jellineger [533] corresponded to type A nuclear inclusions. As these inclusions have been reported in numerous viral and degenerative conditions, as well as a result of abnormal mitoses, they most probably represent non-specific changes.

Astrocytosis presents a substantial part of neuropathological picture of experimental CJD. In the first reported transmission experiment, Beck et al. [69] found moderate to severe astrocytosis in both biopsy and autopsy specimens of CJD virus-infected chimpanzees. In the cerebral cortex, the hypertrophic astrocytes completely disturbed the neuronal architecture. In contrast to scrapie, where the problem of hypertrophy versus proliferation (hyperplasia) is still discussed, the proliferation of astrocytes was unquestionable. Many astrocytes were of the gemistocytic type, similar to those in human CJD. Severe glial reaction was seen also in the striatum, diencephalon and cerebellar cortex. Beck et al. [69] raised the problem of astrocytes as a primary target for CJD virus; in other words, the location of CJD within a vague spectrum of so called "glial dystrophies". This notion was based primarily on observed discrepancies between the severity of astrocytosis and neuronal damage. While such differences have been unequivocally noted, it must be stressed that in most situations the most severely affected brain regions presented also the highest level of astrocytosis. Ultrastructurally, astrocytes in CJD-affected chimpanzees showed abundant glial filaments and accumulations of glycogen, thus presenting a feature of typical reactive cells. The swelling of astrocytes, probably artefactual in nature, was also seen. The presence of severe astrocytic hypertrophy and true hyperplasia in experimental CJD was confirmed in subsequent reports. Particularly, Manuelidis et al. [550–553, 712–713] found severe astrocytosis in experimental CJD in guinea pigs, hamsters and mice. In CJD affected-hamsters "clusters of these cells appear almost as pure cultures; this collection is far in excess of what classically in human and experimental neuropathology is known as reactive astrocytosis". To further substantiate the notion of a primary involvement of astrocytes in CJD, Manuelidis and Manuelidis reported that astrocytes from CJD-affected brains could be maintained in vivo for a long time [713].

In contrast, those established from uninfected brains all died after a short period of time. This problem was further addressed in a serial killing experiment, in which Liberski, working in Gajdusek's laboratory found by means of electron microscopy that astrocytosis paralleled the spongiform changes in parietal cortex and adjacent corpus callosum of mice infected with the Fujisaki strain of CJD virus [679]. As dilated and swollen astrocytic processes were found occasionally in both CJD virus-infected and sham-inoculated animals these were regarded as a result of local suboptimal fixation and not as part of CJD neuropathology, as reported by others [550–553].

Recently, the neuropathology of spongiform encephalopathies has been reproduced in transgenic mice created by microinjection of the chimeric murine cosmid containing a codon 101 substitution within and PrP ORF [515]. The 101 codon substitution is regarded as equivalent to that found in GSS syndrome [511–512]. Transgenic mice presented severe spongiform changes but rather mild or moderate astrocytosis, except in the cerebellum where severe Bergmann radial gliosis was observed. In conclusion, spongiform changes rather than astrocytosis seems to be the primary neuropathological phenomenon.

While astrocytic gliosis is a prominent and ubiquitous finding in CJD, kuru, scrapie and bovine spongiform encephalopathy, it still remains a controversial issue in GSS syndrome. Hudson et al. [516] found mild astrocytosis in 3 cases of GSS syndrome and this was associated mostly with amyloid plaques. In a case reported by Kuzuhara et al. [630] moderate astrocytosis of the cerebellar white matter was found, while severe astrocytic reaction was seen in the inferior colliculus. Vinters et al. [1030] reported astrocytic gliosis through the neocortex while, in contrast, Tateishi et al. [1002] explicitly found astrocytosis only in an area of concomitant infarct. Similarly, Ghetti et al. [409] Nochlin et al. [814] and Pearlman et al. [848] reported severe gliosis only in areas where numerous plaques and neuronal loss were also seen. In 3 cases of GSS syndrome observed by the author at the National Institutes of Health in Bethesda and in the Neurological Institute of the University of Vienna astrocytosis was found throughout cerebral and cerebellar cortex but never approached such a degree as found in CJD cases. Particularly, gemistocytic astrocytes were never seen in these cases. However, in a recent cases from the original Austrian Gerstmann-Sträussler-Scheinker family, the astrocytosis in cerebral cortex approached that in CJD brains and innumerable gemistocytic astrocytes were seen. Thus, the diversity of neuropathology of GSS is perhaps of the same magnitude as that of CJD. Astrocytes in GSS syndrome were characteristically elongated and slender reminiscent of pilocytic astrocytes. Unfortunately, no report on the ultrastructure of astrocytic reaction of GSS syndrome has been published.

4.4.2.2.3. The involvement of astrocytes in formation of amyloid plaques
Amyloid plaques are a neuropathological hallmark of slow virus diseases
and the involvement of astrocytes in plaque morphogenesis has been
discussed recently. Kuru plaques and multicentric plaques are characteristic features of kuru (or CJD) and GSS syndrome, respectively. Cortical
kuru (unicentric) plaques of GSS syndrome consist of amyloid fibrils
within a narrow extracellular space between distended astrocytic processes [101–104]. Amyloid fibrils invaginated "deeply the surrounding
profiles of astrocytes so that the filaments sometimes seemed to be
intracellular". Furthermore, the labyrinth-like formation of basement
membranes not connected to the vessel walls was reported. Such peripheral accumulations of astrocytic processes in close connection with the
amyloid fibrils was noted even in the earliest amyloid plaques [101–104].
This intimate association of amyloid and astrocytes in GSS syndrome led
Boellaard et al. [102] to coin the term *glial plaques*. Glial plaques are
plaques of slow virus disorders or at least GSS syndrome in contrast to
neuritic plaques of Alzheimer's disease (AD) [1056–1058]. As
dystrophic neurites are regarded as sites of amyloid processing and
accumulation in AD [1056] the surface of astrocytes may serve a similar
role in GSS syndrome and, possible in other slow virus diseases. However, as dystrophic neurites are encountered in plaques of scrapie, kuru,
CJD [362] and even GSS syndrome [1002–1004] it is difficult to reconcile
the last hypothesis based on the observation of the relative lack of
dystrophic neurites in glial plaques with their very presence in these
disorders.

4.4.2.4. Scrapie bovine spongiform encephalopathy (BSE), and chronic wasting disease

Generally, the degenerative brain pathology of scrapie consists of
neuronal loss, spongiform changes and astrocytosis (Fig. 27–29). All
three phenomena are encountered in different proportions in different
scrapie models, irrespective of whether they be natural or experimental.
The true nature of astrocytic changes is only poorly understood [325] and
the most pertinent problem in studying glial responses in scrapie is
whether astrocytic proliferation (hyperplasia) or only astrocytic hypertrophy or both are responsible for apparent increase in the number of
astrocytes observed under the light microscope.

In two earlier studies performed in goats with natural [472] and
experimental scrapie in rats infected with the "Chandler" strain of
scrapie virus [841] the problem of hypertrophy versus hyperplasia was
not settled. Hadlow [472], studying scrapie-infected dairy goats, reported
that both hypertrophy and hyperplasia with a predominance of the
former brought about an overall increase in the apparent number of

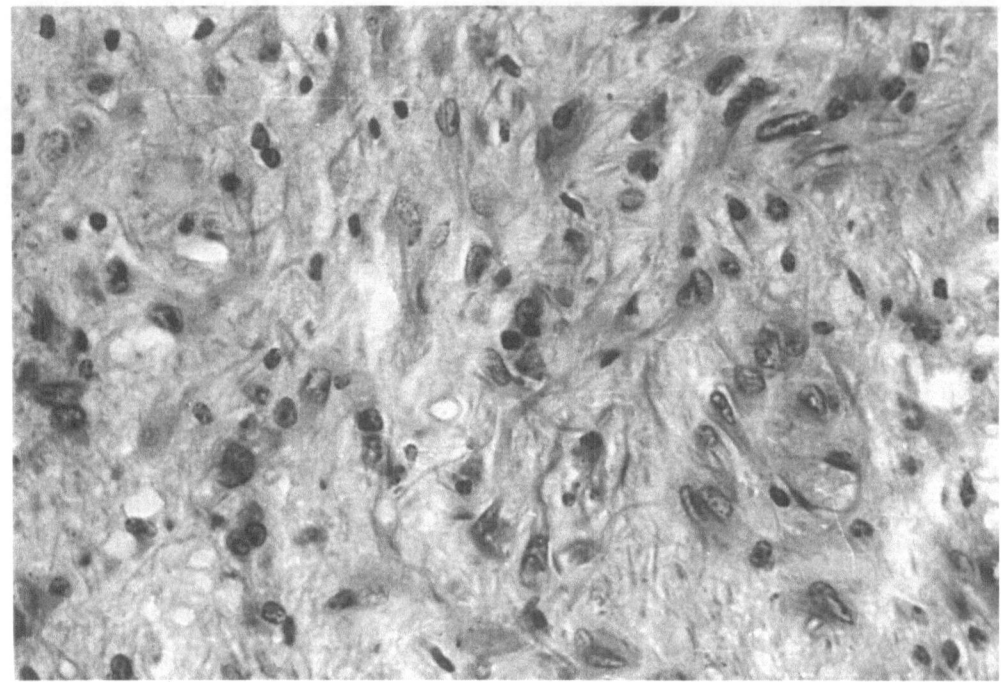

Fig. 27. Severe astrocytosis (gliocytosis) in the brain stem of a hamster infected with the 263K strain of scrapie virus. H & E. Original magnification, ×400

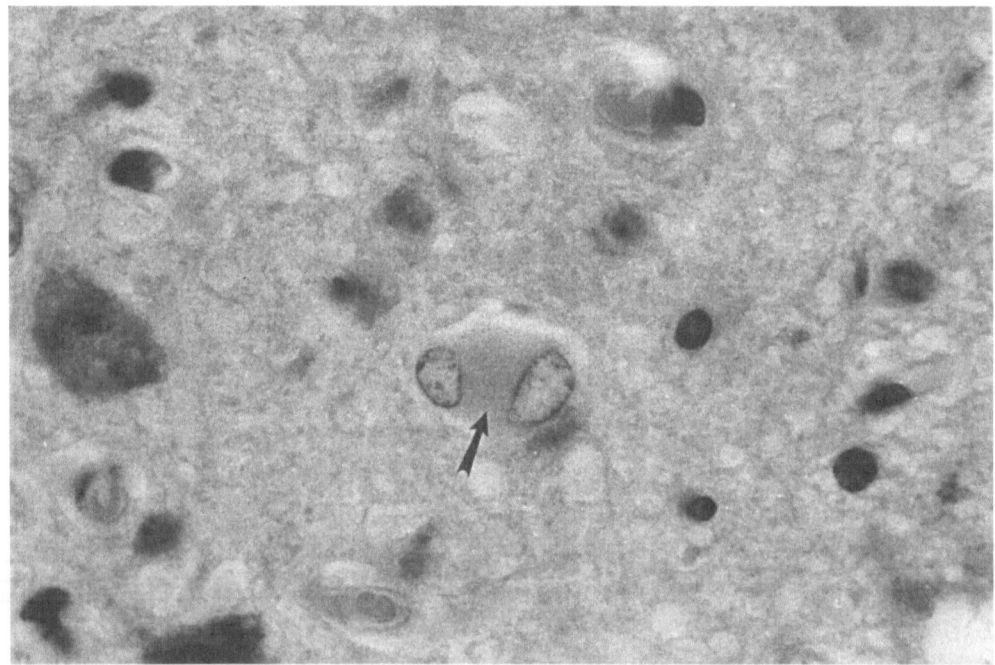

Fig. 28. Dividing astrocyte (arrow) in hippocampus formation of hamster brain infected with the 263K strain of scrapie virus. H & E. Original magnification, ×400

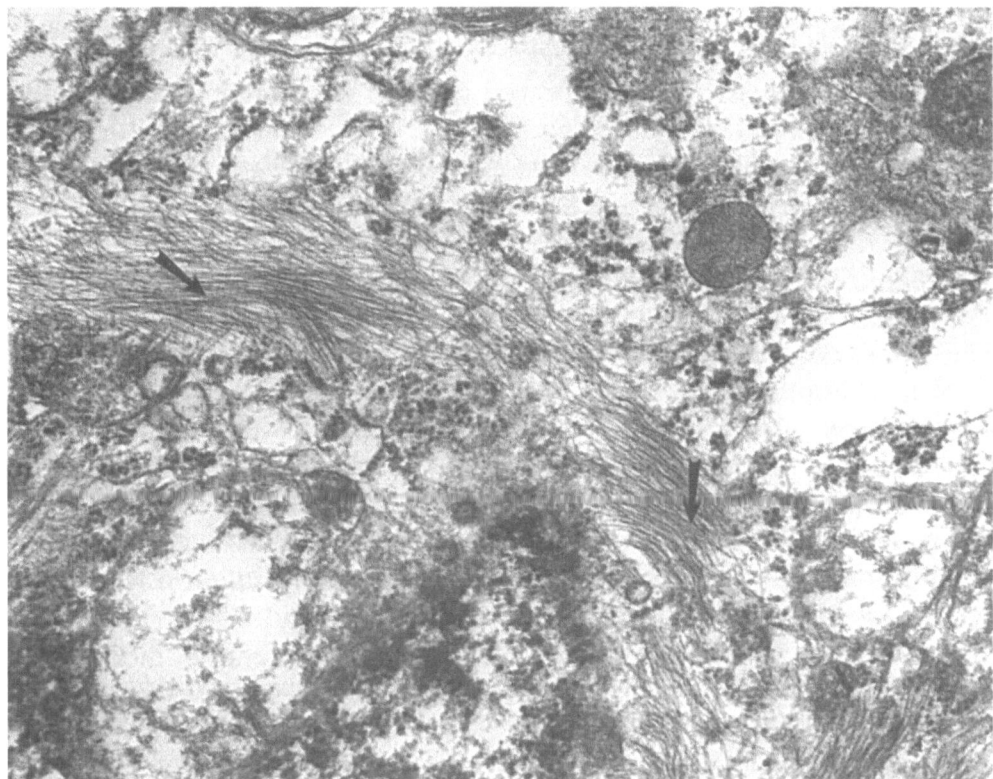

Fig. 29. Hypertrophic astrocyte from hamster brain infected with the 263K strain of scrapie virus. Note numerous glial filaments (arrows). Original magnification, ×12 000

astrocytes seen in sections stained with azure-eosinate and Cajal gold sublimate method. The estimation of the number of astrocytes was difficult, however, as the Cajal method also stains different proportions of astrocytes in normal goat brains. The "scrapie" astrocytes were not always easily distinguished from pleomorphic "normal" astrocytes, but typically they measured up to 14 um in diameter and contained a few chromatin granules. Particularly, the presence of kidney-shaped, elongated and irregularly lobulated nuclei, frequently clustered as 3 or 4 cells reminiscent of those in Alzheimer II cells or "naked nuclei" of Alzheimer has been reported. Furthermore, in the subpial regions of colliculi, mammillary bodies, subependymal parts of diencephalon and parts of the brain stem the pigmented astrocytes characterized by the presence of nuclei surrounded by round and elongated granules were seen. Hypertrophy and proliferation of astrocytes were confined to the affected (vacuolated) gray matter and paralleled that of neuronal loss. The adjacent white matter was involved only occasionally. However, in the midbrain and several thalamic nuclei astrocytosis outnumbered that of vacuolation. Furthermore, cerebral cortex characterized by minimal spongiform changes presented disproportionately severe astrocytosis.

Hypertrophic astrocytes appeared early in the course of disease and regressive changes in these cells has not been seen. The tendency to form glial fibers was scanty as demonstrated with Holzer stain.

Topographically, different brain regions were involved to different degree. The lesions were bilateraly symmetrical and the boundaries between affected and unaffected regions were remarkable sharp. Dense astrocytosis was observed in the pallidum, septal nuclei, and diencephalon. The last structure was most regularly affected. Moderate astrocytosis was seen in striatum and the brain stem where hypertrophy prevailed above proliferation. The minimal astrocytic hypertrophy with slight proliferation was seen in deeper layers of otherwise mostly unaffected cerebral cortex, and cerebellar cortex. In the last area radially arranged glial processes and clusters of Bergmann glial constituted distinctive features. Only in one section of cerebral cortex the site of inoculation was found consisting of small cluster of hypertrophic astrocytes and lipid-laden macrophages. In strict contrast to some murine scrapie models the hippocampus formation was unaffected but diffuse astrocytosis was evident mostly between pyramidal cells and alveus. Hypertrophy and "undoubted" proliferation has been also detected in natural scrapie in goats [474]. Topographically, both experimental and natural caprine scrapie were alike, except that the striatum, pallidum and septal nuclei were only slightly affected in the latter. On the contrary, Pattison and Jones [841] explicitly stated that hypertrophy but not proliferation was a feature of rats infected with the Chandler strain of scrapie virus. Astrocytosis mostly paralleled the vacuolation and was greater after intracerebral than after intraperitoneal inoculation. Furthermore, astrocytosis preceeded the vacuolation by 14 days. Astrocytic end-plates were hypertrophic, and in the later stages of disease "capillaries appeared to be embedded in swollen, darkly staining astrocytic cytoplasm".

This problem was settled in part by Fraser [343, 945] in several models of murine scrapie and by Liberski [659–660, 663, 669] in hamsters infected with the 263K strain of scrapie virus. Fraser [343] introduced the term *gliocytosis* to denote proliferation of astrocytes accompanied by some changes in their morphology and substantial proliferation of rod-like microglial cells. In murine scrapie gliocytosis, encountered in hippocampus and thalamus, is an extremely rare phenomenon found in approximately 3% of ten thousand scrapie-affected brains examined. Gliocytosis occurs in a wide range of scrapie isolates passaged in different strains of mice but almost exlusively after intracerebral inoculation (256 examples after intracerebral inoculation out of 260 overal studied [343]). In more detailed studies of gliocytosis (sclerosis) of the hippocampus formation, Scott and Fraser [945] found its presence paralleled that of severe vacuolation in LM mice and F_1 cross between C57 and VM mice inoculated intracerebrally with the ME7 strain of scrapie virus. In contrast to the data of Pattison and Jones [841] vacuolation always preceeded the appearance of hippocampal sclerosis.

Liberski found gliocytosis in hamsters infected with the 263K strain of scrapie in much higher proportion than that of murine scrapie models [659–660, 663, 669]. Both astrocytic hypertophy and proliferation were observed. Astrocytosis apparently correlated with spongiform changes but not with neuronal loss. In the hippocampus formation, astrocytic changes were seen in both the pyramidal cell layer and the granular cell layer of fascia dentata. Astrocytic hyperplasia was evident and different stages of mitoses were recognized. Many astrocytes were similar to "naked nuclei" of Alzheimer II cells, others contained lobulated and bizarre nuclei more reminiscent to those of hypetrophic reactive astrocytes or astrocytes encountered in multifocal leukoencephalopathy. The presence of glial fibers and Rosenthal fibers, regarded as products of gliofilament condensation and degeneration, were frequently noted. Proliferation of astrocytes was accompanied by rod-like microglial cells.

A few ultrastructural studies provided a consensus concerning glial changes [663, 195, 201–202, 669]. Astrocytes did not show any features which discriminated them from reactive astrocytes of a plethora of neurodegenerative disorders.

The differences in the proportion of abundant glial reaction (gliocytosis) between hamster and murine scrapie obviously needs further experimentation. The 263K strain of scrapie virus emerged in a process of strain selection over the first two or three passages in hamsters [580, 582]. Because of this the description of the pathology of earlier passages were not typical of the 263K strain although the presence of severe astrocytosis was stressed [574, 722]. Whether the strain of scrapie virus or the host (hamster vs. mouse) contributed to the high incidence of gliocytois in scrapie-infected hamsters, as compared to murine scrapie models, is yet to be established.

A few "serial killing" experiments performed so far provide conflicting data on whether astrocytosis appears before or after vacuolation. Marsh and Kimberlin found hypertrophic astrocytes in scrapie-infected hamsters 9 weeks after intracerebral inoculation and preceeded vacuolation by 2 weeks [722]. This initial astrocytic hypertrophy was first observed at the pia-arachnoidoid surfaces and adjacent to the ventricles. In contrast, Liberski and Alwasiak [667] demonstrated that astrocytosis actually followed vacuolation in hamsters infected with the 263K strain of scrapie virus. However, most of the "early" vacuoles appeared to be nonspecific as revealed by means of an electron microscope [669] but the number of reactive astrocytes increased slightly 5 to 6 weeks after intracerebral inoculation. By the same token, unquestionable scrapie-specific vacuoles appeared 8 weeks postinoculation while at that time astrocytosis unequivocally surpassed vacuolation. Masters et al. [735] found astrocytosis detectable at week 7 or 5 by means of routine neuropathological staining or indirect immunofluorescence with monoclonal antibodies C10D5, respectively. Unequivocal spongiform changes appeared in this model 7–8 weeks after inoculation. While the spongiform changes stabilized in intensity at week 9–10 postinoculation, the number of astrocytes increased steadily until the clinical phase of disease and thus paralleled the steady increase in the infectivity titer. This correlation may suggest astrocytes as a target for virus replication and not merely a passively reactive cells. Unfortunately, both experimental studies suffered from the obvious weakness of the use of "poorly vacuolated" models. Thus, the problem whether astrocytosis is a reaction toward the destruction of neuronal elements or astrocytes undergo primary proliferation and hypertrophy cannot be settled.

The data on astrocytic reaction in bovine spongiform encephalopathy (BSE) and chronic wasting disease (CWD) are very limited. In the first report of BSE by Wells et al. [1041] mild gliosis was noted in BSE-affected cattle brains. Furthermore, these data are difficult to interpret as normal cattle brain is characterized by very abundant fibrous astrocytes. Liberski et al. (unpublished observations) found numerous hypertrophic astrocytes, not infrequently bi-nucleated and containing abundant glial filaments, accompanied the neuronal degeneration. Analogously, in chronic wasting disease (CWD) in mule deer, hybrids of mule deer and white tailed deer and Rocky Mountain elk, numerous hypertrophic astrocytes were noticed (Guiroy, Liberski, Yanagihara and Gajdusek, unpublished observations).

4.4.2.5. The blood-brain-barrier in slow virus diseases

The blood brain barier (BBB) consists of tight junctions between the capillary endothelial cells. However, the functioning of BBB is also determined by the astrocytic endfeet [743] which justifies an inclusion in this chapter data concerning BBB impairment in scrapie.

Wisniewski et al. [1051] reported a focal leakage of horseradish peroxidase in scrapie-affected mouse brain. This was particularly prominent in areas where amyloid plaques were present. The "leaking" endothelial cells did not differ ultrastructurally from those unaffected but increased numbers of transport vesicles ere noted. To further characterize the capillary endothelial cells several enzymes were studied cytochemically [1032]. While in control mice alkaline phosphatase was distributed in the form of small patches at the luminal side of the vessel, in scrapie-affected animals it shifted also to the abluminal side. The 5'-nucleotidase was found not only at the luminal side of the vessel wall but also within the endothelial cytoplasm, on the surface of myelin sheaths and the membranes of glial cells. This enzyme was also re-distributed to the abluminal side of the capillary vessel in scrapie-affected mice. In contrast, the topography of NDPase and TPPase was only slightly changed in scrapie-affected animals. The biological significance of the positional shift of these enzymes is uncertain, but it may reflect the impaired BBB as demonstrated by HRP leakage. Whether and how these phenomena contribute to scrapie-related neuropathology is totally unknown.

4.4.2.6. The particular forms of astrocytic reaction in unconventional slow virus diseases

SSVE are polioencephalopathies and corresponding fine structural changes of the grey matter are relatively well described. Recently, however, the panencephalopathic type of CJD, characterized by predominant involvement of the white matter, has been reported (*vide supra*) and axonal and myelin pathology at the ultrastructural level has been described [679, 681, 686]. In mice infected with the Fujisaki strain of CJD virus, myelinated axons present various pathological changes (Fig. 30–33). Initially the myelin sheath is separated by cytoplasmic tonques of astrocytes and macrophages (microglial celles) into several concentric bands with increased myelin periodicity. Cellular processes penetrate between layers of myelin and lift away the outermost lamellae. Then a complicated labyrinth of concentric cellular processes, clearly belonging to either astrocytes or macrophages, invest myelinated axons. In the terminal stages, axons completely denuded of myelin are seen in the center of a concentric network of cellular processes. The myelin fragments penetrate into astrocytes or macrophages where they undergo digestion.

Interestingly, the ultrastructural picture of intramyclin vacuoles is most compatible with that produced in organotypic cultures of mouse spinal cord with tumor necrosis factor (TNF)α [953] (*vide infra*), a lymphokine released from activated macrophages and astrocytes [127, 909, 952–953]. This section, however, concerns the mechanism by which the myelin sheath is removed from the axon and then engulfed by the astrocytes and macrophages.

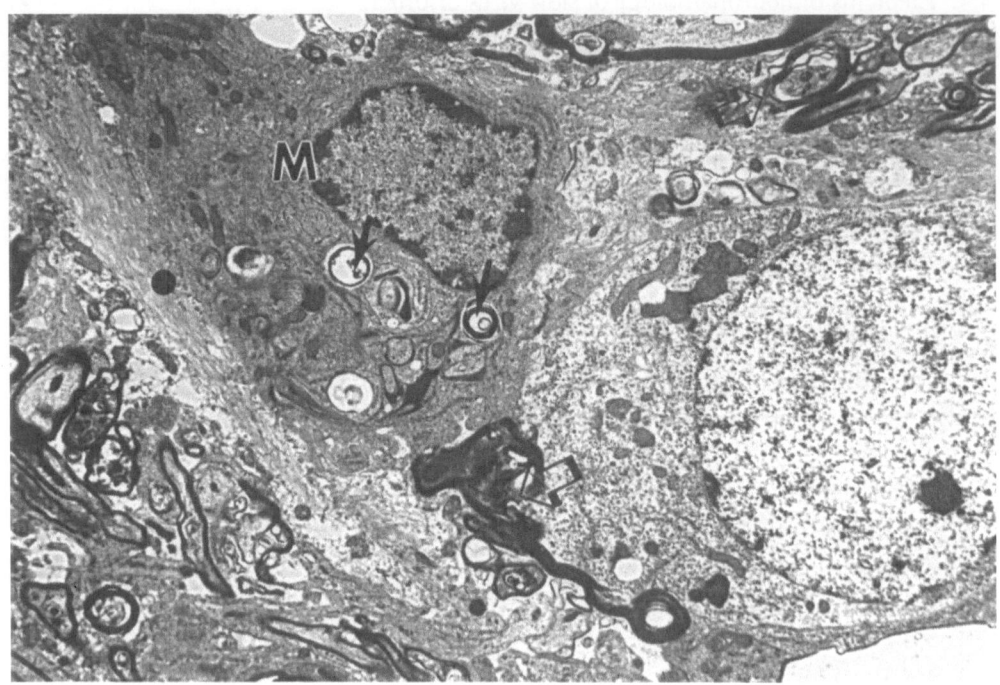

Fig. 30. Cellular reaction in corpus callosum of mouse infected with the Fujisaki strain of CJD virus. Note macrophage (*M*) containing myelin debris (arrows), glial process (circle) and degenerating myelinated fibers (open arrows). Unpublished work of P.P. Liberski, R. Yanagihara, C.J. Gibbs, Jr, and D.C. Gajdusek, the National Institutes of Health, Bethesda, USA. Original magnification, ×12 000

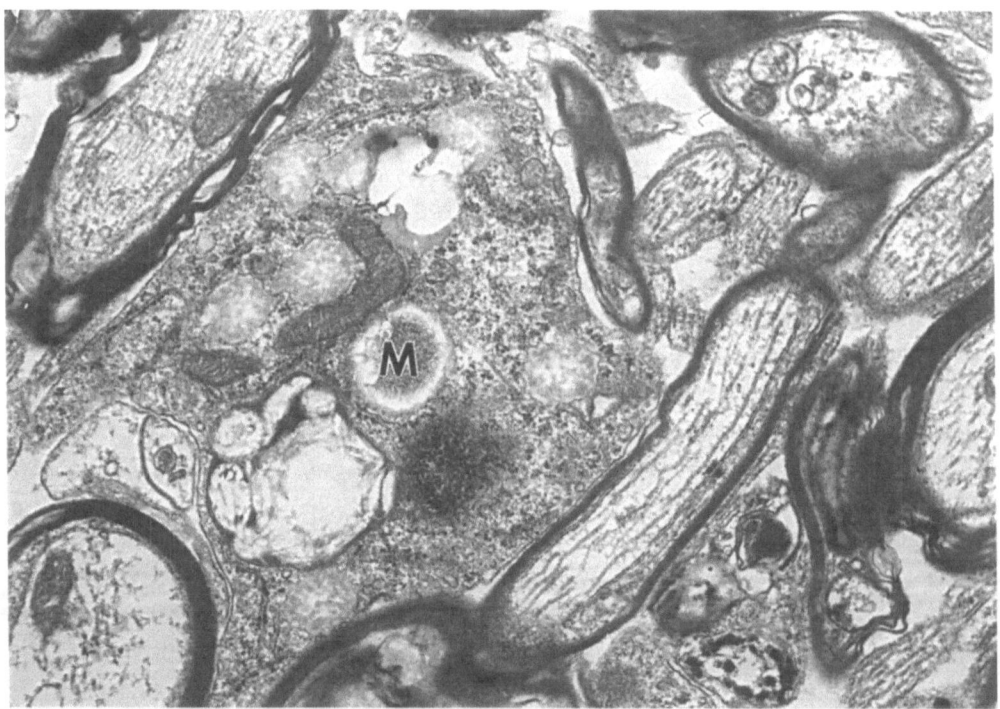

Fig. 31. A macrophage (*M*) in close contact with several degenerating myelinated fibers in corpus callosum of mouse infected with the Fujisaki strain of CJD virus. Unpublished work of P.P. Liberski, R. Yanagihara, C.J. Gibbs, Jr, and D.C. Gajdusek, the National Institutes of Health, Bethesda, USA. Original magnification, ×12 000

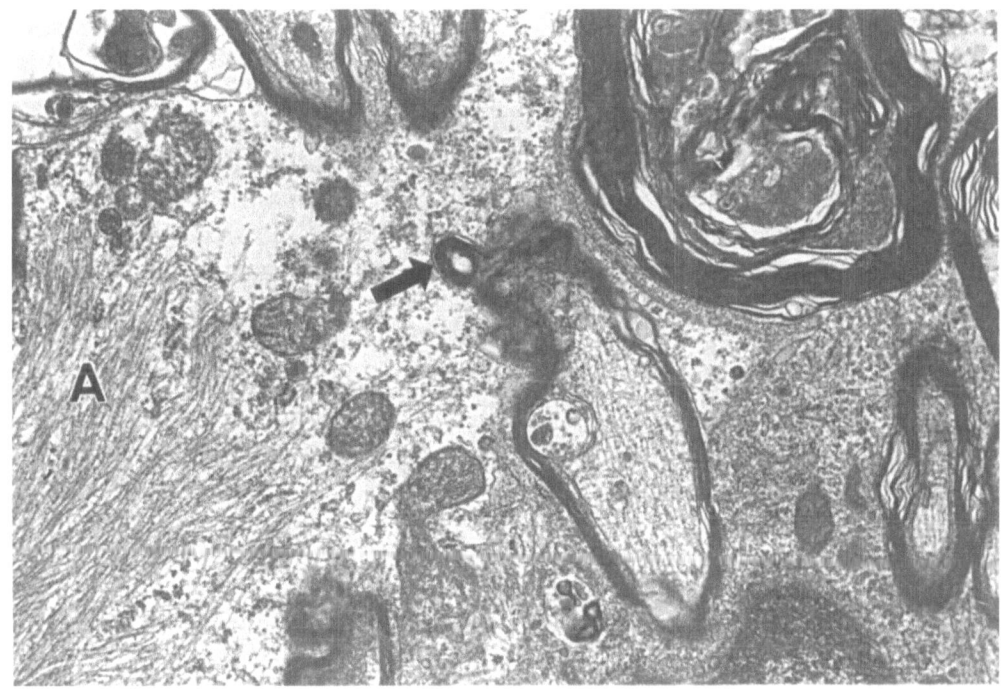

Fig. 32. A myelinated fiber (arrow) enveloped by an astrocytic process (*A*) to be later digested within this cell. Unpublished work of P.P. Liberski, R. Yanagihara, C.J. Gibbs, Jr, and D.C. Gajdusek, the National Institutes of Health, Bethesda, USA. Original magnification, ×12 000

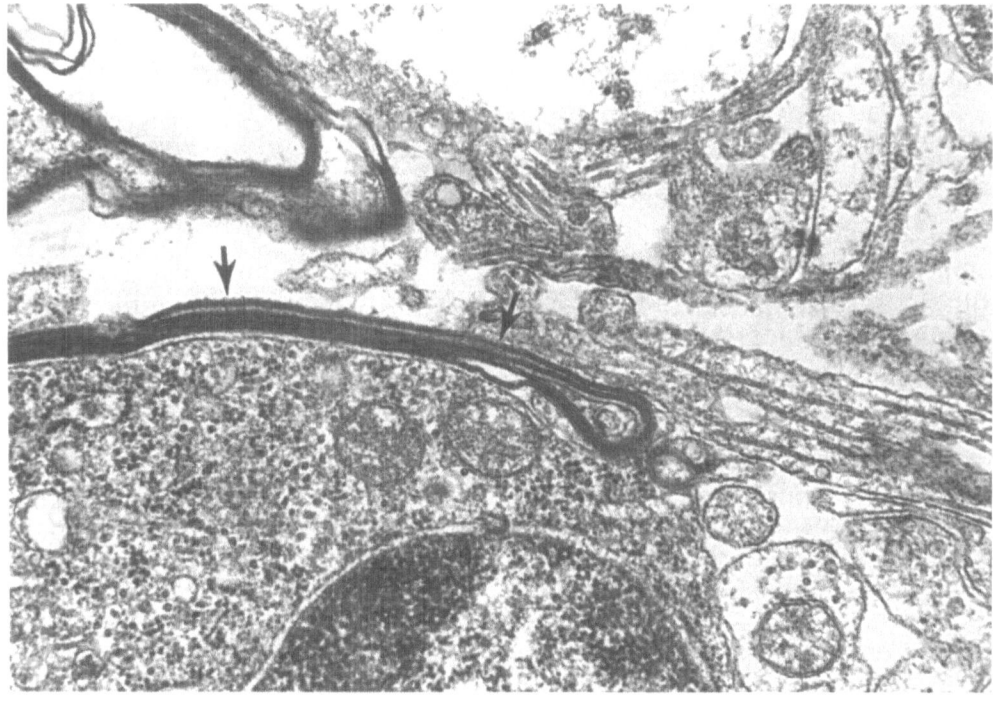

Fig. 33. Collapsed myelin (arrows) in close contact with a macrophage of mouse brain infected with the Fujisaki strain of CJD virus. Unpublished work of P.P. Liberski, R. Yanagihara, C.J. Gibbs, Jr, and D.C. Gajdusek, the National Institutes of Health, Bethesda, USA. Original magnification, ×12 000

The sequence of pathologic events described above is remarkably similar to that involved in the pathogenesis of multiple sclerosis (MS) [858] and experimental allergic encephalomyelitis (EAE) [892-898], except for the obvious lack of lymphocytes and plasma cells.

The invasion of myelin sheath by mononuclear cells originated from the blood was first demonstrated for EAE by Lampert [638-639] and for MS by Sluga [969]. The myelin lamellae were stripped away and later digested by macrophages. This mechanism of demyelination was confirmed in other experimental models of EAE [859], canine distemper, [1055] and visna [406]. Thus, this phenomenon is commonly encountered in different and unrelated demyelinating conditions. At the chronic stage of demyelinating lesions the active phagocytosis terminates, leaving axons invested by processes of fibrous astrocytes [433, 896, 1054-1055]. Furthermore, axons invested by astrocytic processes were observed during serum-induced demyelination in vitro [629, 898] or even in certain heredodegenerative processes [497]. These axons enveloped with cellular processes are basically identical to those encoutered in experimental CJD, except that axons in purely demyelinating conditions were mostly totally naked. Thus, it may implicate that the process of myelin damage in CJD is either incomplete or abortive. Basically the same conclusion of abortive host reaction against the scrapie virus has been recently reached on the basis of increased expression of beta-2-macroglobulin gene by molecular cloning rechniques [300]. The concentration of beta-2-macroglobulin, associated with the major histocompatibility (MHC) class I antigen, is increased in several malignant and infectious processes. It therefore appear, that SSVE evoke an active response "which is not apparent histologically" [300]. Thus, a basically similar mechanism is operative via different cellular components.

As mentioned above, vacuolation of myelinated axons in experimental CJD [679, 681, 686, 931], scrapie [331, 669] and kuru [640-641] closely resembles that induced by tumor necrosis factor-alpha (TNF-alpha) in mouse spinal cord cultures [953] and that found in demyelinating diseases. To determine whether lymphokines released from activated macrophages and astrocytes are involved in the pathogenesis of myelin dilatations in experimental CJD, brain tissues from CJD virus-infected mice were examined by the avidin biotin immunohistochemical technique, using a polyclonal rabbit antiserum raised against murine TNF-alpha (Liberski, Nerurkar, Yanagihara, Gajdusek, unpublished data).

Marked astrocytosis was first noted 14 weeks following intracerebral inoculation, coincident with the onset of myelin dilatation. All astrocytes stained intensely with antiserum against GFAP (vide supra), but only hypertrophic, gemistocytic bizzare astrocytes in areas of extensive myelin sheath vacuolation exhibited robust immunoreactivity with the antibody against TNF-alpha.

Tumor necrosis factor alfa/cachectin is a lymphokine mediating several responses to infection and cancer [359, 1016–1018]. In the central nervous system TNF alfa is released from activated astrocytes and macrophages/monocytes [909] and the role of TNF in demyelinating diseases as multiple sclerosis [485] and AIDS-related vacuolar myelopathy [458, 740, 1017] has recently been suggested.

The functional significance of TNF alpha within astrocytes is not clearly understood. Cultured astrocytes released multilineage hematopoietic colony-stimulating activity as recently demonstrated in murine bone marrow colony-stimulating assays [359]. Furthermore, Massa et al. [727] found TNF alpha-mediated amplification of class II (I^a) MHC antigens on astrocytes induced by interferon gamma (IFN) or infection with measles virus [687]. Bacterial adjuvants as lipopolysaccharide and muramyl dipeptide induced I^a antigen on astrocytes, and this induction was, in turn, amplified by alpha-TNF. Lipopolysaccharides and neurotropic viruses stimulate production of TNF-alpha in astrocytes [687]. All these data suggest astrocytes to participate in a complicated network of immunological responses within the brain.

4.4.2.7. Expression of glial fibrillary acidic protein (GFAP)

First isolated from old multiple sclerosis plaques [310–311] glial fibrillary acidic protein (GFAP) is a major protein component of glial filaments, a class of intermediate filaments specific for astrocytes. GFAP is a 49 kDa protein accompanying on polyacrylamide gel electrophoresis (PAGE) by products of a proteolytic cleavage of apparent molecular weigth down to 40 kDa [310–311]. It is regarded as a useful marker for normal, hypertrophic and neoplastic astrocytes.

Gliosis is a prominent feature in almost all models of subacute spongiform virus encephalopathies (*vide supra*), however, the use of GFAP as a marker for gliosis has only recently been recognized. Mackenzie studied Compton mice infected with the "Chandler" (139A) strain of scrapie virus by means of immunohistochemistry and antibodies against GFAP [699]. The use of GFAP as an astrocytic marker proved to be extremely useful, particularly for quantitative estimation of astrocytosis, previously so complicated by the insensitivity of routine H&E staining [659, 663] and capriciousness of Cajal metal impregnation technique [343]. Abundant GFAP-immunopositive astrocytes were seen in corpus callosum, hippocampus, cerebellum and spinal cord. This location was disease-specific, as a different pattern of GFAP-immuno-reactivity was observed in mice infected with Semliki forest virus or mice intoxicated with cuprizone. Furthermore, GFAP-immunoreactive astrocytes were easily detected in scrapie-affected sheep. Noteworthy, there was no correlation between clinical signs of scrapie and GFAP-immunoreactivity in brain stem, either between the distribution of spongiform vacuoles and GFAP-immunoreactity. The hyperproduction of GFAP was recently confirmed in mice infected with the C506 strain of scrapie virus and in scrapie-affected sheep [298].

Such astrocytic reaction marked by robust GFAP immunostaining was called "hypergliotic" [254] as being out of proportion to the degree of neuronal damage. Furthermore, the regional distribution of GFAP-immunoreactive astrocytes paralleled that of PrP [254–255, 257]. GFAP concentration, measured in homogenates of whole scrapie-affected hamster brain, increased 20 to 30 days following intracerebral inocula-tion [254]. The initial rise was slow and accelerated some 60 days postinoculation, when the first signs of clinical scrapie were also observed. PrP 27–30 (a proteinase K-cleavage product of the precursor protein PrP 33–35^{sc}) was first detectable approximately 45 days postinoculation; thus, the accelerated increase of GFAP concentration clearly followed that of PrP. A similar rise of PrP followed by that of GFAP was observed in selected brain regions. For instance, in the thalamus GFAP concentration increased 50 to 55 days postinoculation preceded by increased concentration of PrP 33–35^{sc} by approximately 5 to 10 days. It thus seemed that PrP induces reactive gliosis. To test this hypothesis directly the influence of PrP on astrocytic growth *in vitro* has been studied [254]. Primary astrocytic cultures (more than 90% of astrocytes) exhibited a numerical increased of astrocytes in the third day of the presence of PrP. Furthermore, a dramatic increase in GFAP-immunopositive glial filaments was observed following PrP supplementa-tion of the culture medium. Such an increase in GFAP concentration paralleled that of an increase of GFAP mRNA, as detected by a cDNA probe encoding GFAP. In conclusion, PrP was found to be a potent growth factor for astrocytes. It has been suggested that PrP released into extracellular spaces induces reactive astrocytic gliosis. Subsequently, PrP^{Sc} has been localized also to astrocytes [130]. Following formic acid pretreatment PrP^{sc} was immunolocalized to astrocytes in scrapie-infected but not in control animals. In serial experiments, PrP^{sc} was detected in astrocytes 8 weeks following inoculation, then increased to the stage when it was detetctable diffusely within neuropil. The accumulation of PrP^{sc} within astrocytes preceded astrocytosis first observed 12 weeks following inoculation and the appearance of scrapie amyloid, 16 weeks postinoculation. From the above discussion it seems that PrP, GFAP, and astrocytosis are somehow related, and that PrP is the growth factor for astrocytes.

4.4.3. The neuropathology of amyloid plaque

4.4.3.1. Kuru

The general neuropathology of kuru, the first human slow virus disorder discovered in 1957 by Gajdusek and Zigas [389–390] and transmitted to nonhuman primates by Gajdusek, Gibbs and Alpers [388], was discussed in a previous pragaraph. In this chapter the neuropathology of amyloid plaques, a hallmark of the whole group of slow virus disorders, will be discussed in detail.

The amyloid plaques encountered in kuru, currently known as kuru plaques, were first described in 1959 by Klatzo and Gajdusek [612–613] as plaque-like structures. Plaques, measuring 20 to 60 um in diameter, were found most frequently in the cerebellum and occasionally in the basal ganglia. In the cerebellum, they were most numerous in the granular layer and sparsely distributed in the molecular layer and in the

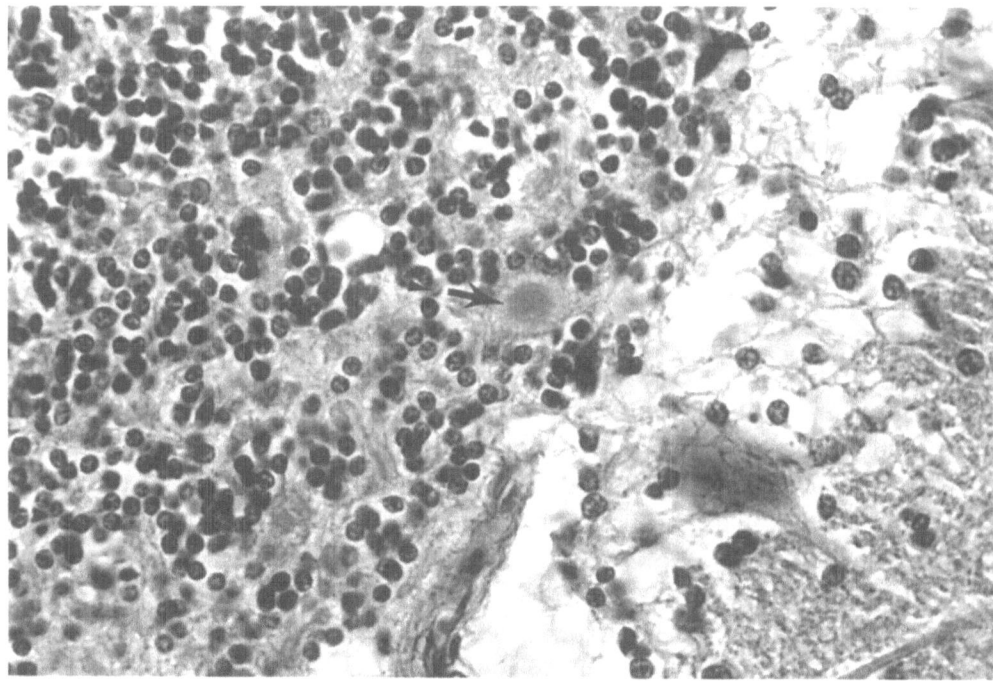

Fig. 34. Stellate plaque, "kuru plaque" (arrow) from a case of kuru (Kupenota). Courtesy of Dr. D.C. Gajdusek and Dr. C.J. Gibbs, Jr, the National Institutes of Health, Bethesda, USA. H & E. Original magnification, ×160

white matter [612–613, 809, 958]. In a more recent study [948], however, plaques were also seen in the occipital and motor cortices; thus, plaques are more widespread then previously recognized.

Kuru plaques are spherical structures (Fig. 34), containing a dark core surrounded by a delicate fibrillary border. Plaques are weakly eosinophilic, moderately argentophilic, stained according to Bielschowsky, Holmes and Hortega's methods, and they are PAS positive, stain for amyloid with Congo red and they are birefringent under polarized light. In contrast to classical neuritic plaques of "cocarde" type of Alzheimer's disease (AD) / senile dementia of Alzheimer type (SDAT) [1057–1058] the peripheral "halo" composed of dystrophic neurites have not been found in kuru plaques by means of Von Braunmuhl staining [612–613]. However, glial cells, mostly microglial cells, and occasionally astrocytes are frequently observed surrounding these plaques.

4.4.3.2. Creutzfeldt-Jakob disease (CJD)

Overall, more than 75% of kuru patients exhibit amyloid plaques [61–63, 612–613]. In comparison, only a small number of CJD patients of European or American descent show plaques [729].

Chou and Martin [213] were the first to describe kuru plaques in a CJD case. PAS-positive, congophilic plaques, measuring 4 to 40 um in diameter, were seen in the granular layer, cerebellar white matter and molecular layer, in descending frequency. Ultrastructurally, they were composed of fibrils measuring 10 to 13 nm in diameter, interwoven at the center and arranged radially at the periphery.

Amyloid plaques have been detected in a small subsequent number of cases [5, 499, 626, 673, 1064–1065, 1073]. The morphology and tinctorial properties of these plaques do not differ from those described by Chou and Martin [213] in a CJD case or from those encountered in kuru (Fig. 35–36). The presence of glial cells around plaques have been noted [626]. Plaques in CJD and kuru are mostly spherical structures or represent "a more intricate branching structure" when reconstructed from serial consecutive sections [626]. Interestingly, small plaques have a tendency to cluster together [5, 626] and this phenomenon is reminiscent of multicentric plaques of Gerstmann-Sträussler-Scheinker (GSS) syndrome (vide infra). Whether CJD cases with multiple small plaques clustered together (multicentric plaques?) should be re-classified as GSS syndrome cases is yet to be established.

Ultrastructurally, amyloid plaques are composed of fibrils densely interwoven to form a core surrounded by a peripheral lighter zone consisting of fibrils radially oriented [626]. The similarity of fibrils from

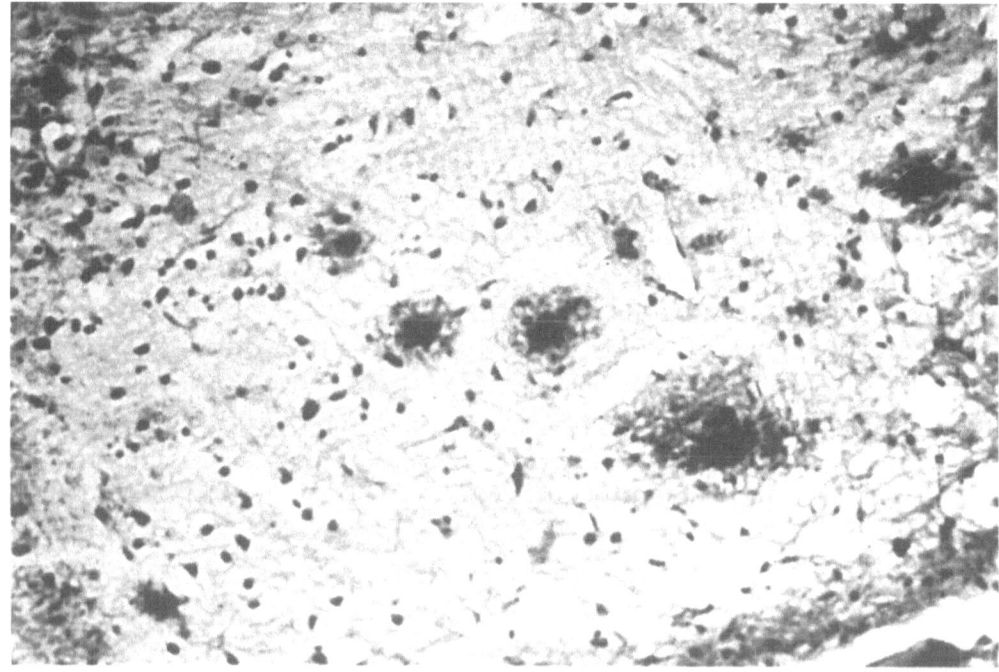

Fig. 35. Numerous kuru plaques. Original figure reproduced from the paper by Gerstmann, Sträussler and Scheinker [407]

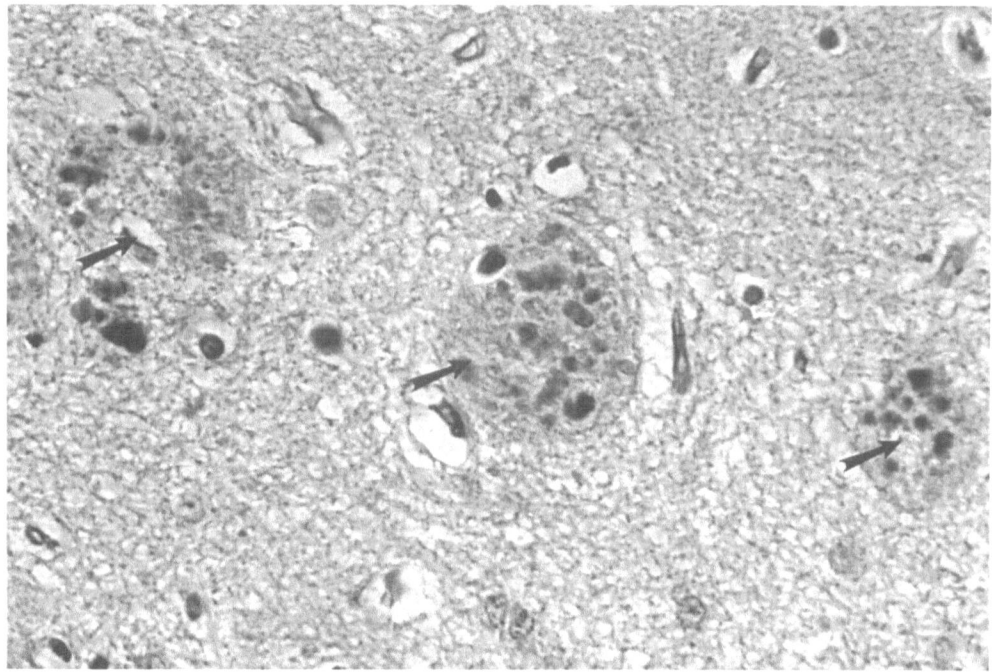

Fig. 36. Numerous multicentric plaques (arrows) from a case of GSS syndrome. Material courtesy of Dr. D.C. Gajdusek, the National Institutes of Health, Bethesda, USA. PAS. Original magnification, ×400

these plaques to those of senile plaques of AD/SDAT have been noticed [510]. Lattices consisting of parallel arrays of delicate fibrils or lamellae were seen at the plaque periphery [626]. As in kuru, dystrophic neurites were not detected with silver staining or by thin-section electron microscopy [510]. Interestingly, a familial CJD case characterized not only by plaques but also by congophilic angiopathy, a phenomenon highly characteristic of AD/SDAT has recently been reported [546].

As mentioned previously, Japanese CJD patients with abundant plaques represent a separate subgroup due to their long clinical course, severe involvement of white matter and frequent absence of typical periodic synchronous discharge on EEG [604–608, 1000, 1004, 1006, 1064–1065]. In contrast to American and European CJD patients in whom plaques are mostly confined to the cerebellum, Japanese patients, show widespread distribution of plaques in cerebrum, cerebellum, brain stem, and spinal cord [1064–1065]. Furthermore, a wide spectrum of plaque types have been observed: primitive plaques without a noticeable core, unicentric plaques of "cocarde" type, multicentric plaques, and kuru plaques. This plaque pleomorphism is characteristic for GSS syndrome but it seems that within a subgroup of Japanese CJD cases with plaques GSS syndrome is overrepresented as evidenced by a high proportion of CJD cases with the 102 codon mutation of the PRNP gene [297]. This mutation is regarded as specific for GSS syndrome [511–152].

Amyloid plaques have been reproduced experimentally but only at the first passage of CJD in mice [1003]. As plaques are observed mostly in the CJD cases of long duration and since the incubation period in primary transmission is much longer than that of subsequent passages, the time factor may clearly contribute to the presence of amyloid plaques in the primary but not in serial transmissions. Totally, 30 out of 31 CJD inocula produced plaques [607] and, suprisingly, plaques were not seen in experimental GSS syndrome (3 cases). However, plaques were observed following primary passage from a case belonging to the GSS syndrome family from which the Fujisaki strain of CJD virus has been isolated [1002]. Recently, brain tissues obtained from the CJD-affected older brother of the proband was also transmitted to mice with the production of amyloid plaques [1002]. Thus, despite the widespread plaque formation in human GSS syndrome the production of plaques in experimental GSS syndrome is as variable as in experimental CJD in which experimental plaque production is independent of their presence in human brain used as inoculum.

In experimental CJD, spherical plaques, measuring 10 to 30 um in diameter, develop in areas adjacent to the lateral ventricles: corpus callosum, hippocampus, and parietal white matter [1003]. As in human CJD, plaques are weakly eosinophilic, are PAS-positive, stain green with Masson's trichrome technique, are weakly argentophilic, and are birefringent after Congo red or thioflavin T staining.

Ultrastructurally, they consist of 7 to 10 nm amyloid fibrils, interwoven at the center and radiating at the periphery. A few astrocytes and microglial cells are visible at the periphery, but dystrophic neurites are rare or totally absent. It seems that macrophages either play an active role in plaque formation or they are reactive toward the presence of plaques [495]. Plaques clustered in one area are surrounded by macrophages. The plasma membranes between macrophages and the plaque itself are frequently fragmented and amyloid bundles seem to protrude outside of the Golgi apparatus or rough endoplasmic reticulum. At this stage, macrophages apparently leave the amyloid plaque and amyloid plaques transform to appear as "felt-like structures" [495]. This mechanism of plaque formation is reminiscent to that postulated for AD/ SDAT [1057–1058]. Activated macrophages/microglia are also observed at the periphery of amyloid plaques in IM mice infected with the 87V strain of scrapie virus [1053]. Furthermore, nucleoside diphosphatase (NDPase) activity, as detected by immunoelectron microscopy, is associated with amyloid fibers within microglial cells but not with smooth endoplasmic reticulum and cell membranes as in control microglial cells. Thus, such intimate association of microglial cells and amyloid seems to be a common phenomenon for the whole SSVE group.

A separate group represent CJD cases with abundant and otherwise typical senile (neuritic) plaques, and thus represent a narrow "gray" area between CJD and SDAT/AD (Fig. 37–38) [131, 134, 145, 463, 498, 668, 675–676]. Whether such cases are a distinct pathological process with overlapping features of both CJD and SDAT/AD, or both disorders coexisting in the same patients is yet to be established. The presence of typical senile plaques in otherwise definite CJD cases is a rare phenomenon; among approximately 1200 CJD cases referred to Gajdusek's laboratory only 22 cases show senile plaques in excess to those found in age-matched control senile brains.

Hirano et al. [498] were perhaps the first to describe neuritic plaques in an otherwise typical CJD case. Ultrastructurally plaques were composed of numerous dystrophic neurites with an admixture of amyloid. In a case of familial neurodegenerative disease (classified as atypical GSS syndrome or AD/SDAT due to the presence of ataxia and dementia), de Courten-Myers and Mandybur [258] found numerous plaques virtually identical to those encountered in AD. Furthermore, abundant NFT and spongiform changes were observed in this case. Similarly two CJD cases from a large kindred have been described by Azzarelli et al. [31]. These two brothers were also characterized by NFT and numerous pleomorphic plaques including cerebellar kuru-type plaques. The chemical composi-

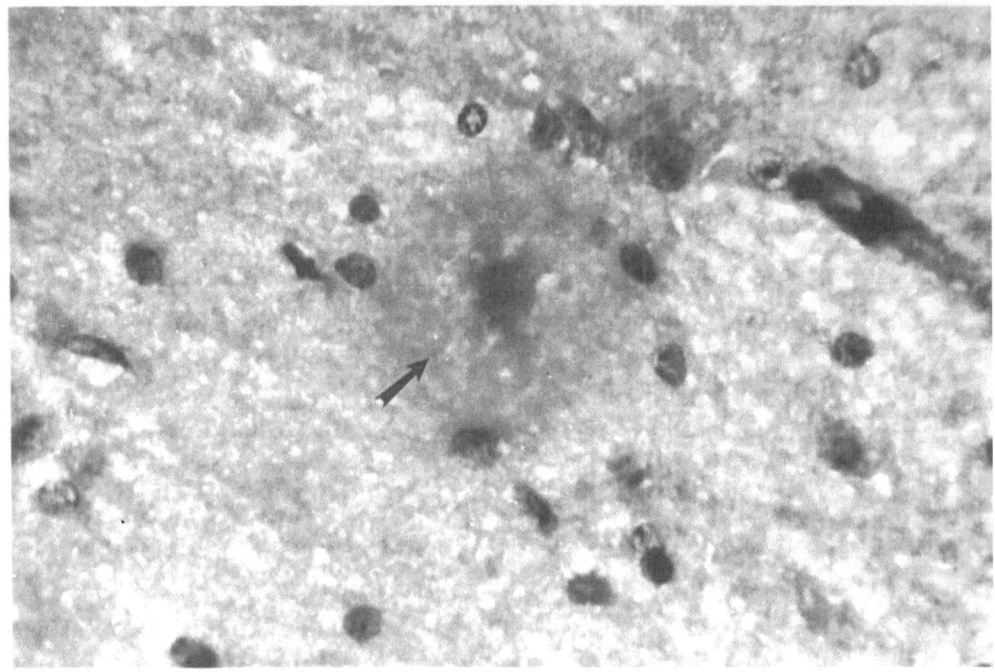

Fig. 37. Amyloid (neuritic) plaque (arrow) from an atypical case of CJD with plaques and paired helical filaments. Unpublished work of P.P. Liberski, J. Alwasiak and W. Papierz. These plaques were later proved to be A4-immunopositive [463]. A & E. Original magnification, ×400

tion of these plaques remained uncertain until immunohistochemistry showed them to be A4-positive in de Courten and Mandybur case [409]. The plaques in Hirano et al. [498] and Azzarelli et al. [31] cases have never been characterized immunohistochemically. Further, such cases were recently reported by Liberski, Papierz and Alwasiak [668, 675–676] and Brown et al. [131–134]. The definite CJD nature of Brown's case was further corroborated by isolation of scrapie-associated fibrils (SAF) and successful transmission. Transmission experiments of Liberski's case were still in progress at the time of this writing (Liberski and Brown, unpublished data).

The biological significance of these findings is uncertain and several hypotheses may be proposed. First, the coexistence of CJD and AD/SDAT may be purely coincidental. However, taking into account the rarity of CJD (but obviously not AD), and more so the rarity of CJD with plaques, this explanation sounds implausible. More tantalizing is the hypothesis that the very presence of one cerebral amyloidosis [367, 373, 378–379, 385] predisposes the development of another. Recently, Guiroy and Gajdusek [461–462] presented a hypothesis that viruses causing subacute spongiform encephalopathies act as so-called amyloid

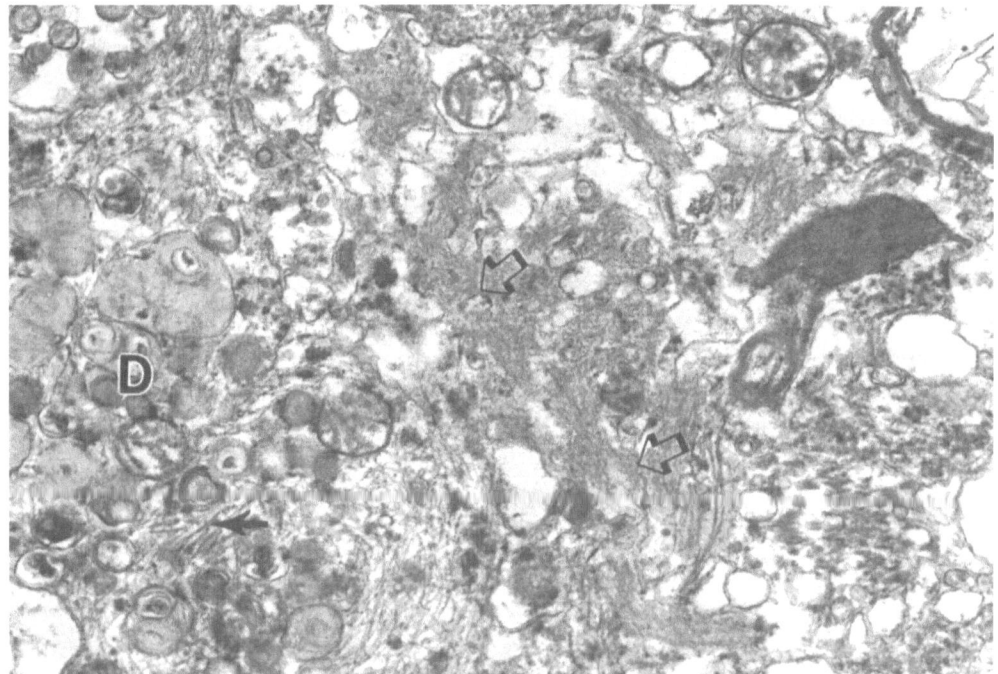

Fig. 38. Neuritic plaque from an atypical case of CJD with plaques and paired helical filaments. Unpublished work of P.P. Liberski, J. Alwasiak and W. Papierz. Note dystrophic neurite (*D*) containing paired helical filaments (arrow) and amyloid fibers (open arrows). Original magnification, $\times 10\,000$

enhancing factors, low molecular weight polypeptides, serving as nucleants in accelerating their own production and the formation of amyloid. It is plausible that CJD virus, irrespective of its molecular nature, starts transformation of a cellular isoform of prion protein (PrP^c) to its CJD isoform (PrP^{CJD}) acting similarly to "crystal template directing its own crystalization or crystal lattice from a source of pre-synthesized host precursor proteins, and an inorganic cation receptor nucleus" [378–378]. Such a self-perpetuating process may influence and enhance another process of a transformation of beta amyloid precursor protein into A4 amyloid of AD/SDAT. If the above hypothesis is valid, then the coexistence of CJD and AD/SDAT reflects the entanglement of both pathogenetic processes responsible for the ultimate deposition of molecularly different forms of amyloid.

4.4.3.3. Gerstmann-Sträussler-Scheinker (GSS) syndrome

GSS syndrome is an extremely rare familial disorder discovered in 1936 by Austrian investigators [407]. Neuropathologically, the syndrome is characterized by widespread plaque formation already illustrated in the original description (Fig. 35 and 39) [407]. The molecular layer of the cerebellum is most frequently involved (116 plaques/mm^2), following by the granular cell layer (19 plaques/mm^2), parahippocampal gyrus (15.9 plaques/mm^2), insular gyrus (11.5 plaques/mm^2), and superior frontal gyrus (10.6 plaques/mm^2) [409]. The position of GSS syndrome within the spectrum of CJD was established by transmission to nonhuman primates [729]. However, the close similarity to kuru based on a common neuropathology has been pointed out by Seitelberger [949] even before the slow virus era was initiated.

Plaques in GSS syndrome are pleomorphic (Fig. 39–43) [1, 100–106, 516, 630, 848, 850, 934, 936, 941]. In addition to kuru plaques, primitive plaques and senile plaques of "cocarde" type (typical for AD/SDAT), multicentric plaques (compound plaques) have been described in GSS syndrome [101–106], and these are regarded as a hallmark for this syndrome. Multicentric plaques consist of several small plaques clustered together or a central core of amyloid surrounded by many smaller amyloid globules. Noteworthy, while kuru plaques retain their individual form even when clustered together, multicentric plaques grow together, at least at the central region. It has been suggested that this is the reason

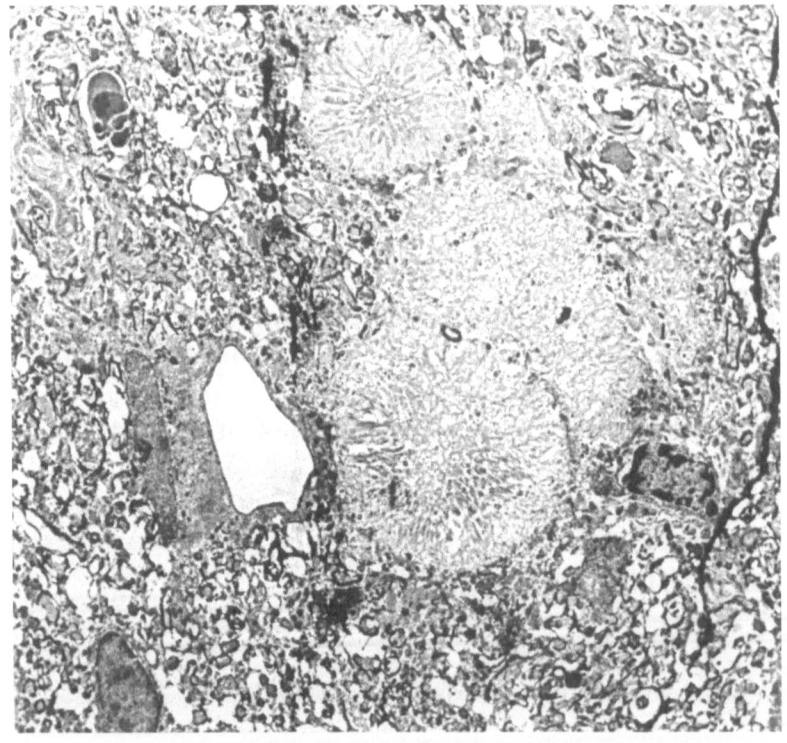

Fig. 39. Three kuru plaques from a case of GSS syndrome. Courtesy of Dr. J. Boellaard. Original magnification, ×3500

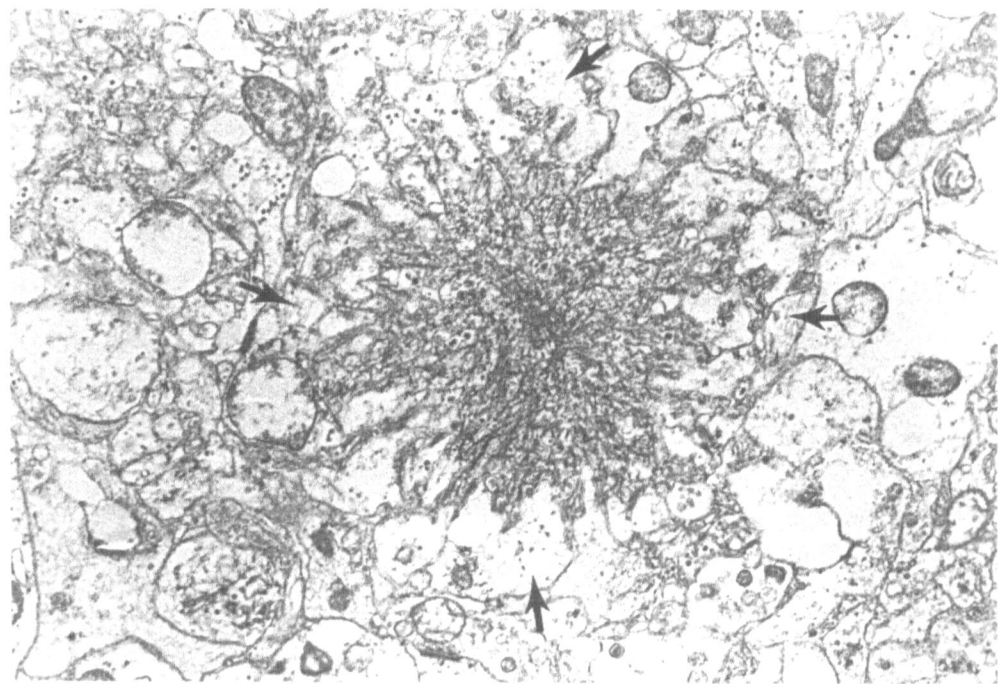

Fig. 40. Stellate kuru plaque from a case of GSS syndrome. Courtesy of Dr. J. Boellaard. Note numerous astrocytic processes (arrows). Original magnification, ×5000

Fig. 41. A margin of a multicentric plaque from a case of GSS syndrome. Material courtesy of Dr. J. Boellaard. Note numerous astrocytic processes (arrows). Original magnification, ×30 000

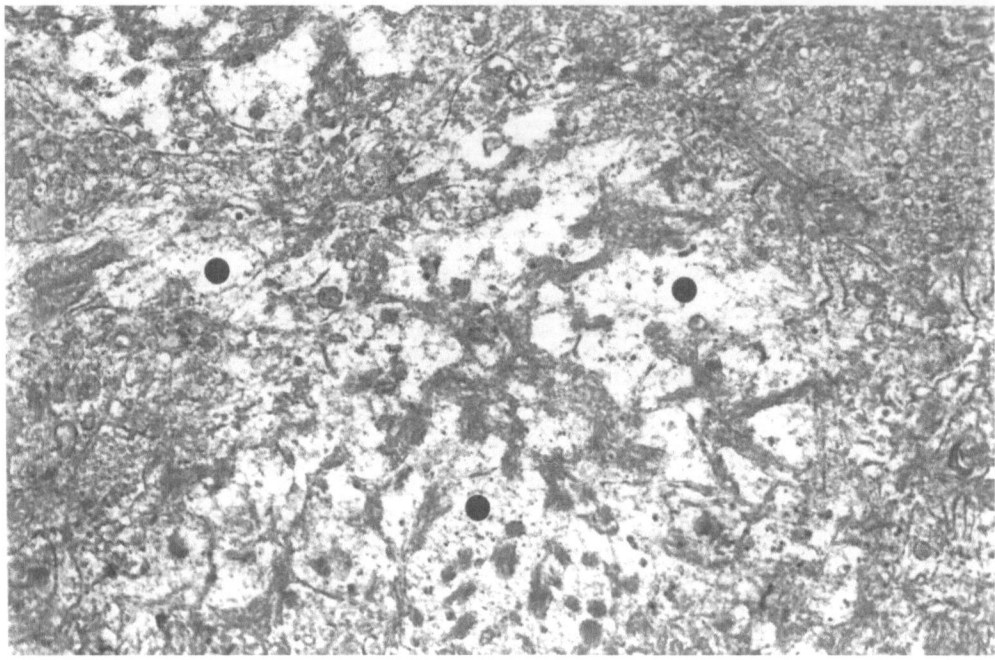

Fig. 42. Multicentric plaque from a case of GSS syndrome. Material courtesy of Dr. J. Boellaard. Note numerous astrocytic processes (circles). Original magnification, ×7000

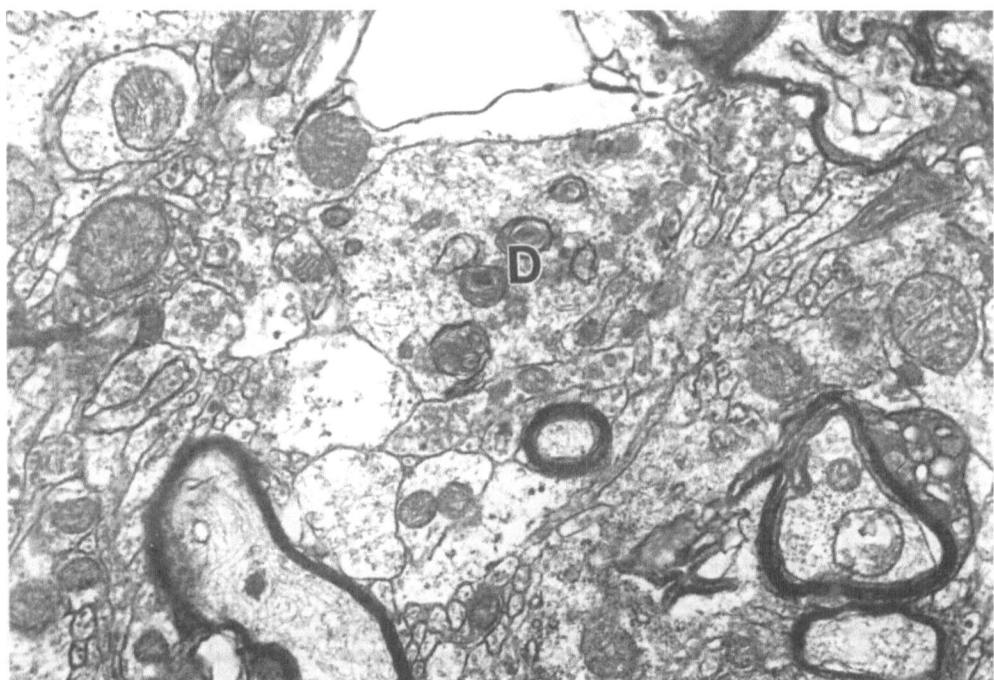

Fig. 43. A margin of a multicentric plaque from a case of GSS syndrome. Material courtesy of Dr. J. Boellaard. Note dystrophic neurite (*D*). Original magnification, ×12 000

for the relatively large size of multicentric plaques (Boellaard, personal communication). Multicentric plaques exhibit tinctorial properties virtually identical to that of kuru plaques of CJD.

Ultrastructurally, different types of plaques were observed (Fig. 40–43). Primitive plaques consist of numerous enlarged dystrophic neurites [936] indistinguishable from neuritic plaques of AD/SDAT or CJD [682, 1057]. Amyloid plaques consist of "dark filamentous and amorphous deposits" [936] composed of fibrils measuring 10 to 15 nm in diameter [630, 936]. Characteristically, these plaques were enveloped by numerous astrocytic processes [101–106, 936]. Labyrinth-like formations of basement membranes not associated with blood vessels were occasionally seen [936]. Plaques evolved from very small deposits to fully developed multicentric plaques. Kuru plaques did not differ from those previously described in kuru and CJD but, in contrast to both, dystrophic neurites were frequently seen at the periphery of them as well as at the periphery of typical multicentric plaques [101–106, 630]. Dystrophic neurites found in GSS syndrome are indistinguishable from those of AD/SDAT [634, 636, 682], but in contrast to them they have never been observed to contain paired helical filaments (PHF) (Fig. 41–43). Furthermore, dystrophic neurites in GSS syndrome are ubiquitinated [768]. Immunohistochemical staining with antibodies against ubiquitin revealed numerous immunopositive "dot-like deposits" at the periphery of plaques, corresponding to dystrophic neurites. Not well characterized structures, "dense osmiophilic bodies" (autophagic vacuoles?), "uncoated amorphous densities", and a network consisting of short filaments measuring 10–15 nm in diameter were immunodecorated by means of immunoelectron microscopic techniques. Noteworthy, antibodies against synaptophysin did not label any particular structures and the interpretation of the last phenomenon is lacking. It must be noted, however, that cerebellar kuru plaques may consist of only stellate amyloid cores without dystrophic neurites [630].

Typical GSS syndrome neuropathology may coexist with neurofibrillary tangles (NFT) [409, 411, 814], not regarded as part of CJD/GSS syndrome neuropathology. In Ghetti's case [409] dystrophic neurites clearly participated in the formation of plaques in cerebral but not cerebellar cortex. Furthermore, axons and dendrites in the proximity of the plaque area presented bending rather than normal straight course.

4.4.3.4. Scrapie and bovine spongiform encephalopathy and chronic wasting disease

PAS-positive amyloid plaques were first reported in natural scrapie by Beck and Daniel [62]. Furthermore, Bruce found amyloid plaques as occasional findings during a routine histopathological survey of 200 brains of sheep and goats experimentally inoculated with scrapie [156]. In contrast, cerebrovascular amyloidosis in scrapie was reported only recently [432], in part because scrapie-associated changes are most advanced in the brain stem, while cerebrovascular amyloidosis is observed rather in forebrain structures. Congophilic angiopathy was found in 11 of 20 scrapie-affected sheep brains and none of sheep brains affected with other disorders. Routine H & E staining revealed congophilic angiopathy only in 4 brains, the remaining seven cases were detected by polarized light. In routine sections, congophilic angiopathy appears as pale pink thickening of a vessel wall and its surroundings. Cerebrovascular amyloid was typically congophilic, PAS positive and showed bright green birefringence. The incidence of the lesion was low, congophilic angiopathy was found only at 1–2 locations per section and thus easily missed from routine histopathological examination. Affected vessels were found mostly within gray matter and within the gray/white matter junction. In the cerebellum they were observed in molecular and granular layers.

Fraser and Bruce were the first to report plaques in experimental scrapie [348, 156]. Using the 87A, 125A, and 138A strains of scrapie virus passaged in C57Bl and VM mice, they reported several distinct types of amyloid plaques. It is noteworthy, that many intermediate types (transitional stages in a yet to be discovered morphogenesis of amyloid plaques?) are also observed. All plaques, irrespective of the type, are weakly eosinophilic. They are PAS positive. Stained with Masson trichrome technique, they are argentophilic, and birefringent after staining with Congo red or thioflavin S. Plaques are observed in both gray and white matter, particularly in the corpus callosum, thalamus and cerebral cortex. They are seen less frequently in hypothalamus, hippocampus, brain stem, and cerebellar granule cell layer. On the basis of their morphology Bruce and Fraser [156] singled out the following types of amyloid plaques: 1, Shadowy plaques – measured approximately 12 um in diameter. They are weakly stained with Masson trichrome, PAS, Congo red and thioflavin S; in H & E-stained sections they are frequently not visible. 2, Stellate plaques, roughly identical to kuru plaques mentioned above. These plaques are particularly well stained with Masson trichrome. 3, Amorphous plaques, clearly defined areas measured 10–20 um in diameter. Amyloid filaments at the periphery are well delineated. 4, Giant plaques – measured up to 100 um in diameter, these plaques are structureless in appearance. 5, Diffuse amyloid deposits – mostly associated with the needle track scars in the thalamus, occasionally these plaques are formed by a coalescence of several discrete plaques of other type. 6, Perivascular amyloid deposits analogous to congophilic angiopathy. Astrocytic processes and microglial cells are occasionally seen at the periphery of plaques.

Ultrastructurally, only the appearance of the stellate type of plaques was reported [156]. Generally the ultrastructure of these plaques did not differ much from that of kuru plaques of kuru, CJD, and GSS syndrome. In contrast to kuru plaques, numerous typical dystrophic neurites containing pleomorphic inclusions and reactive microglial cells are visible at the periphery. Paired helical filaments, typical to dystrophic neurites of Alzheimer disease have never been seen.

In bovine spongiform encephalopathy, plaques have been reported only in experimental disease in mice [43] but not in natural disease in cattle [1041]. In contrast, plaques are frequently seen in chronic wasting disease in captive mule deer and elk [467–470].

4.4.3.5. The biology of amyloid plaques

4.4.3.5.1. The biology of amyloid plaques;
the correlation between the number of amyloid plaques and strain of virus,
inoculation route and aging
The only available data on the biology of amyloid plaques are based on experiments in mice infected with different strains of scrapie virus. The number of amyloid plaques is the passagable characteristic of a given strain of scrapie virus in a particular genotype of host [157, 178]. The incidence of cerebral amyloidosis in different experimental models varied from 0 to 100% but the intracerebral route always produced the highest number of plaques [153]. 87A and 87V strains of scrapie virus produced plaques in all studied genotypes of mice, ME7 and 22A produced moderate number of plaques, while 22C, 79A and 79A produced no plaques at all [157]. The passage history, in terms of mouse genotype, did not influence the overall plaque-producing ability of the particular strain. The effect of host genotype was also noticeable but only for 22A and ME7 strains. It seemed that the plaque-producing ability of any given strain of scrapie virus is dependent on the genotype of the host. The incidence of cerebral amyloidosis strictly reflects the duration of the incubation period [157]. However, this characteristic is not merely a simple consequence of a prolonged incubation period, as the range of incubation periods of plaque-producing strains is virtually identical to those of non plaque-producing strains. Furthermore, the incidence of amyloidosis in mice of different genotypes infected with the 87A and 87V strains of scrapie virus is the same, while the lengths of incubation periods are completely different. Moreover, age of mice at the time of inoculation did not influence the incidence of cerebral amyloidosis [160].

Thus, the incidence of cerebral amyloidosis is not the consequence of aging.

Smilar observations were made in several different strains of mice infected with two differest isolates of CJD virus [610]. The Fujisaki (Fukuoka-1) strain of CJD (latter reclassified as GSS syndrome) virus produced PrP plaques in 4.5% of infected mice while Fukuoka-2 strain in 90.1% plaques. It is noteworthy, that the proportions of mice with plaques did not reflect their number in original human cases (large number of plaques in a case from which the Fujisaki strain was isolated and only a few plaques in a case from which Fukuoka-2 strain was derived). Furthermore, it seems that this proportion corresponded to the virus titer ($10^{6.7-7.4}$ LD_{50} for the Fujisaki strain and $10^{8.2}$ LD_{50} for the Fukuoka-2 strain) and the amount of PrP^{CJD} (30 ug/g of brain tissue for the Fujisaki strain and 90 ug/g of brain tissue for the Fukuoka-2 strain).

The topography of plaques is dependent on the strain of scrapie virus, the genotype of the host and the route of injection. In VM mice infected with the 87V strain of scrapie virus, plaques appeared 100 days before spongiform changes could be detected. Similar results were obtained with LM and SM mice infected with the 87A strain of scrapie virus. After intracerebral inoculation, plaques appeared in the periventricular area and spread to regions remote from the ventricles only later, when spongiform changes were observed. In contrast, in mice infected intraperitoneally, plaques appeared mostly in the thalamus and superior colliculus being completely absent from the periventricular area of the corpus callosum and hippocampus [153]. Noteworthy, the number of plaques corresponded with the density of white matter tracts [424]. The number of plaques was higher on the ipsilateral side of inoculation [160] while following midline inoculation equal numbers of plaques were found on each side. Plaques clustered along the needle track [160, 1053] and numerous macrophages were associated with plaques [1053, 1058]

4.4.3.5.2. The immunohistochemistry of amyloid plaques

Prion protein, a protein of apparent molecular weight 27–30 kilodaltons (kDa) (PrP27–30) [820]; also known as SAF protein [503] or scrapie amyloid [925, 378]), is a product of limited proteolysis of a precursor protein of apparent molecular weight 33–35 kDa (PrP 33–35) [766, 820]), or a scrapie (or CJD) isoform of a prion protein (PrP 33–35^{sc} or PrP 33–35^{CJD}) [820]. PrP33–35 sc(or PrPCJD) is encoded by a cellular gene mapped to chromosome 20 in humans and 2 in mice [52]. The product of this gene is a cellular isoform of prion protein (PrPc) trans-formed to PrP 33–35sc (CJD) by an unknown process, presumably posttranslational modification (the problem is discussed in detail in chapter 2).

The purification of PrP 27–30 from scrapie-infected hamster brain enabled the development of polyclonal and monoclonal antibodies [74, 48–51] and immunolocalization of this protein [254–257].

In uninfected hamsters, PrP-immunoreactivity was localized mostly to the neuronal perikarya (except nuclei) and the proximal portion of dendrites [257]. The neuropil was diffusely and weakly stained. In contrast, in hamsters infected with the 263K strain of scrapie virus, strong PrP-immunoreactivity was localized to the neuropil. Different anatomical regions stained differently. Dorsal hippocampus, particularly the hilus of the dentate gyrus, including the CA4 region, the stratum oriens of the CA1 region, thalamus and septal nuclei, stained most intensely. The most intense staining was observed in the neuropil of regions which were most affected with spongiform changes. Smaller number of neurons was stained, and if so, the entire cell body except intraneuronal vacuoles was immunopositive. PrP immunopositive plaque-like structures, measuring 10–20 um in diameter, were observed scattered throughout the neuropil, most frequently in the efferent pathway of hippocampus, some grey matter regions, and subependymal areas of the third and lateral ventricles and subpially in cerebral hemispheres. PrP-immunoreactivity has also been studied in several models of murine scrapie: MB/Dk mice infected with the 87V strain of scrapie virus and BSC/Dk or VM/Dk mice infected with the ME7 strain of scrapie virus, respectively [161]. Technically, only periodate-lysine-paraformaldehyde or p-benzoquinone fixation, followed by formic acid pretreatment [161, 257, 604; Liberski, unpublished data], enabled localization of PrP immunoreactivity outside plaques. In uninfected mice, PrP immunoreactivity was limited to neuronal perikarya. In contrast, diffuse PrP immunoreactivity was detected in the neuropil of scrapie virus-infected mice. Noteworthy, neuronal perikarya remained largely unstained and were entirely delineated from the background of densely stained neuropil. However, some other neurons along with their processes (Bruce et al. [161] mentioned only axons, but dendrites are clearly visible on published photomicrographs) exhibited strong granular PrP immunoreactivity. The overall pattern of PrP immunostaining was different for both scrapie models and reflected, in turn, a different topographic pattern of vacuolation. In mice infected with the ME7 strain of scrapie both PrP immunopositivity and vacuolation were of diffuse type. However, subtle differences in the intensity of PrP immunostaining were detectable. In contrast, in mice infected with the 87V strain of scrapie virus, both PrP immunostaining and vacuolation were restricted to several regions; namely, ventral thalamic nucleus, dorsal lateral geniculate nucleus, CA2 region of hippocampus, ventral pallidum, substantia nigra, the medial habenular nucleus, inferior olives, and cochlear nucleus. Furthermore, in a temporal study

PrP immunoreactivity initially appeared at 140 days postinoculation. At that time PrP immunoreactivity was confined to dystrophic neurites and became diffuse in the neuropil at 210 days postinoculation and preceeded the appearance of vacuolation by several weeks. Overall, such precise targeting of both PrP immunoreactivity and spongiform vacuolation to the same neuroanatomical structures clearly suggest its central role in SSVE pathogenesis.

Amyloid plaques were clearly immunostained following formic acid pretreatment. Furthermore, several circular structures, measuring 2–5 um in diameter, were strongly immunostained. These structures were localized either at the periphery of amyloid plaques or as single or clustered structures in regions with no PrP immunostaining. Their localization made the interpretations of dystrophic neurites plausible.

The only other structure exhibiting PrP immunopositivity was cerebrovascular amyloid (congiophilic angiopathy) in natural and experimental scrapie in sheep, and in several models of scrapie in mice [22]. Noteworthy, PrP-immunopositive amyloid plaques adjacent to vessels with PrP-immunopositive congophilic angiopathy were readily detected. The immunoreactivity of cerebrovascular amyloid was highly enhanced following formic acid pretreatment, and such a procedure revealed numerous small (5–10 um) plaque-like structures in the subventricular regions.

The relationship between PrP immunoreactivity and scrapie pathology was also studied [254–257]. The localization of glial fibrillary acidic protein (GFAP)-immunopositive reactive astrocytes paralleled that of PrP immunoreactivity. For instance, PrP immunopositivity and reactive astrocytosis were absent in septal nuclei. In contrast, both PrP immunopositivity and intense reactive gliosis were detected in the medial-ventricular border of the caudate nucleus. Such immunohistochemical observations were confirmed by semi-quantitative Western blot of PrP. The thalamus exhibited the highest concentration of PrP by Western blot which paralleled that of strong diffuse PrP immunopositivity by immunohistochemistry. On the contrary, the caudate nucleus, presenting the lowest concentration of PrP, exhibited almost negligible PrP immunostaining (except for the medial-ventricular border). Using, peroxidase immunoelectron microscopy, DeArmond et al. [256] found that PrP27–30 antiserum detected filaments located within distended extracellular space of the subependymal region. Filaments measured 16 nm in diamater and were up to 1000 nm in length. They narrowed every 8 nm and such narrowing may represent twists. Overall, these "prion" filaments met generally accepted criteria for amyloid. Plaques were easily detected in semi-thin sections of brain tissues from hamsters infected with the 263K strain of scrapie virus as "skyblue material" in perivascular, subependymal and subpial regions [1046]. Ultrastructurally, dystrophic neurites containing electron dense bodies and extracellularly located filaments, measuring 7–9 to 12–17 nm in diameter, were seen in these regions. Furthermore, cells with invaginated cytoplasm, probably macrophages/microglial cells, were seen in the vicinity of the amyloid deposits and amyloid fibers seemed to project from the interior of these cells into the extracellular space. The last phenomenon resembled that found in experimental CJD and scrapie in mice. Both extracellular and intracellular amyloid filaments were immunodecorated using antisera against PrP27–30 or synthetic peptides encompassing the N-terminus of PrP27–30 [1046]. In summary, there is no doubt that PrP is indeed the component of amyloid fibers in plaques.

In mice infected with different strains of scrapie virus, PrP immunoreactivity have been found exclusively in plaques [742]. Furthermore, the congophilic angiopathy of natural scrapie in sheep was PrP immunopositive only after formic acid pretreatment [742], the common method for unmasking amyloid epitopes [604]. PrP-immunopositive plaques are also stained with antibodies against paired helical filaments (PHF) and tau [123]. Anti-PHF and anti-tau antibodies also labelled yet to be identified structures at the periphery of plaques. Such immunolocalization of PHF and tau suggested dystrophic neurites. Furthermore, as PHF are not detected in dystrophic neurites in scrapie [429, 661–662, 679, 681–683] such PHF or tau immunoreactivity was ascribed to the unpolymerized form of tau protein (i.e. unformed abnormal fibrils).

In the first published report [608] equivocal staining of plaques in CJD and GSS syndrome was obtained. After limited proteolysis with trypsin or proteinase K, PrP-immunopositive plaques were unequivocally detected. Suprisingly, plaques in GSS syndrome were immunolabeled less intensely than plaques in CJD. Furthermore, PrP-immunopositive plaques were detected in mice infected experimentally with CJD virus [608]. When the peroxidase-antiperoxidase (PAP) method was compared with the avidin-biotin complex (ABC) technique, the former produced significantly less background staining. The same conclusions were reached by the author (Liberski, unpublished data).

PrP-immunopositive plaques have been found in CJD in a few subsequent reports. Using antibodies raised against amyloid plaque cores purified from a case of GSS syndrome and against a synthetic peptide encompassing 15 amino acids of the N-terminus of human PrP, Kitamoto and Tateishi [605] found PrP immunoreactivity in all GSS syndrome cases and in 20 out of 34 CJD cases. Furthermore, when CJD cases were re-analysed according to the duration of disease, 95% of cases with clinical courses longer than 11 months showed PrP plaques but only one case of short clinical course presented PrP-immunopositive plaques. This study confirmed earlier observations that kuru plaques are associated with CJD of longer clinical course [729] even when PrP immunohisto-chemistry detected plaques invisible with classical neuropathological staining. Liberski, Budka, Kitamoto, Link, Tateishi, Brown (670; manuscript in preparation) using four different antisera against purified PrP or synthetic peptides found PrP-immunopositive plaques in only one out of 31 consecutive CJD cases and this particular case exhibited protracted clinical course. The same conclusion was recently reached by Liberski et al. [673] based on PrP immunohistochemistry of two series of CJD cases from Poland and the Netherlands, respectively. Thus, it seems that the high percentage of Japanese CJD with PrP-immuno-positive plaques reflects the differences in virus strain or host genetic makeup, or both, in turn reflected by differences in the clinical course. By the same token, the proportion of cases of GSS syndrome within Japanese series may be overrepresented [297]. Recently, PrP-immuno-positive plaques have been detected in familial CJD of French origin [44].

Kuru [852; Liberski, unpublished data] and GSS syndrome plaques are PrP immunopositive and all types of plaques (primitive, kuru, uni- and multicentric plaques) are equally immunoreactive. The PrP amyloid is distinguishable from that of the A4 type by means of electron micro-scopy [102]. A4 amyloid consists of bundles of fibrils measuring 6 nm in diameter, often with close contact with microglial cells. In contrast, kuru plaques in GSS syndrome consist of fibrils measuring 12 nm in diameter and in close contact to astrocytes. Amyloid fibrils of multicentric plaques measure 6 to 11 nm in diameter and they are embedded in amorphous electron-dense material. Close contact with astrocytes have been reported [102]. As the plaque morphology is not sensitive enough to differentiate between different types of amyloid in CJD/GSS syndrome or AD/SDAT, the classification of plaques based rather on their chemical composition has been recently suggested [907]. The beta-plaque, com-posed of A4 or beta amyloid is found in AD/SDAT, Down's syndrome, and aging, while PrP plaques are composed of PrPs in CJD, GSS syn-drome, kuru, scrapie and chronic wasting disease [464, 466–470].

PrP-immunopositive plaques have been detected in the first passage of experimental CJD in mice in 30 of 33 transmission experiments [607]. In contrast, spongiform changes were produced in mice in 23 of 33 transmission experiments. Thus, the presence of kuru plaques may be more sensitive indicator of a positive primary isolation than spongiform changes. The rabbit anti-human PrP antibodies recognized both human and murine plaques while mouse anti-human PrP antibodies recognized only human kuru plaques. Therefore, murine kuru plaques are com-posed of murine but not human PrP [607]. Anti-PrP immunostaining detects not only kuru plaques but also diffuse and small PrP-immuno-positive deposits, otherwise undetectable with Congo red staining.

4.4.3.5.3. *The relation of amyloid plaques to neuroaxonal dystrophy*

In a discussion of amyloid plaques and neuroaxonal dystrophy (NAD) one of the most pertinent question is the relationship between these two phenomena during the morphogenesis of amyloid plaques. Moreover, NAD is the ultrastructural link between subacute spongiform virus enccphalopathies (SSVE) and senile dementia of the Alzheimer type (SDAT) [636, 682]. In the first ultrastructural studies of SDAT, dystrophic neurites (DN) and not neuronal perikarya were found to constitute the senile plaques already known from classic neuropathology and silver staining [445, 625, 696, 1008].

Regardless of the animal species and chemical composition of amyloid, the amyloid plaques are composed of various proportions of three basic elements: DN, amyloid fibrils, and reactive glial (mostly microglial) cells [424–426, 1052, 1056–1058]. In three types of amyloid plaques separated by Wisniewski and Terry [1057], the primitive plaque is composed exclusively of DN without amyloid fibers. On the contrary, the burnt-out plaque is completely devoid of DN. The classical (or senile or cocarde) plaque possesses both structures. It should be remembered however, that the interpretation of the ultrastructural image is highly dependent on the plane of the section; thus the possibility that the neuritic plaque comprises DN (and analogously, the primitive plaque, amyloid fibrils) which are outside the given plane of section cannot be completely ruled out. Furthermore, NAD in SDAT is widespread much beyond that associated with plaques. The common morphogenesis of primitive and senile plaques has also been doubted [426–427]. As the number of primitive plaques decreases with age the number of senile plaques increases, thus their different topography and age-related behavior may reflect different morphogenesis accomplished by a different role of DN.

In SSVE and SDAT the amyloids are modified neuronal membrane proteins of unknown functions but DN in both conditions exhibit virtually the same ultrastructural pattern, except that those in SDAT additionally contain paired helical filaments [636, 682]. Whether amyloid or DN appears first during the morphogenesis of neuritic plaques has yet to be determined. However, the electron microscopic observations that in rodents infected with plaque-producing and "non plaque-producing" strains of scrapie [362, 429] the earliest plaque is composed exclusively of DN without admixture of amyloid, along with a finding of abundant NAD in SDAT suggests that DN may form niduses on which amyloid precursor proteins are processed and deposited.

4.4.4. Tubulovesicular structures

4.4.4.1. Introduction

Attempts, using thin-section transmission electron microscopy, to define the structure of the infectious virus of scrapie (and Creutzfeldt-Jakob disease, kuru, and, bovine spongiform encephalopathy) have been fraught with failures, false starts, and misinterpretations.

In this chapter we shall start with a discussion of the structures which have been reported, but obviously are irrelevant to the disease itself or are clearly of artefactual nature. Then we shall concentrate on the "tubulovesicular structures" (TVS) or "scrapie-associated particles". These structures of unknown chemical composition and unresolved biological significance are the only ones unique to this group of disorders at the level of thin-section transmission electron microscopy. Whether TVS represent a part of or aggregate of the infectious scrapie virus or a pathological product of the disease is yet to be established.

4.4.4.2. "Virus-like particles" of unknown significance

Perhaps the best known example of misinterpretation of electron microscopic data is the report of the "isolation of 14-nm virus-like particles from mouse brain infected with scrapie agent" by Cho and Greig [211]. The extraction procedure used by these investigators consisted of repeated fluorocarbon (Freon 113, trichloro-trifluoroethane) treatment followed by CsCl density gradient ultracentrifugation. Negative-stain electron microscopy of $1.29-1.30\,\mathrm{g\,cm^{-3}}$ and $1.33-1.34\,\mathrm{g\,cm^{-3}}$ fractions showed spherical particles measuring 14-nm in diameter. The authors claimed that "if the 14-nm particles isolated from scrapie-infected mouse brain are indeed a virus, they represent the smallest one ever reported". However, in subsequent studies these 14-nm particles were demonstrated not only in scrapie-infected but also in control mice [212]. Furthermore, as the 14-nm "virus-like particles" were aggregated by anti-mouse ferritin antisera they probably represented nothing more than ferritin.

The highest number of irrelevant or misinterpreted structures have been reported by Narang and co-workers in different scrapie models [330, 798–799, 803]. In natural scrapie in sheep the following "virus-like" structures have been reported: 1) "barred structures" [799] representing lamellar cytoplasmic bodies – normal cellular organelles of unknown significance [497, 665, 672]; 2) "virus-like particles" measuring 15–26 nm in diameter and 75 nm in length [799, 803].4 As these "particles" are clearly formed by "separating itself from plasma membrane" they represent "vesicular or curly fragments of plasma membrane" already reported by Lampert et al. [644] (vide supra); 3) "tubular structures" measuring 20 nm in diameter associated with electron-lucent vesicles 25 to 70 in diameter and "elongated dumb-bell shaped tubules" 44 nm in diameter. Other reported tubules measured 12 nm in diameter with a suggestion of cross striation [799]. Similar tubular particles have been found in experimental scrapie in rat [330]. These electron-lucent tubules and vesicles are most consistent with "branching tubules" – abnormal subcellular organelles characteristic of neuroaxonal dystrophy – a well characterized type of neuronal degeneration (vide infra).

Different types of "virus-like particles" have been reported in Creutzfeldt-Jakob disease. Bots et al. [117] reported hexagonal "virus-like particles" measuring 80–115 nm in diameter and containing electron dense cores. These particles were observed to adhere to the cell membranes. The 90-nm particles containing electron-dense cores accompanied by "nucleocapsid-like" tubules measuring 15 nm in diameter were reported by Narang [802] and particles measuring 70–110 nm in diameter by Sibayama et al. [955]. Overall, these particles are vaguely similar to type C of retrovirus particles. Since retroviruses are not implicated as the causative agent of the subacute spongiform virus encephalopathies the significance of these findings, if any, is completely unknown.

4.4.4.3. Tubulovesicular structures

Tubulovesicular structures (TVS) (also known as the scrapie – associated particles) are the only structures unique to all of the subacute spongiform virus encephalopathies (Fig. 44–48). TVS were first described by David-Ferreira et al. [249] in NIH Swiss mice inoculated intracerebrally with the Chandler strain of scrapie virus. TVS were described as "particles and rods ranging in diameter from 320 to 360A". Interestingly, rods were covered with spikes. It must be stressed that it is very clear from the above-mentioned description and subsequent discussion that the reported diameter of TVS have been slightly different when presented by different investigators, but factors inherent to electron microscopic techniques (namely swelling or dehydratation) must be taken into account as plausible explanations.

In natural scrapie in sheep, TVS appeared as membrane-bound accumulations of round particles measuring 35 nm in diameter. Electron-dense cores could be demonstrated in some of them [90]; (Fig. 2, inset). It seems that "cucumber-shaped bodies" measuring approximately 20 nm

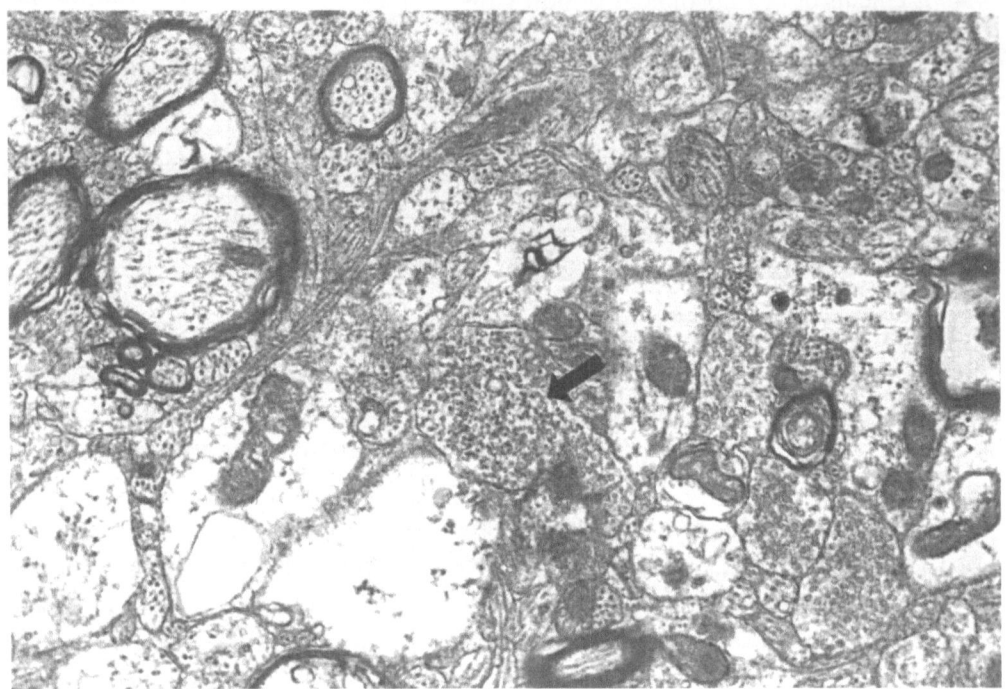

Fig. 44. Low power electron micrograph of neuronal process containing tubulovesicular structures (arrow) in hamster brain infected with the 263K strain of scrapie virus. Original magnification, ×7000

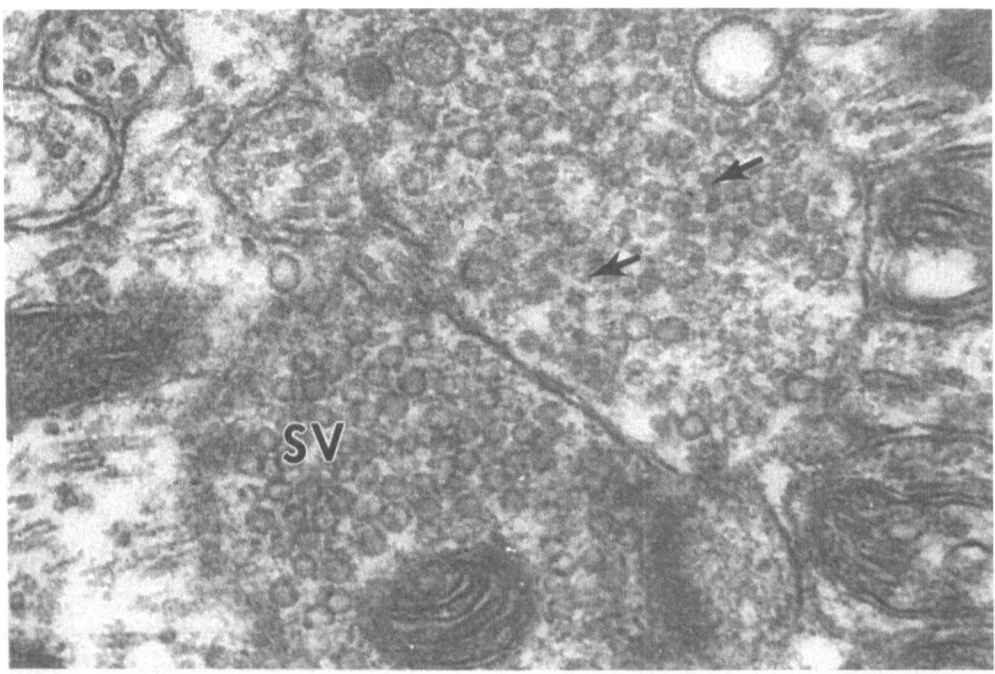

Fig. 45. Tubulovesicular structures (TVS) within axonal terminal of hamster brain infected with the 263K strain of scrapie virus. Note that the size of TVS (arrows) is smaller than that of synaptic vesicles (SV). Original magnification, ×30 000

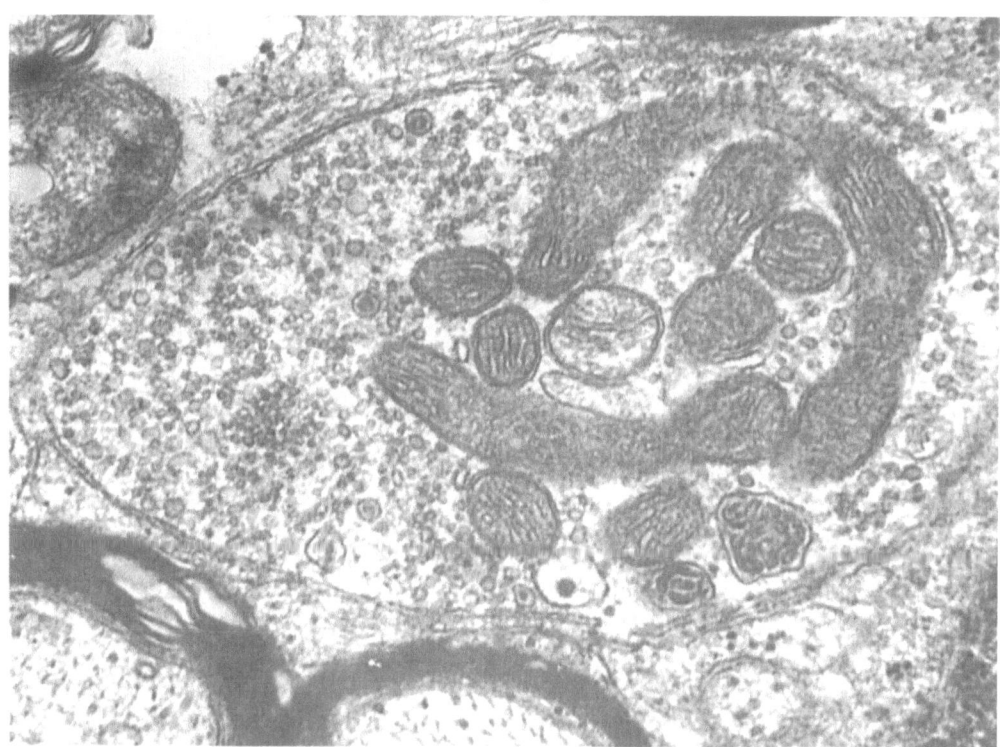

Fig. 46. Tubulovesicular structures (TVS) within axonal terminal of hamster brain infected with the 263K strain of scrapie virus. Note that the size of TVS is smaller than that of synaptic vesicles. Original magnification, ×20 000

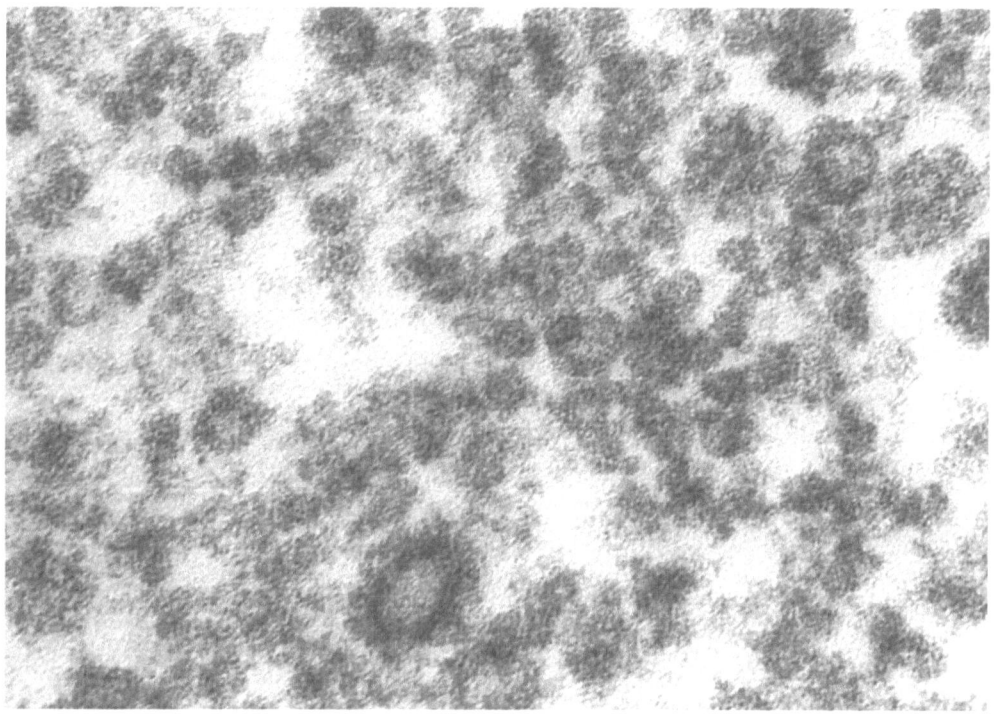

Fig. 47. Details of previous figure. Note pleomorphism of TVS. Original magnification, ×140 000

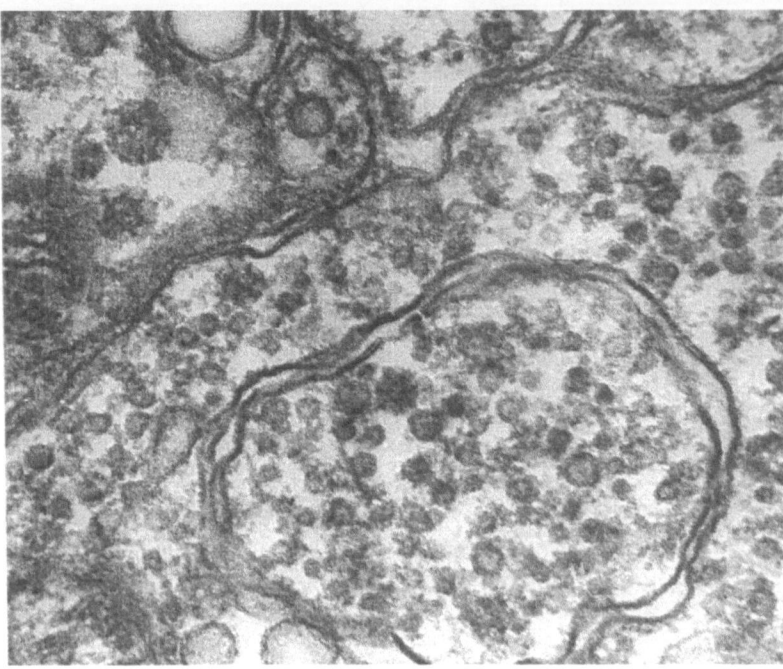

Fig. 48. High power electron micrograph of TVS from a brain biopsy of a human CJD case. Material courtesy of Prof. H. Budka and Prof. E. Sluga, the Neurological Institute of University of Vienna, Austria. Original magnification, ×85000

in diameter and 60 nm in length reported by Narang [799] represent TVS despite differences in size.

TVS have been reported in nearly all models of scrapie in rodents studied to date. Lamar et al. [633] found TVS in postsynaptic processes (dendrites) of ICR mice inoculated intracerebrally or subcutaneously with the Klenck strain of scrapie virus isolated from a Suffolk ram. Spherical TVS reported by these investigators measured 30–35 nm in diameter; some TVS, however, appeared as long rods. Interestingly TVS could not be found in spleens taken from scrapie-affected mice or in brain cell cultures. It is noteworthy that TVS were not found in neuroblastoma cultures infected with scrapie virus (Liberski, unpublished observations). A possible explanation for this phenomenon based on an estimation of the infectivity titer is provided later in the text. In Sprague-Dawley rats inoculated intracerebrally with the Chandler strain of scrapie virus TVS appeared as tubular "cucumber-shaped particles" measuring approximately 20 nm in diameter and 60 nm in length and containing 4-nm electron dense cores [330]. Contrary to other reports, these structures were found not only in neuronal processes but also in neuronal perikarya. Narang et al. [807] extended these observations, reporting TVS in BSVS (Compton White) or LACG mice inoculated intracerebrally with the Chandler strain of scrapie virus. In these models TVS were seen only in postsynaptic terminals and measured 26 nm in diameter. When stained with ruthenium red, TVS appeared larger, approaching 33 nm in diameter [798, 800]. In Swiss mice inoculated with the Chandler strain of scrapie virus TVS were found in enlarged post-synaptic processes (dendrites) or in unidentified neuronal processes [39–41]. Electron-dense spherical TVS measured 23 nm in diameter and frequently formed "paracrystalline" tubular or vermicellar arrays [41]. A tilting analysis of such arrays revealed a clear change in the apparent axis of these tubular aggregates, suggesting that they resulted from over-lapping spherical profiles. Baringer et al. [41] constructed a topological model of TVS arrays which, when tilted, gave an image consistent with that observed by electron microscopy and supported the notion that tubular arrays consisted of round structures. Interestingly, when hamsters infected with the 263K strain of scrapie were studied, TVS could not be detected [40, 42]. These negative data were used to support the hypothesis that TVS could not represent a significant ultrastructural finding (much less the infectious virus or an aggregate of it) as they could not be found in the scrapie model representing the highest titer of infectivity [149]. Recently, however, three groups of investigators reported TVS in hamsters infected with the 263K strain of scrapie virus. Narang et al. [801, 804] reported TVS measuring 22–24 nm in diameter in postsynaptic terminals (and dendrites!) accompanied by larger vesicular profiles measuring 100–110 nm in diameter and "smaller number of tubulofilamentous profiles" which were 200 nm in length. As the latter structures were neither shown nor characterized it is impossible to judge their significance. Liberski et al. [685] identified TVS in the same hamster model, not only in postsynaptic but also presynaptic terminals. In the latter, they were mixed with, but easily differentiated from synaptic vesicles. Affected processes also contained normal-appearing mitochondria, large vesicles (measuring 100–150 nm in diameter), multivesicular bodies and large electron lucent cisterns sur-rounded by "fuzzy" membranes. TVS measured 20 to 50 in diameter (mean, approximately 35 nm). Interestingly, when electron micrographs were re-examined for the presence of TVS apparently not present at the time of publishing [660] they were easily identified. Thus, what is quite typical for electron microscopy, when a new finding is reported it appeared that the phenomenon has already been present but it had not been recognized. Gibson and Doughty [428] reported the most extensive survey of TVS in different experimental scrapie models in rodents and sheep. In agreement with Liberski et al. [660] these investigators found TVS in both pre- and postsynaptic terminals, but postsynaptic terminals predominated (41 profiles out of 109 in postsynaptic, compared to 12 in presynaptic terminals). Gibson and Doughty [428] reported the presence of dense "paracrystalline" arrays apparently absent in hamsters infected with the 263K strain of scrapie. TVS were most numerous in VL mice infected with the ME7 strain of scrapie virus followed by C3H mice infected with the 22C and 79A strains of scrapie virus. Furthermore, when VM mice infected with the 87V strain of scrapie (a model producing a large quantity of amyloid deposits) were examined, numerous synaptic terminals containing TVS were found in the vicinity of amyloid plaques. TVS were also found in Cheviot sheep inoculated with the ME7 strain of scrapie virus, but the number of affected processes was the lowest among reported models.

In CJD, TVS were reported infrequently. In natural CJD the presence of TVS was cited but never shown by Narang [801]. However, TVS were found recently to be a consistent finding in well-fixed brain biopsies of patients with CJD [671; Liberski, Budka, Sluga, Barcikowska, Kwiecinski, manuscript in preparation]. In chimpanzees inoculated either with human or chimpanzee CJD isolates, TVS were reported in one animal as spherical and elongated structures measuring 30–40 nm in diameter [642].

TVS have been reported in NIH Swiss mice inoculated intracerebrally or intraocularly with the Fujisaki strain of CJD virus [680, 685]. They were of the same diameter and localization as in hamsters infected with the 263K strain of scrapie virus. It seemed only that TVS in experimental CJD were more densely clustered than those in experimental scrapie in hamsters.

The exact topology of TVS is not entirely clear. In most published electron micrographs TVS appeared as spheres measuring between 20 and 40 nm in dimater. Some investigators stressed that the tubular "arrays" of TVS are spurious, brought about by overlapping spherical profiles [40–42]. Recently, however, Liberski et al. [680] clearly demonstrated short tubular forms of TVS; thus it is evident that TVS are pleomorphic structures existing in at least two forms – spheres and short tubules. It is also possible that TVS circular profiles correspond to those of short tubules cut in transverse section. Some investigators strongly believe that this pleomorphism of TVS preclude them to represent a part of or aggregate of the infectious scrapie virus. While we do not prejudge the true nature of TVS, it is noteworthy to remember the pleomorphism of other viruses. Virions of hepatitis delta virus, for example, exist as spheres and short tubules.

The nature and significance of TVS is unknown. Only limited data are available concerning the appearance of TVS through the incubation time. In hamsters inoculated with the 263K strain of scrapie virus, TVS were initially observed as early as 3 weeks after intracerebral inoculation, but their numbers increased only with the onset of disease at 9 to 10 weeks after inoculation [680]. Noteworthy, vacuolation and astrocytosis were detected at 8 weeks postinoculation and, thus, followed the appearance of TVS. In Webster-Swiss mice, TVS were found 12 weeks and 16 weeks after intracerebral inoculation and intradermal (footpads) inoculation, respectively [797]. In NIH Swiss mice infected with the Fujisaki strain of CJD virus, TVS were first seen 13 weeks after intracerebral inoculation and their numbers increased dramatically at 18 weeks after inoculation, when the first signs of clinical disease were noted. Vacuolation and astrocytosis were detected at the same time, and the increase in the number of processes containing TVS paralleled the increasing intensity of vacuolation and astrocytosis. TVS were approximately twice as abundant in terminally ill mice following intraocular inoculation as in mice following intracerebral inoculation. In contrast, the intensity of vacuolation and astrocytosis was not dependent on the route of inoculation. Several conclusions can be drawn from these studies. TVS appeared early in the incubation period, preceeding the onset of clinical disease. Furthermore, in scrapie-infected hamsters TVS preceeded the appearance of other neuropathological changes. The approximately 1000-fold lower infectivity titer of the Fujisaki strain of CJD virus, compared to the 263K strain of scrapie virus, may have accounted for the delayed appearance of TVS in experimental CJD. However, the number and density of TVS appear to correlate with the duration of the incubation period. TVS were more abundant in mice with experimental CJD, which has a longer incubation period than in scrapie in hamsters characterized by a short incubation period. When the incubation period of experimental CJD was further prolonged by means of intraocular injection, the number and density of TVS increased even more, while the intensity of astrocytosis and vacuolation did not. The apparent correlation between the number of neuronal processes containing TVS and infectivity titer may explain why in cell cultures infected with scrapie virus (the titer: 1 LD_{50} per 631 to 7943 neuroblastoma cells [191–192]) TVS could not be found and why their number in experimental scrapie in sheep was so low [428]. Furthermore, it should be remembered that virus particles cannot be detected when the virus titer is lower than 106 infectious particles per ml.

The chemical composition of TVS is unknown. Earlier studies reported that ruthenium red enhances contrast of TVS [798, 800]. These staining properties of TVS may be interpreted as an evidence for the presence of glycosyl residues within TVS. Furthermore, PrP33–35^{sc}, a protein of central role in scrapie pathogenesis as recently evidenced by molecular studies is a sialoglycoprotein [111]. Thus, it is tempting to speculate that PrP33–35^{sc} may be a component of TVS. Obviously, immunoelectron microscopy should be undertaken to solve this problem as published immunogold electron microscopic studies did not reveal so far any TVS staining.

The isolation and purification of TVS has been unsuccessful [801]. However, when homogenataes of scrapie-affected brains and spleens were analysed by ultrafiltration followed by zonal centrifugation, spherical particles measuring 30–60 nm in diameter were detected by thin-section and negative-staining transmission electron microscopy [961]. Occasionally, 30–60 nm spheres were accompanied by smaller spherical structures measuring 8–10 and 1, any correlation between TVS and particles described by Siakatos et al. [960] has yet to be established.

In conclusion, TVS appear to be the only ultrastructural marker for the whole group of SSVE and thus, even despite complete knowledge on their biological significance, are worth of further investigation.

4.4.5. Neuroaxonal dystrophy

4.4.5.1. Neuroaxonal dystrophy: a common form of neuronal degeneration

Neuroaxonal dystrophy (NAD) is a well known form of neuronal degeneration (for review: [536–537, 950–951]). The first ultrastructural description of NAD was provided by Lampert [634] who separated several classes of dystrophic neurites (DN): reactive, degenerative, and *bona fide* dystrophic. Reactive neurites contained mitochondria, dense bodies (heterogeneous and amorphous) and vesicular elements accompanied by increased amounts of neurofilaments. Degenerative axons additionally contained axonal debris probably originating from neurofilaments. Dystrophic axons contained numerous subcellular organelles, particularly mitochondria, electron-dense material, multi-granular bodies, electron dense bodies and neurofilaments. Rc-evaluation of Lampert's classic paper [634] suggests that the differences between these three different classes of DN are only quantitative. Numerous further studies proved the intrinsic heterogeneity of DN in almost all models so far examined [93] but a few investigators favor the narrower definition of DN as structures containing predominantly, if not ex-clusively, cisterns and arranged loops of endoplasmic reticulum, believing that other classes of abnormal neurites represent reactive phenomena [950–951].

NAD may occur in apparently "normal" brains. Dayan found NAD in all 47 species of vertebrates except birds [253]. Rees observed DN in biopsies of apparently normal human brains and their numbers did not correlate with age, except in those araes containing neuritic plaques [900–901]. In contrast, Jellinger [537] reported such a correlation in humans: 85% of individuals below the age of 10 years were free of neuritic changes, as compared to 40% of those between 10 and 20 years of age, 10% in the fourth decade of life and only 2% above 40 year of life. The intensity of NAD strongly depends on brain topography [361, 537, 538]. Thus, the reported difference of correlations of the density of DN with age may reflect the intrinsic vulnerability (pathoclisis) of a particular brain area for the development of NAD. Furthermore, Sotelo and Palay hypothesized that the presence of NAD in "normal" brains merely reflects the structural remodeling processes as part of the general "plasticity" of the brain [979].

It is thus evident that the presence of NAD is not par excellence a pathological process but it becomes pathological only if found in increased density.

NAD has been described in the central and peripheral nervous system in plethora of naturally occurring and experimentally induced pathological conditions. Ultrastructurally, NAD has been reported in humans with familial intrahepatic choleostasis [926], subacute sclerosing panencephalitis [676], and in several cases of human neuroaxonal dystrophy (HNAD) including Seitelberger's and Hallervorden Spatz diseases, respectively [28, 168, 231, 536–537, 702, 795, 929, 950–951, 956, 1009, 1049, 1059, 1063]. Furthermore, DN have been found in monkeys with subacute combined degeneration [7], sheep with alpha mannosidosis [523], goats with beta-mannosidosis [539], dogs with hereditary striato-nigral and cerebello-olivary degenerations [787], canine neuroaxonal dystrophy [230] and spinal cord transection [541], guinea pigs as a consequence of a long term demyelination [899], and cats with hereditary neuroaxonal dystrophy [1060], unilateral rubrospinal tractotomy [46], and interruption of afferent fibers to the cuneate nucleus [1033]. NAD has been observed in rats and mice with aging of autonomic ganglia [938], thalamus degeneration [47], nerve ligation [262], optic nerve crush [935], transection of the corpus callosum [264], impaired regeneration of sciatic nerve [937], diabetes mellitus [689, 1062], hypertension [614], vitamin E-deficiency [637, 851] and several intoxications [93, 215, 554, 990]. Analogously to HNAD, Mukoyama et al. [791] reported gracile axonal dystrophy mouse – a mutant mice characterized by abundant DN. At the end, DN were seen in vitamin E-deficient chicken [1071]. Ultrastructurally, DN in all these conditions contain electron-dense bodies and, less frequently, neurofilaments, clefts and layered loops and cisterns [683]. Thus, the pattern of neuronal degeneration is nearly identical, irrespective of the cause.

The presence of dystrophic changes within otherwise normal synaptic terminals led Gonatas [444] and Gonatas and Moss [447] to speculate on the "synapse diseases". In a sense, CJD and Alzheimer's disease [636] are also classic examples of "synapse disease" along with human neuro-axonal dystrophy (HNAD), Tay-Sachs disease [444, 447], vitamin-E deficiency [637, 851] and remodeling processes [979].

The mechanism of DN formation is unclear and the impairment of slow axonal transport has been postulated recently to be a common pathogenetic pathway in SSVE, Alzheimer's disease and several neuro-degenerative disorders [370–373]. Interference with axoplasmic transport (anterograde, retrograde or turn-around) is directly responsible for DN formation in the aging nucleus gracilis, ligated sciatic nerve, acrylamide neuropathy and transected spinal cord axons [456, 541, 853, 1069]. Under these cirumstances, the ultrastructural pictures of DN are nearly identical to that encountered in SSVE. Thus, the same mechanism may be assumed to operate in SSVE and other neurodegenerative disorders.

4.4.5.2. Neuroaxonal dystrophy in subacute spongiform virus encephalopathies

It has been shown recently that NAD is an important ultrastructural marker for SSVE (Fig. 49–55) [429, 660–662, 678, 682–683]. This is noteworthy in light of the fact that vacuolation in some models may be significantly negligible [726].

Irrespective of disease (Creutzfeldt-Jakob disease, GSS syndrome, scrapie, BSE and CWD) and the affected species, DN are ovoid in shape and usually observed as isolated structures lying against a background of devastated neuropil. Occasionally, DN form small clusters which have been interpreted as premature or abortive neuritic plaques [362, 660–662, 678, 682–683]. If observed in longitudinal sections larger DN form bullous swellings in continuity with normal-appearing axons.

The internal structure of DN is similar irrespective of their origin and their classification as dendrites or axons (Fig. 49–55). DN contain mitochondria, some of which exhibit features of degeneration, and numerous pleomorphic, membrane-bound inclusions. Ultrastructurally these inclusions could be classified as (1) electron-dense homogeneous structures, (2) multigranular, (3) multivesicular, (4) multilamellar, (5) electron-lucent cisterns and arranged, frequently tubular profiles of endoplasmic reticulum. In CJD, scrapie and CWD DN containing electron-dense bodies predominate, and in BSE those containing neuro-filamentous masses are abundant. In BSE, dense-bodies and mitochondria are entrapped among neurofilamentous masses. It should be stressed,

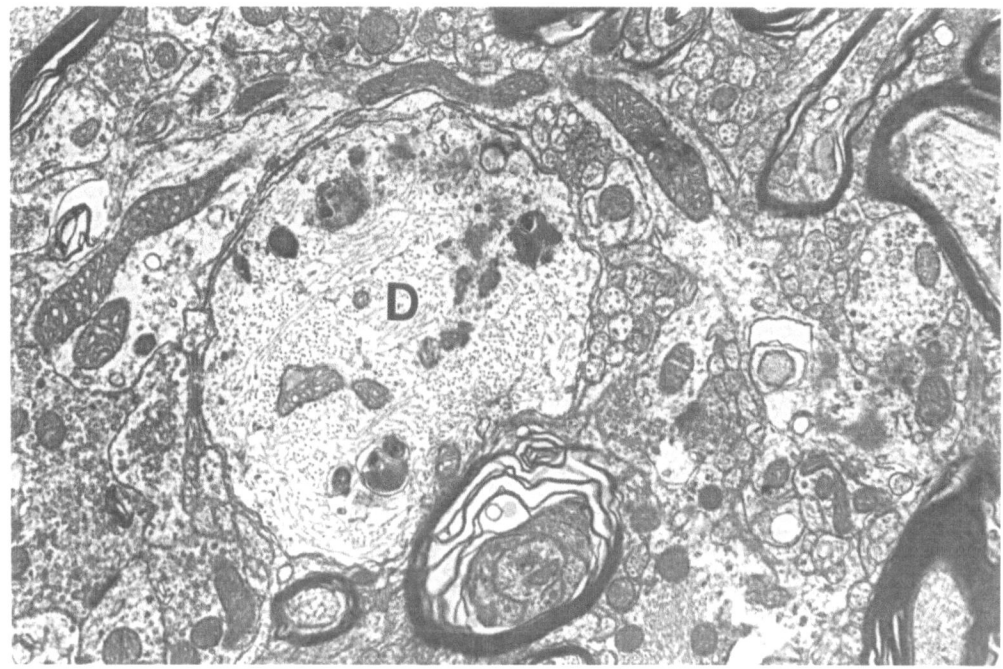

Fig. 49. Dystrophic neurite (D) containing neurofilaments and electron-dense inclusions from the brain of a spider monkey infected with kuru virus. Unpublished work of P.P. Liberski, P. Brown and D.C. Gajdusek, the National Institutes of Health, Bethesda, USA. Original magnification, ×12 000

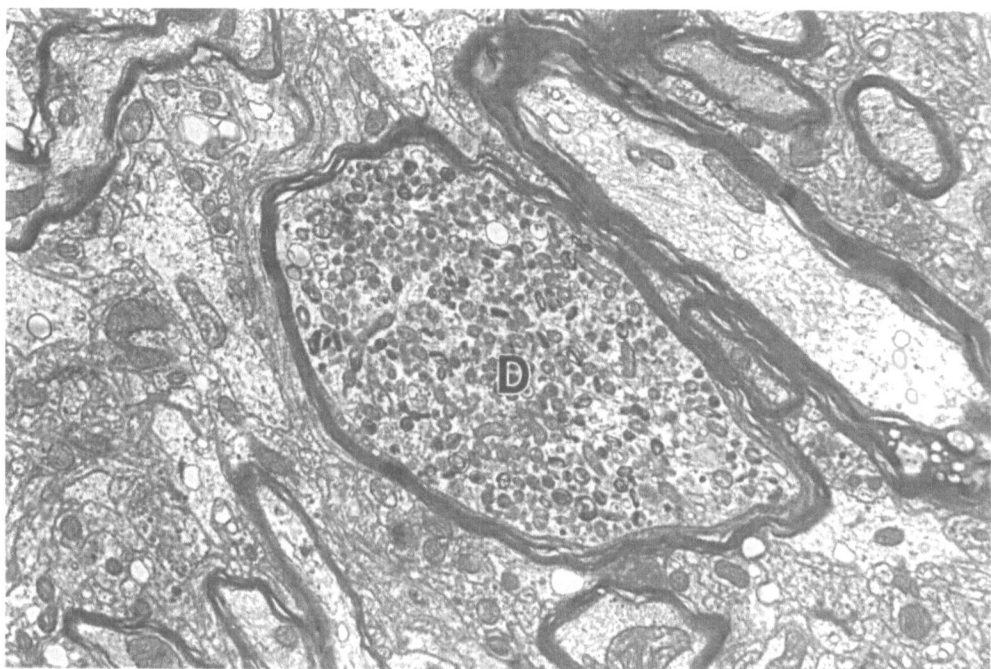

Fig. 50. Dystrophic neurite (D) containing numerous degenerating mitochondria and electron-dense inclusions from a brain biopsy of a human CJD case. Material courtesy of Prof. H. Budka and Prof. E. Sluga, the Neurological Institute of University of Vienna, Austria. Original magnification, ×7000

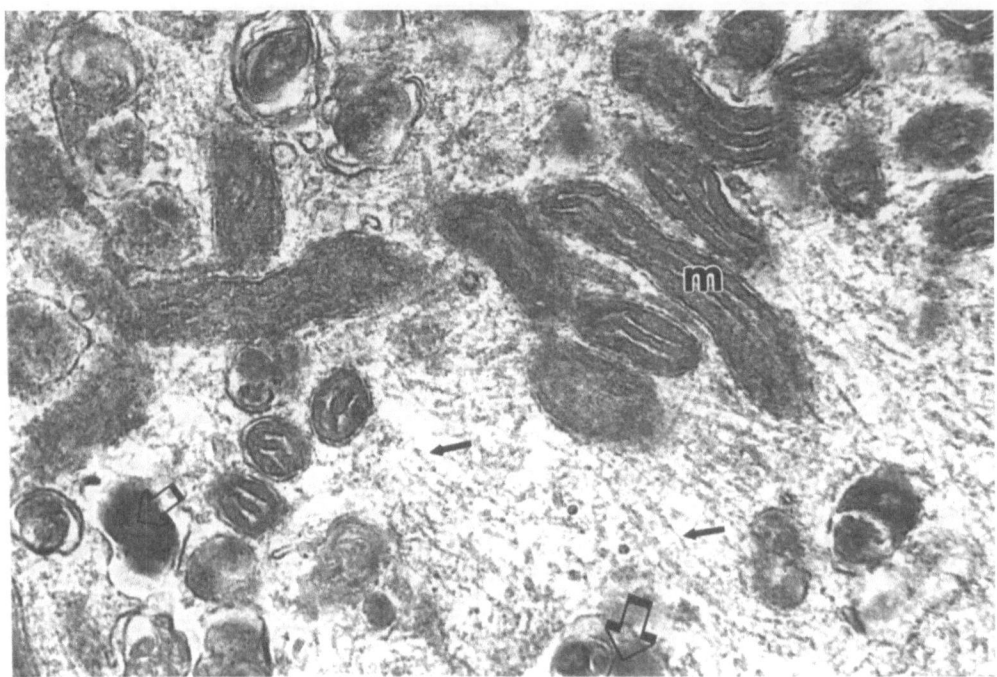

Fig. 51. High power electron micrograph of a dystrophic neurite containing numerous degenerating mitochondria (*M*), neurofilaments (arrows) and electron-dense inclusions (open arrows) from a brain biopsy of a human CJD case. Material courtesy of Prof. H. Budka and Prof. E. Sluga, the Neurological Institute of University of Vienna, Austria. Original magnification, ×30 000

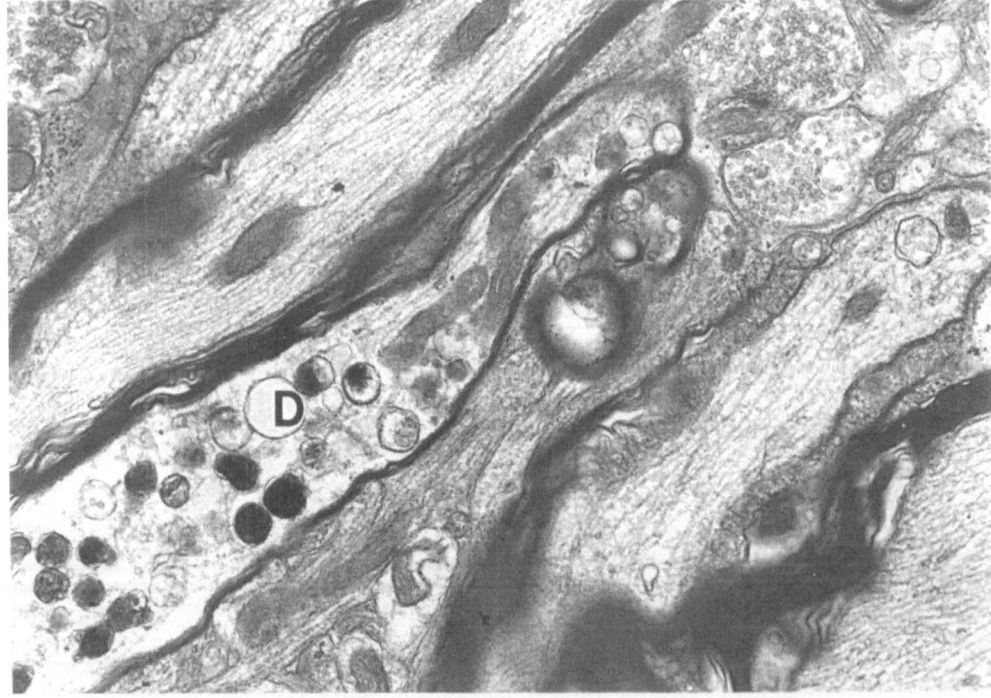

Fig. 52. Longitudinal section of dystrophic neurite (arrow) from hamster brain infected with the 263K strain of scrapie virus. Original magnification, ×10 000

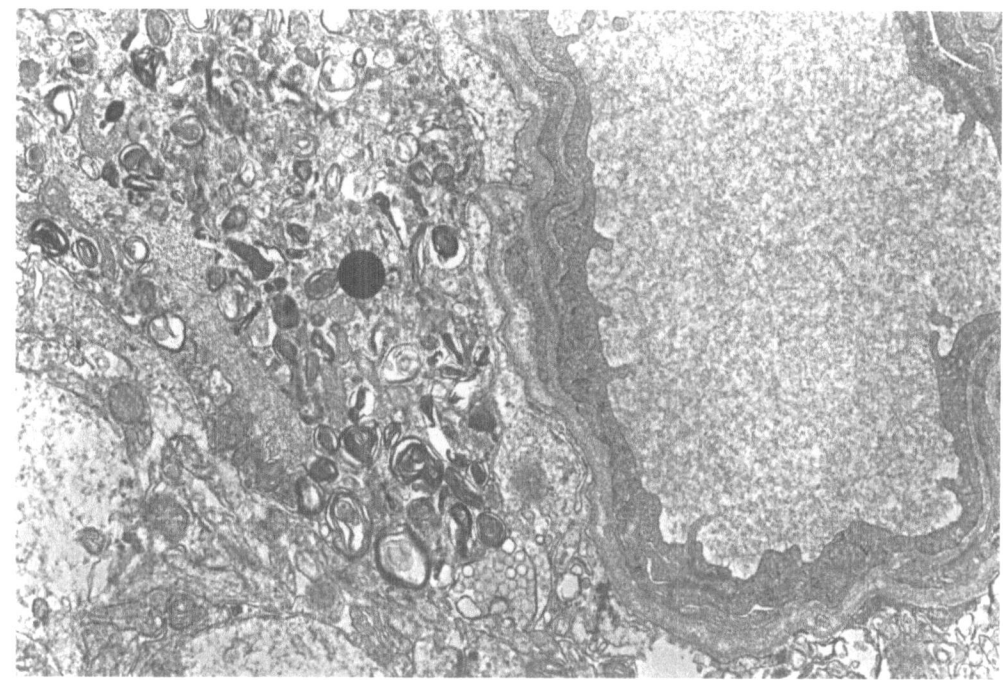

Fig. 53. Large dystrophic neurite (circle) in close contact with blood vessel from brain of BSE-affected cow. Material courtesy of Dr. Gerald A.H. Wells, Central Veterinary Laboratory, Ministry of Agriculture, Fisheries and Food, Weybridge, U.K. Original magnification, ×10 000

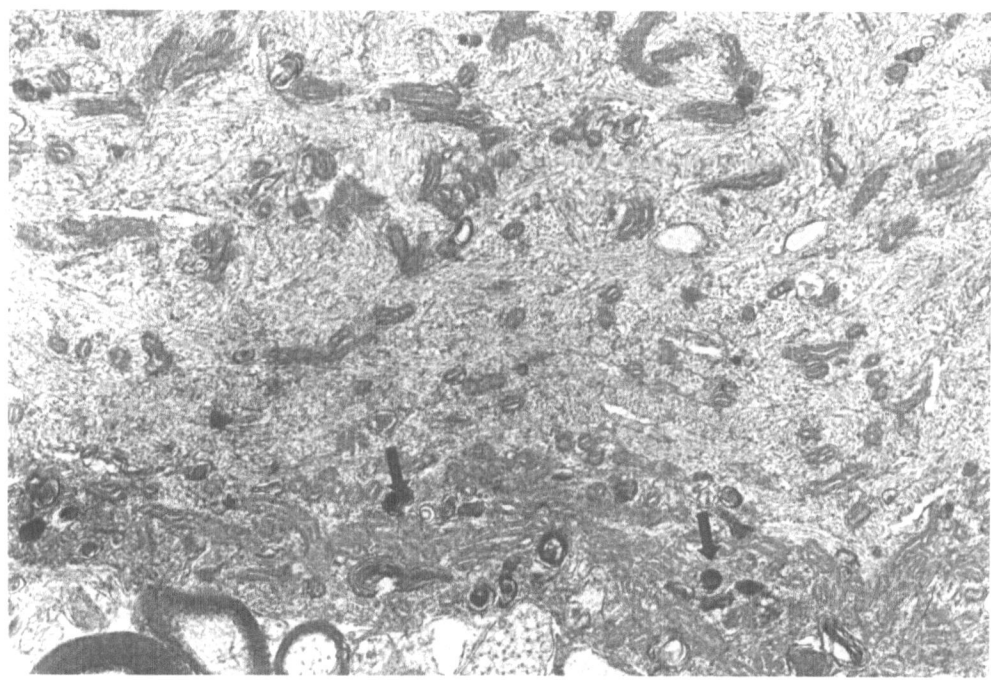

Fig. 54. A fragment of dystrophic neurite containing innumerable neurofilaments and electron-dense bodies (arrows) from the brain of a BSE-affected cow. Material courtesy of Dr. Gerald A.H. Wells, Central Veterinary Laboratory, Ministry of Agriculture, Fisheries and Food, Weybridge, U.K. Original magnification, ×10 000

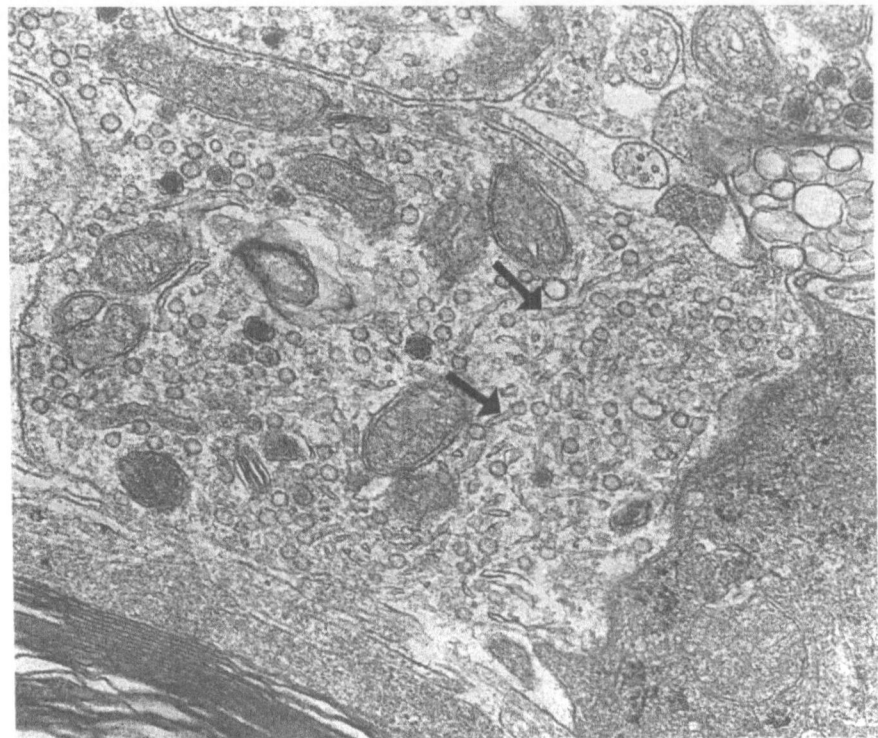

Fig. 55. Dystrophic neurite containing "branching cisterns" (arrows) from the brain of a BSE-affected cow. Material courtesy of Dr. Gerald A.H. Wells, Central Veterinary Laboratory, Ministry of Agriculture, Fisheries and Food, Weybridge, U.K. Original magnification, ×20 000

however, that all types of DN may be found in any slow virus disease. Noteworthy, the electron-lucent cisterns and stacked membranes regarded as pathognomic for HNAD have recently been reported in scrapie and BSE [666, 669].

Gonatas et al. [448] were the first to report DN in brain biopsies from two CJD cases. The most affected processes were unidentified, but one was defined as axonal preterminal. Lampert et al. [640–643] reported DN containing mitochondria, vesicles, electron-dense, and lamellar inclusions in experimental kuru in chimpanzees and Spider monkeys but not in experimental CJD in the same species of nonhuman primates. However, the presence of DN containing altered subcellular organelles or neurofilamentous masses was subsequently confirmed, albeit regarded as a nonspecific finding, in experimental CJD [643]. Lampert et al. [640–643] stressed the presence of NAD in neurodegenerative disorders other than SSVE and suggested the common pathogenetic mechanism. Ribadeau-Dumas and Escourolle [903–904] described a few neuronal processes containing altered subcellular organelles regarded as unspecific alterations reflecting secondary degenerating phenomena. DN have been subsequently reported in natural and experimental CJD in guinea pigs and hamsters, respectively [550–553, 967, 712–713].

DN, erroneously interpreted as astrocytic processes, were first seen in experimantal scrapie in mice by Field and Raine [331]. Chandler [195, 201–202], studying scrapie-infected gerbils and mice, found electron-dense lamellar subcellular organelles within axons, axonal terminals and dendrites. Noteworthy, he has shown the neuronal cytoplasm containing electron-dense lamellar bodies. The occasional presence of electron-dense bodies within a neuronal somata corresponds with those observations of Yaima and Suzuki [1066], Cummings et al. [243], and Gibson and Liberski [429], supporting the notion of a neuronal origin of dense bodies. Furthermore, DN have been observed in natural scrapie in sheep [88, 92]. In the latter situation, NAD has been widespread and occurred in several brain stem nuclei and the cerebellum. Interestingly, Bignami and Parry [92], noticed the similarities of DN and "Herring bodies" – accumulations of neurosecretory granules in the infundibulum nerve fibers undergoing degeneration [88]. Furthermore, the possibility of "disturbances in nerve flow" was considered as a possible pathomechanism responsible for development of NAD in scrapie-affected sheep.

The significance of NAD in SSVE is unclear. NAD is widespread in SSVE and appears relatively early in the incubation period. The increase in DN may preceed vacuolation, as in C3H mice infected with the ME7 strain of scrapie virus and in hamsters inoculated with the 263K strain of scrapie virus [429, 669, 682]. By contrast, DN increased in number concurrent with or following the onset of vacuolation in C3H mice infected with the 22C or 79A strains of scrapie virus and in NIH Swiss mice infected with the Fujisaki strain of CJD virus [429, 682]. NAD and vacuolation probably evolve independently, and it has been suggested that NAD may be an ultrastructural marker for the depopulation of neurons.

As mentioned previously the pathomechanism by of which NAD develops is yet to be elucidated, but most investigators agree that the disturbances of both anterograde and retrograde axoplasmic flow contribute to the formation of DN. Recently, Machado-Salas [697], in an elegant study based on combined Golgi and ultrastructural techniques applied to a brain biopsy of a CJD patient, showed multiple spherules along the longitudinal axes of neurites which ultrastructurally corresponded to typical DN. Analogous distensions have also been reported in hamsters infected with the 263K strain of scrapie [501]. Furthermore, Machado-Salas has suggested that the stagnation of axoplamic flow blocked between such spherules led to a decrease of cytoskeleton elements within them and, subsequently, to their emptiness followed by formation of spongiform vacuoles [697]. Such an hypothesis, although speculative, having the potential power to explain both NAD and the formation of spongiform vacuoles, should certainly be tested.

The problem of the patogenesis of NAD in SSVE has been recently addressed by immunocytochemical studies on the expression of the 200 kDa neurofilament protein in CJD brains (Liberski and Budka, manuscript in preparation). Immunohistochemistry showed numerous structures with a strong immunoreactivity for the 200 kDa neurofilament protein. In both control (brains with other unrelated pathologies) and CJD brains innumerable axons of different diameters were immunopositive. In control material the only other consistently observed 200 kDa neurofilament protein-immunopositive structures were spheroids of the medulla. These spheroids corresponded to those DN encountered in small numbers in apparently normal brains.

Three types of apparently abnormal structures were observed in CJD brains.

1. Neurons accumulating 200 kDa neurofilament protein were detected in 22 of 31 CJD brains. Some of these cells had normal appearance, particularly large cortical pyramidal neurons. Their apical dendrites and/or initial axonal segments were also immunopositive and could be traced for some distance. However, other neurons appeared distended with excentrically placed nuclei and homogenously stained cytoplasm. Intranuclear vacuoles, characteristic for SSVE, were frequently obserwed within these neurons.

2. Neuritic distensions (neuritic swellings, DN) were observed mainly in the white matter, but also at the junction between gray and white matter. They were ovoid or elongated in shape. Most were connected to one end of an axon, but many appeared as bullous swellings along the course of axon. In several axons 3 or 4 such distensions could be traced along the visible part of the axonal segment. Noteworthy, such immunostained neurites adapted to other pre-existing structures such as vessels and, particularly, spongiform vacuoles. In such locations they were distorted and bent. Frequently, the immunostaining was more intense at the periphery and distensions had the appearance of "an eye of the needle". Still other distensions clearly encompassed typical CJD vacuoles. In such situations vacuoles were surrounded by a narrow ring of strong immunopositivity. Furthermore, several neurons were observed with vacuoles distending immunopositive branching dendrites. Occasionally, such swellings were of proximal localization and the close proximity of it to the neuronal cell body was easily recognized. The smaller blebs and globules were frequently seen and these probably corresponded to DN of smaller diameters cut in transverse planes.

3. Spheroids were observed mostly within medulla predominantly but not exclusively in anterior horns. They were larger than neuritic swellings and strongly immunopositive for the 200 kDa neurofilament protein. Immunopositivity, sometimes located at the periphery, was of diffuse or granular type and small vacuoles were frequently seen within larger spheroids. Larger spheroids, in turn, were often surrounded by smaller ones or by clusters of immunopositive blebs and globules.

The accumulations of the 200 kDa neurofilament protein within neurons and DN deserves further comments. Few 200 kDa neurofilament protein-immunopositive neuritic distensions were seen in control material and this fact presumambly reflected their limited presence in apparently normal brains. Indeed, Smith [971] and Clark et al. [217] found increased number of neurofilamentous swellings in human spinal cord up to approximately 20 years of age. After that age the plateau of the number of axonal swellings was reached. In CJD brains the average score of axonal swellings was always higher than in control brains and strong 200 kDa neurofilament protein immunoreactivity presumambly reflected the increased number of neurofilaments as detected by transmission electron microscopy.

Neurofilament proteins are synthesized within neuronal perikarya and then are transported down the axons with slow axoplasmic transport [932]. Thus, the increased accumulation of neurofilament proteins may be expected when the impairment of slow axoplasmic transport occurs. Such increased neurofilament accumulations has been predicted by Gajdusek [371–373] to be a common pathogenetic mechanism for SSVE, Alzheimer's disease and motor neuron disease. Furthermore, an analogous pattern of 200 kDa neurofilament protein accumulation was recently found in mice infected with the Fujisaki strain of CJD virus and in hamsters infected with the 263K and 22CH strain of scrapie virus (Liberski, Budka, Yanagihara, Gajdusek – unpublished observations). Thus, it seems sufficiently established that accumulations of neurofilament protein(s) within abnormal neuritic distensions is an intrinsic feature of CJD and scrapie neuropathology, in both human and laboratory animals.

The anti-200 kDa neurofilament protein antibody used in this study recognized mostly phosphorylated epitopes. Thus, the presence of immunopositive neurons probably reflect abnormal distribution of phosphorylated neurofilaments which, under normal conditions, are present on axons and not in neuronal perikarya [932]. The abnormal distribution of phosphorylated neurofilament protein within neurons has recently been reported in motor neurons disease [706], Pick's disease, corticonigral degeneration, Alzheimer's disease [216], hereditary canine spinal muscular atrophy, disease of Shaker calves, motor neuron disease in rabbits and pigs and "swayback" and zebra myelopathy [230]. Further-more, neurons containing phosphorylated neurofilament epitopes have been recently reported in CJD [796]. The significance of abnormal distribution of phosphorylated neurofilament proteins is uncertain. As the phosphorylation of neurofilament proteins is associated with the diminished transport of them from neuronal somata to the periphery [932]. It is thus tempting to speculate that such abnormal distribution reflects the general impairment of slow axonal transport in CJD, and thus is directly responsible for neuronal degeneration.

4.4.6. Different forms of neuronal degeneration in slow virus disorders

4.4.6.1. The degeneration and neuronal loss

All subacute spongiform virus encephalopathies (SSVE) are purely neurodegenerative processes. The classical triad of spongiform changes, astrocytosis and neuronal loss is regarded as the neuropathological hall-mark for the whole group. However, the study of neuronal degeneration and neuronal loss has been almost totally neglected in the literature and only limited data have been published so far.

The Golgi impregnation method was proven to be a useful tool to evaluate neuronal changes in a variety of naturally occurring and ex-perimentally induced neurological disorders. In human CJD several abnormalities have been detected in cerebral cortical neurons [645]. Pyramidal neurons exhibits dendrites of reduced size and irregular shape. The number of spines seem to be reduced and numerous dendritic swellings of spherical or ovoid shapes are seen. Several dendrites show regular constrictions and dilatations along their courses. Such changes seemed to originate distally and proceed proximally with an effect that proximal dendritic segments or apical dendrites are less affected. Swellings are seen mostly in distal dendritic segments and they are observed proximally only when the total number of swellings are highly increased. In a Golgi impregnation study of a brain biopsy taken from a CJD patient, a reduced number of dendrites along with their smaller caliber was reported [322]. Particularly, large pyramidal neurons seemed to lose 30 to 50% (even up to 90%) of their dendrites. Concomitant loss of dendritic spines was observed. Several axons and dendrites of stellate and pyramidal neurons presented globular and ovoid swelling. In con-trast to previous data [322], swellings observed by Machado-Salas [967] were located in all dendritic segments irrespective of whether within an initial portion, or a primary or a secondary dendritic branches. Swellings were several times larger than the original neuronal processes, 3–5 times larger than dendrites and 8–10 larger than axons, respectively. It seems that swellings impregnated by the Golgi method corresponded to several types of ultrastructural findings, however, the exact comparison still awaits the combined Golgi-electron microscopic study. Several swellings contained electron-dense inclusion bodies and thus corresponded to those dystrophic neurites of neuroaxonal dystrophy. Other swellings, whether in axons or dendrites, contained electron-lucent cytoplasm with a smaller than normally number of microtubules and/or neurofilaments. Still others, showed distended cell membranes (or the outhermost lamellae of the myelin sheath) surrounded totally "empty" space. These observations on natural CJD has been confirmed in hamsters infected experimentally with CJD virus [550]. Cortical pyramidal neurons presented loss of dendrites and dendritic spines. Focal distensions were frequently seen along neuronal processes, and these showed no predilection for any particular segment of a dendrite or an axon. Analogous changes have been reported in hamsters infected with the 263K strain of scrapie virus [501]. While lower numbers of cortical neurons were stained by the Golgi method in scrapie-affected hamsters, loss of spines was evident, particularly on the apical and oblique dendritic systems. Furthermore, numerous dendritic and axonal distensions (varicosities) were seen in several brain regions. They were ovoid or spherial in shape but occasionally showed bubble-like appearance eccentric to the shaft. Noteworthy, no correlation between number of distensions and the degree of spongiform changes could be detected.

The biological significance of this loss of dendrites and spines, together with their distorted shapes, are unclear. It is plausible, however, that dendritic and spine alterations are associated with or even responsible for the typical periodic synchronous discharges observed by EEG in most CJD patients [1019, 1079]. Moreover, synaptic alterations were observed early in the course of experimental CJD in guinea pigs [552]. Long, undulating synaptic membranes with irregular focal thickenings of synaptic densities were seen as early as 4 weeks postinuculation. Whether such changes correspond to the dendritic or spine alteration as detected by Golgi method is yet to be established.

Neuronal loss is more easily estimated in highly organized structures, thus hippocampus has been studied in some detail for neuronal degeneration in several models of murine scrapie [945]. However, in most models of murine scrapie, neuronal loss is not readily apparent, but degeneration of the whole pyramidal cell layer of the hippocampus formation, accompanied by a robust glial reaction (gliocytosis), is prominent in selected models [347]. The degree of hippocampal pyramidal cell layer degeneration varied between different scrapie models, from subtle vacuolation to the severe cell loss against the background of gliotic hippocampus. With minimal cell loss, only isolated pyknotic neurons were observed. When the process progressed, the entire CA1 region of the hippocampus appeared devoid of neurons. Then, the whole CA1 region appeared thinner leaving only a sparse band of remaining neurons. In the most affected lesions, the pyramidal cells were totally replaced by abundant glial cells ("sclerosed"). The severity of hippocampal involvement varied between scrapie models, it was observed only in LM mice (and F1 crosses between VM and C57BL) infected with the ME7 strain of the scrapie virus and predominantly if not exclusively after the intracerebral route of inoculation. Furthermore, the degree of hippocampus involvement was highly correlated with the degree of vacuolation in the affected regions. It seems that certain level of vacuolation must be approached for the hippocampal pyramidal cells to degenerate. However, this particular threshold was not the same for all models examined. Such experimental observations were recently confirmed in natural CJD [783], in which hippocampus involvement is not regarded as frequently observed. In 5 out of 6 CJD cases severe spongiform changes, accompanied by gliosis and mild neuronal loss, were observed from the prosubiculum to the CA1 to CA3 regions of the hippocampus formation. In contrast to murine scrapie models, however, severe neuronal loss was not observed.

4.4.6.2. Intranuclear eosinophilic inclusions

Intranuclear eosinophilic inclusions have been reported in neurons of the substantia nigra in a series of CJD cases from Chile [181]. These inclusions were stained according to the Van Gieson's technique, were PAS negative, and did not stain by the Nissl technique. Their association with CJD is unknown.

4.4.6.3. Spiroplasma-like inclusions

First reported in human CJD [56, 452, 902], spiroplasma-like inclusions measuring 375 to 1000 nm in length and 30 to 115 nm in diameter, were limited to axons or presynaptic terminals. Occasionally, the continuity between spiroplasma-like inclusions and the cell membranes was evident. Furthermore, in hamsters infected with the 263K strain of scrapie virus, spiroplasma-like inclusions were found frequent frequently [664]. However, the close proximity of spirals and neurotubules was demonstrated and, frequently, spiroplasma-like inclusions resolved into aggregations of coated vesicles attached to a neurotubule. Thus, spiroplasma-like inclusions have no connection to real spiroplasma, the cell-wall free procaryotes taxonomically belonging to the class Mollicutes [1022]. The last conclusion is strongly supported by the lack of any antibodies directed against several spiroplasma species in CJD patients [654]. The recently reported cross-reactivity of scrapie associated-fibrils (SAF) and Spiroplasma mirum fibril protein [57] is yet to be explained but the *claim that spiroplama is the causative agent* for the whole group of subacute spongiform virus encephalopathies seems totally implausible.

4.4.6.4. Cerebellar lamellar bodies

In 1970 Peat and Field [849] reported "unusual barred structures" in kuru-affected brains. Such structures have been further reported in both scrapie virus-affected mice and hamsters and sham-inoculated control animals [665, 672]. Profiles of Cytoplasmic lamellar bodies (CLB) were seen in dendrites of granular cell layers in C3H mice infected with the 22C or 79A strain of scrapie virus and sham-inoculated controls, and dendrites and somata of Purkinje cells of hamsters infected with the 263K strain of scrapie virus, respectively. The CLB were dispersed rather randomly in affected cytoplasm. However, the close apposition of the CLB either to Golgi apparatus or endoplasmic reticulum was frequently observed, but because an abundance of these structures was always seen, even this particular location was interpreted as fortuitous.

Typically, the CLB were composed of stacks of electron-lucent cisterns separated by bands of higher electron density, several dense and light bars were observed in each CLB. The width of electron-dense bands was approximately 46 nm, while electron-lucent cisterns varied in width from approximately 50 nm to three or four times of that of dense bands.

The presence of the CLB in both scrapie-affected and sham-inoculated animals suggest that these structures are of no connection with any subacute spongiform virus encephalopathies. Indeed, they were reported in numerous naturally occurring and experimentally induced conditions [497]. The origin of the CLB is uncertain. Rough endoplasmic reticulum is a subcellular component from which the CLB might have originated. However, the comparable electron densities of both mitochondria and the CLB suggest that the CLB may originate from these subcellular organelles.

4.4.6.5. Intraneuronal inclusions in bovine spongiform encephalopathy

Bovine spongiform encephalopathy (BSE), a novel disease recently discovered among Friesian/Holstein cattle in England [1041], pathologically resembles other subacute spongiform virus encephalopathies (SSVE). However, two types of intraneuronal inclusions, not reported in other types of SSVE, have been reported in BSE [666]. The first type, observed also in brain from control cattle, was located in neuronal perikarya and appeared to be lysosomal in origin, consisting of electron-dense, compact lamellae arranged in concentric or parallel arrays. Occasionally, these inclusions exhibited a central area of lower electron density or a globule of low electron density budding from one end. The exact localization of the low electron-density component seemed to be dependent on a plane of section.

The second type of inclusion, located mainly in neuronal processes, was composed of a dense network of approximately 10 nm tubules and circular profiles within a structureless matrix and surrounded by a common membrane. The circular profiles were suggestive of tubular structures on cross sections. Occasionally the subsurface cistern was noticed in involved processes which also contained synaptic vesicles, mitochondria, dense bodies and electron-lucent cisterns.

Neuronal inclusions probably reflect the neurodegenerative nature of BSE and the limited repertoire of the central nervous system. Such inclusions have never been reported in other SSVE, but membrane-bound inclusions composed of 10 nm tubules vaguely resembled the "type d" inclusions found in rats experimentally infected with scrapie [798].

4.4.6.6. Autophagic vacuoles in Creutzfeldt-Jakob disease and scrapie

In this paragraph we discuss the autophagic vacuoles (Fig. 56) in scrapie and CJD, suggesting that autophagic vacuoles may contribute to the neuronal degeneration leading to neuronal loss.

Autophagic vacuoles (AV) were observed in both experimental CJD in mice and scrapie in hamsters. They were more abundant in hamsters infected with the 263K strain of scrapie virus than in mice affected with CJD. AV were observed only in neuronal perikarya and their processes but not in glial cells. They were composed of areas of the cytoplasm sequestrated with single, double, or multiple membranes. Sequestrated cytoplasm contained ribosomes, occasionally mitochondria, small secondary vacuoles with vesicles, or had a homogenously dense appearance, presumably reflecting degenerative changes. It seems that AV developed when otherwise normal area of the cytoplasm was surrounded by proliferating membranes originating from the endoplasmic reticulum. Large AV contained membrane-bound compartments with

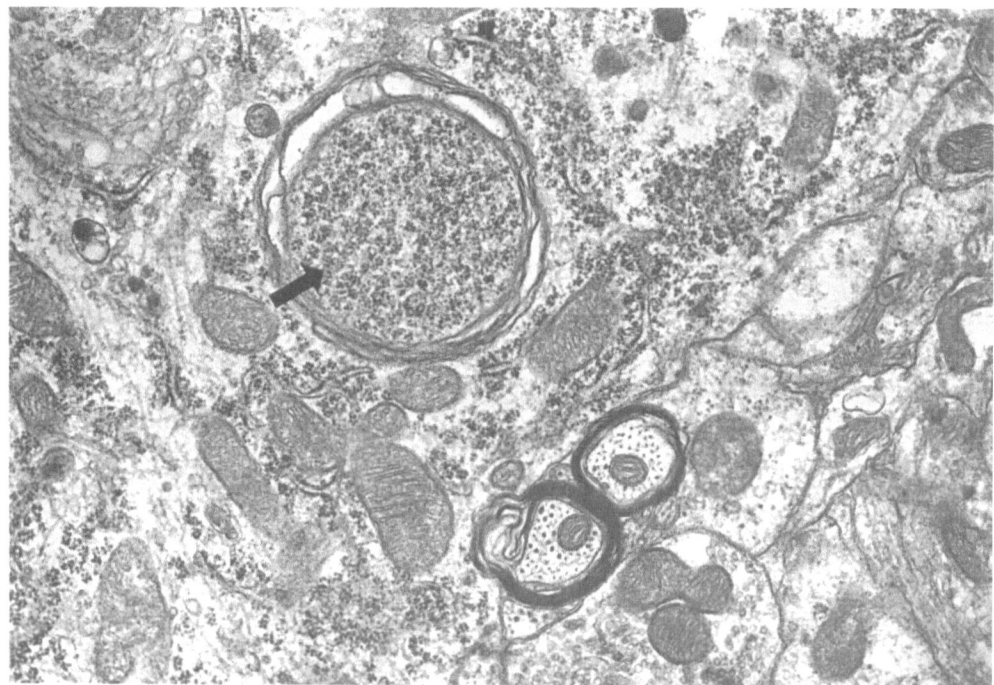

Fig. 56. Autophagic vacuole (arrow) within neuronal cytoplasm of hamster brain infected with the 263K strain of scrapie virus. Original magnification, ×10 000

electron dense cellular detritus. Furthermore, AV were detected occasionally in dystrophic neurites and in these structures they separated the dystrophic part from the rest of the neurite.

AV in experimental CJD [100–106] and scrapie (100; Liberski, Yanagihara, Gibbs and Gajdusek, manuscript in preparation) have only recently been reported. Interestingly, AV have been illustrated but erroneously designated "vesiculated structures" in scrapie-infected gerbils [199] or "whorls" in scrapie-infected hamsters [669].

Neuronal loss belongs to the classical triad of neuropathological phenomena, regarded as pathognomonic for subacute spongiform virus encephalopathies. While neuronal loss has been recognized almost from the beginning of slow virus research [88] the exact pathomechanism by which neurons degenerate has yet to be elucidated. Two forms of neuronal degenerations are known to exist in SSVE: neuroaxonal dystrophy and intraneuronal vacuolation. While the former is almost never detected in neuronal perikarya (Gibson and Liberski – unpublished observations) and could be responsible for neuronal death only through a "dying back" process [534], the latter is clearly separated from severe neuronal loss in certain experimental scrapie models [945]. Thus, the other mechanism should be searched for and the formation of AV is a good candidate for it.

Almost from the onset of modern SSVE research scrapie and CJD have been regarded as disorders of membranes [299, 644]. This concept was supported by a finding that, prion protein 33–35 (PrP 33–35^{sc}), a host-encoded protein with transmembrane domains, is crucial for scrapie pathogenesis. As pointed out above, PrP 33–35 is expressed almost exclusively in neurons and, during the course of the disease it is processed by unknown mechanisms to form intra- and extracellular amyloid aggregates: scrapie-associated fibrils (prion rods) and cores of amyloid plaques. PrP 33–35^{sc} has already been implicated in the vacuole formation [257], and it is tempting to speculate that abnormal processing of PrP 33–35^{sc} may lead to locally originated degeneration, initially sequestrated within the AV and then, when diffuse, causing neuronal death and neuronal depopulation.

5. Final conclusions

"Skepticism is like a microscope whose magnification is constantly increased: the sharp image that one begins with finally dissolves, because it is not possible to see ultimate things: their existence is only to be inferred"

Stanislaw Lem "His Masters Voice". Translated from the Polish
by M. Kandel, A Helen and Kurt Wolff Book,
Harcourt Brace Jovanovich, Publishers,
San Diego–New York – London 1983

While the ultimate answers in the field of the slow unconventional viruses have not been provided yet, at least new powerful tools were constructed which, potentially, may furnish such answers. Investigators from the teams of Prusiner and Weissmann teams constructed mice without PrP [3]. These mice were produced using homologous recombination with a 4.8 kb Prn-p gene fragment in which codons 4 to 187 were replaced by a neomycin phosphotransferase (*neo*) gene under the control of the herpes simplex virus thymidine kinases promotor. PrP-lacking mice, observed by a period of up to 7 months, surprisingly did not show any obvious abnormality. The last finding may suggest that PrP^c, despite its ubiquitous presence, is not a protein with a function which is necessary for an integrity of an organism. However, as function of PrP is even not suspected (besides the postulated role in lymphocyte activation: see paragraph 2.2.4) these results may simply reflect the absence of knowledge where to search for.

When the mice without PrP were inoculated with scrapie they would provide the ultimate answer of the nature of the scrapie virus. If mice inoculated with scrapie developed disease, the answer would be that PrP is a component necessary neither for pathogenesis nor for the virus. The author believes that such an answer is highly improbable. However, if these mice did not develop scrapie but replication of the virus were to be detected by end-point titration it would mean that PrP is crucial for pathogenesis and, by the same token, is not a part of the virus. Only, if inoculated mice developed neither scrapie nor showed replication, it would mean that PrP is the necessary component of the scrapie virus. As the tool is available, it is only a question of time and proper experimentation to solve the puzzle.

For the most neutral observers, the pendulum seems to switch toward "prion hypothesis" mostly because the search for the virus proved to so be completely fruitless. However, careful analysis of accumulated data seems to suggest that the Prusiner group achieved a goal slightly different from that they have pursued, instead to find the nature of the virus they solved the problem of virus pathogenesis and proved that SSVE are the *transmissible cerebral amyloidoses*.

The SSVE recapitulate in many details the pathogenesis of another form of cerebral amyloidosis: the non-transmissible cerebral amyloidoses of Alzheimer's disease, senile dementia of Alzheimer type, Down's syndrome, and hereditary hemorrhage with amyloidosis, Dutch type (HCHWA-A) [2, 6, 13], and this group of disorders will be used here to shed some light on the emerging solution of the puzzle of the pathogenesis of the SSVE. While crucial differences between transmissible and non-transmissible cerebral amyloidoses stem from the very fact of existence of chemically different amyloids, the molecular pathogenesis remains basically the same: the synthesis, processing and final accumulation of appropriate amyloids. Even the neuropathological hallmark in both groups of cerebral amyloidoses is the same: the amyloid plaque. In the SSVE, PrP^c plays a role of amyloid precursor protein, while modified PrP (i.e. PrP^{sc} and PrP 27–30) is the final deposit (*vide supra*). In non-transmissible amyloidoses, the amyloid precursor protein (APP) is analogous to PrP^c while its proteolytic cleavage product, a 39–42 amino acid protein of M_r of 4.2 kDa, designated beta-A4 amyloid, is the final deposit [2, 8, 9, 12, 14, 20]. The processing of APP is much more complicated than that of PrP^c [10, 23]. The alternative splicing of the primary transcript produces at least 6 different isoforms of APP: APP_{365}, APP_{563}, APP_{695}, APP_{714}, APP_{751} i APP_{770} [23]. Noteworthy, APP_{751} and APP_{770} additionally contain inserts (of 56–76 amino acids) homologous to the Kunitz type of serine protease inhibitors. Beta-A4 (the final deposit) is encoded as an internal peptide of APP_{695}, APP_{714}, APP_{751} i APP_{770} but not of APP_{365}, APP_{563} [11–12]. Analogously to PrPs, where both transmembrane and secretory forms are known (at least *in vitro*; compare paragraph: 2.2.2.2), both transmembrane and secretory forms of APP have been discovered [11, 19]. Secretory forms of APP are generated by a proteolytic cleavage of a still undiscovered "secretase" by which the "constitutive" pathway of APP metabolism must be switched to a still obscure "amyloidogenic" pathway. Analogously, a yet to be discovered "constitutive" pathway of PrP^c metabolism must be altered (and this alterations is still unknown) to produce amyloidogenic PrP^{sc}. The final deposit of beta-A4 is the amyloid plaque, a structure analogous to PrP plaque of the SSVE.

It is very clear from the above description that both, *transmissible* and *non-transmissible* beta amyloidoses, follow the same basic pattern of pathogenesis. But the cause of either of them seems to be entirely different. However, even within a group of non-transmissible beta amyloidoses the etiology of the particular disease is different. Neuropathology of Down's syndrome is caused by a simple gene overdose (beta-A4 gene is located on chromosome 21 in humans [9]), the cause of sporadic AD/SDAT (like sporadic CJD) is unknown while the etiology of familial forms of AD/SDAT is probably linked to mutations within the ORF of the APP gene [4, 5, 7, 15, 21]. The last mechanism recapitulates that postulated for familial CJD.

The discovery that several mutations within the PRNP gene are linked to the familial occurrence of CJD and GSS syndrome (see paragraph: 2.3.8 and 2.3.9–2.3.13) were followed by analogous discoveries of point mutations within the APP gene on chromosome 21 [4, 5, 7, 15, 21]. APP gene mutations were tightly linked to the occurrence of familial AD cases. Analogous mutations within appropriate amyloid (cystatin C) was also discovered in another *non-transmissible* cerebral amyloidosis – hereditary haemorrhage with amyloidosis – Icelandic type (HCHWA-I; [16]). The importance of these mutations was further elaborated by use of transgenic mice. However, while transgenic mice constructed with a mutated Prn-p gene (see paragraph: 2.3.8) clearly produced neurodegenerative disease (transgenic mice which harbors Prn-p gene with other CJD-linked mutations are available), transgenic mice produced with *the native* (without known mutations) sequence of APP gene did not develop any neuropathology reminiscent of Alzheimer's disease [1, 17, 18, 22]. Thus, a structural abnormality of amyloid precursor gene must be present to exert its deleterious influence. In this context, the presence of sporadic cases (those without mutations) of both AD and CJD is still unresolved puzzle.

The working hypothesis to present here is that changes in amyloid precursor proteins, irrespective of their chemical nature (APP or PrP^c), are not causes of cerebral amyloidoses but contribute to pathogenesis, i.e. amyloidogenesis responsible for the final amyloid deposit (in the form of beta-A4 or $PrP^{sc(CJD\ or\ GSS)}$). While the cause of non-transmissible cerebral amyloidoses is unknown, the cause of transmissible cerebral amyloidosis is the virus (however unconventional) which is still obscure.

References for "final conclusions"

1. Anonymous (1991) Amyloid deposits: a case of mistaken identity. Science 1991: 1201

2. Beyreuther K, Masters CL (1991) Amyloid precursor protein (APP) and betaA4 amyloid in the etiology of Alzheimer's disease: precursor-product relationships in the derangement of neuronal function. Brain Pathol 1: 241–252

3. Bueler H, Fischer M, Lang Y, Bluthmann H, Lipp H-P, DeArmond SJ, Prusiner SB, Aguet M, Weissmann C (1992) Normal development and behaviour of mice lacking the neuronal cell-surface PrP protein. Nature 356: 577–582

4. Chartier-Harlin M-C, Crawford F, Houlden H, Warren A, Hughes D, Fidani L, Goate A, Rossor M, Roques P, Hardy J, Mullan M (1991) Early-onset Alzheimer's disease caused by mutations at codon 717 of the beta-amyloid precursor protein gene. Nature 353: 844–846

5. Crawford F, Hardy J, Mullan M, Goate A, Hughes D, Fidani L, Roques P, Rossor M, Chartier-Harlin M-C (1991) Sequencing of exons 16 and 17 of the beta-amyloid precursor gene in 14 families with early onset Alzheimer's disease fails to reveal mutations in the beta-amyloid sequence. Neurosci Letters 133: 1–2

6. Glenner GG, Murphy MM (1991) Amyloidosis of the nervous system. J Neurol Sci 94: 1–28

7. Goate A, Chartier-Harlin M-C, Mullan M, Brown J, Crawford F, Fidani L, Giuffra L, Haynes A, Irving N, James L, Mant R, Newton P, Rooke K, Roques P, Tabot C, Pericak-Vance M, Roses A, Williamson R, Rossor M, Owen M, Hardy J (1991) Segregation of a nonsense mutation in the amyloid precursor protein gene with familial Alzheimer's disease. Nature 349: 704–706

8. Goedert M, Spillantini MG, Crowther RA (1991) Tau proteins and neurofibrillary degeneration. Brain Pathol 1: 279–286

9. Goldgaber D, Lehrman MI, McBride O-W, Saffiotti U, Gajdusek DC (1987) Characterization and chromosomal localization of a cDNA encoding brain amyloid of Alzheimer's disease. Science 235: 877–880

10. Hardy J, Allsop D (1991) Amyloid deposition as the central event in the etiology of Alzheimer's disease. Trends Pharmacol Sci 12: 383

11. Ishihura S (1991) Proteolytic cleavage of the Alzheimer's disease amyloid A4 precursor protein. J Neurochem 56: 363–369

12. Kang J, Lemaire H-G, Unterbeck A, Salbaum JM, Masters CL, Grzeschik K-H, Multhaup G, Beyreuther K, Muller-Hill B (1987) The precursor of Alzheimer's disease amyloid A4 protein resembles a cell surface receptor. Nature 325: 733–736

13. Masters CL, Beyreuther K, Alzheimer's disease: molecular basis of structural lesions. Brain Pathol 1: 226–228

14. Masters CL, Simms G, Weinman NA, Multhaup G, McDonald BL, Beyreuther K (1985) Amyloid plaque core protein in Alzheimer disease and Down syndrome. Proc Natl Acad Sci USA 82: 4245–4249

15. Murrell J, Farlow M, Ghetti B, Benson MD (1991) A mutation in the amyloid precursor protein associated with hereditary Alzheimer's disease. Science 254: 97–99

16. Palsdottir A, Abrahamson M, Thorsteinsson L, Arnason A, Olaffson I, Grubb A, Jensson A (1988) Mutation in cystatin C causes hereditary brain hemorrhage. Lancet ii: 603–604

17. Quon D, Wang Y, Marian Scardina J, Murakami K, Codell B (1991) Formation of beta-amyloid protein deposits in brains of transgenic mice. Nature 352: 239–241

18. Robertson M (1991) Alzheimer's disease and amyloid. Nature 356: 103

19. Selkoe DJ (1990) Deciphering Alzheimer's disease: the amyloid precursor protein yield new clues. Science 248: 1058–1060

20. Selkoe DJ, Bell DS, Podlisny MB, Price DL, Cork LC (1987) Conservation of brain amyloid proteins in aged mammals and humans with Alzheimer's disease. Science 235: 873–877
21. St George-Hyslop PH, Tanzi RE, Podlinsky RJ, Haines JL, Nee L, Watkins PC, Myers RH, Feldman RG, Pollen D, Drachman D, Growdon J, Bruni A, Foncin J-F, Salmon D, Frommelt P, Amaducci L, Sorbi S, Piacentini S, Stewart GD, Hobbs WJ, Conneally PM, Gusella JF (1987) The genetic defect causing familial Alzheimer's disease maps on chromosome 21. Science 235: 885–890
22. Wirak DO, Bayney R, Ramabhadran TV, Fracasso RP, Hart JT, Hauer PE, Hsiau P, Pekar SK, Scangos GA, Trapp BD, Unterbeck AJ (1991) Deposits of amyloid beta protein in the central nervous system of transgenic mice. Science 253: 323–325
23. Younkin SG (1991) Processing of the Alzheimer's disease beta-A4 amyloid protein precursor (APP). Brain Pathol 1: 253–262

Addendum

Since the time of submission of this supplement, no major breakthrough has been achieved but the most important data will be reviewed here. The major leitmotiv of the current scrapie research, the irreconcilable views of scrapie agent as "prion" [14] or a "virus" [2], is still at work. Neither formal paper on a positive transmission from mice with spontaneous neurodegenerative disorder (see paragraph: 2.3.8) nor results of mice without PrP (see paragraph: 2.3.8) inoculated with scrapie virus have been published. It was, however, reported that amphotericin B delayed the onset of clinical signs and Prp^{sc} accumulation in hamster brains but not replication of the 263K strain of scrapie virus [19]. Such a clear dissociation of infectivity titer and Prp^{sc} level may indicate the Prp^{sc} is not a part of the virus but, on the same token, it may be more important than replication by itself for the development of disease. Using transgenic mice harboring hamster Prn-p gene (see: paragraph 2.4.2), Hecker et al. [6] reported total block of replication of the murine isolate of scrapie by the 263K (Sc237) strain (hamster) of scrapie virus. In contrast, there was no competition between two hamster strains of scrapie virus inoculated into syrian hamsters, despite clear differences of incubation period, neuropathology and PrP^{sc} distribution between them. Furthermore, the sequence of PrP^{sc} appearance in different brain regions seems to followed neuro-anatomical connection between them [6]. Scott et al. [16] arrived at the same con-clusions using different experimental approach (see: paragraph 3.7) and provided some evidence that neuron-to-neuron spread of the virus is highly restricted by the function of gene *Sinc* (or Prn-p). Furthermore, the transport of PrP^{sc} along the white matter tracts was recently reported [18].

The role of mutations within the PRNP gene was not further clarified. The 178 codon mutation (see: paragraphs 2.3.8–2.3.13) was not only independently found in another case of familial CJD [4] but also in the novel neurodegenerative disorder – fatal familial insomnia [9]. As there are no data on the transmissibility of this peculiar disease, its association with SSVE is at the momemnt conjectural. Another mutation, in codon 198, was found in the peculiar Indiana kindred of GSS syndrome [3]. Furthermore, cases with additional 129$^{Val/Val}$ homozygocity presented earlier onset than those with 129$^{Val/Met}$ heterozygocity. An analogous influence of codon 129 polymorphism on the pattern of clinical signs and the incidence and morphology of PrP plaques was recently reported from Japan [13]. Cases with PRNP codon 129 homozygocity showed longer duration, and neither myoclonus nor typical EEG records. It seems thus, that codon 129 homozygocity is not as previously suggested (see: paragraph 2.3.11) the *conditio sine qua non* to get CJD but it modifies the pattern of incoming disease. Using embryo transfer experiments, it was demonstrated that scrapie is passaged vertically from ewes to lambs, thus clarifying this part of the mode of scrapie transmission in the field and making implications for the BSE epidemic [5].

"Transitional" stage PrP (in part proteinase K resistant like PrP^{CJD} and extra-cellular like PrP^c) was detected in peripheral tissues of Libyan Jews harboring 200 codon mutation [10]. The PrP^{sc} was immunolocalized to lysosomes by immunoelectronmicroscopy *in vitro* [8]. The cellular compartment at which PrP^c is transformed into PrP^{sc} is still conjectural but it seems that PrP^c which later become PrP^{sc} is transported from Golgi to the cell surface where it becomes transiently accessible to PIPLC [1]. This transformation of PrP^c to PrP^{sc} may place in still unidentified compartment between mid-Golgi and lysosomes [17]. Furthermore, the removal of the 67 amino acid residues takes place after PrP^{sc} is synthesized. It is plausible that PrP^c is internalized from the cell membrane, possible by the non-clathrin coated vesicles so called caveolae [1, 17].

Only a few new data on the neuropathology has been published. It was demonstrated that, like in Alzheimer's disease, microglial cells are part of the PrP plaque in GSS syndrome [11; Barcikowska, Liberski, Boellaard, Brown, Gajdusek, Budka, Acta Neuropathol (Berl) *in press*]. Using hydrolytic autoclaving, PrP^{sc} was immunolocalized not only to plaques but also diffusely in the neuropil [7]. Furthermore, PrP^{sc} may coexist in beta-A4 amyloid in the same plaque of GSS syndrome [12]. At the end, alpha-beta crystallin, a protein highly homologous to small heat-shock proteins was immunolocalized to the reactive astrocytes in CJD [15].

1. Borchelt D, Taraboulos A, Prusiner SB (1992) Evidence for synthesis of scrapie prion proteins in the endocytic pathway. J Biol Chem 267: 16188–16199
2. Chesebro B (1992) PrP and the scrapie agent. Nature 356: 560
3. Dlouhy SR, Hsiao K, Foroud T, Conneally MP, Prusiner SB, Hodes MB, Ghett B (1992) Linkage of the Indiana kindred of Gerstmann-Straussler-Scheinker disease to the prion protein gene. Nature Genetics 1: 64–67
4. Fink JK, Warren JT, Drury I, Murman D, Peacock BA (1992) Allele-specific sequencing confirms novel prion gene polymorphism in Creutzfeldt-Jakob disease. Neurology 41: 1647–1650
5. Foster JD, McKelvey WAC, Mylne MJA, Williams A, Hunter N, Hope K, Fraser H (1992) Studies on maternal transmission of scrapie in sheep by embryo transfer. Vet Rec 130: 341–343
6. Hecker R, Taraboulos A, Scott M, Pan K-M, Yang S-L, Torchia M, Jendroska K, DeArmond SJ, Prusiner SB (1992) Replication of distinct scrapie prion isolates is region specific in brains of transgenic mice and hamsters. Genes Develop 6: 1213–1228
7. Kitamoto T, Shin R-W, Doh-ura K, Tomokane M, Miyazano M, Muramoto T, Tateishi J (1992) Abnormal isoform of prion proteins accumulates in the synaptic structures of the central nervous system in patients with Creutzfeldt-Jakob disease. Am J Pathol 140: 1285–1294
8. McKinley MP, Taraboulos A, Kenaga L, Serban D, Stieber A, DeArmond SJ, Prusiner SB, Gonatas N (1991) Ultrastructural localization of scrapie prion proteis in cytoplasmic vesicles of infected cultured cells. Lab Invest 65: 622–628
9. Medori R, Tritschler H-J, LeBlanc A, Villare F, Manetto V, Chen HY, Xue R, Leal S, Montagna P, Cortelli P, Tinuper P, Avoni P, Mochi M, Baruzzi A, Hauw JJ, Ott J, Lugaresi L, Autilio-Gambetti L, Gambetti P (1992) Fatal familial insomnia, a prion disease with a mutation at codon 178 of the prion protein gene. New Eng J Med 326: 444–449
10. Meiner Z, Halimi M, Polakiewicz RD, Prusiner SB, Gabizon R (1992) Presence of prion protein in peripheral tissues of Libyan Jews with Creutzfeldt-Jakob disease. Neurology 42: 1355–1360
11. Miyazono M, Iwaki T, Kitamoto T, Kaneko Y, Doh-ura K, Tateishi J (1992) A comparative immunohistochemical study of kuru and senile plaques with a

special reference to glial reaction at various stages of amyloid plaque formation. Am J Pathol 139: 589–598

12. Miyozano M, Kitamoto T, Iwaki T, Tatieshi J (1992) Colocalization of prion protein and beta protein in the same amyloid plaques in patients with Gerstmann-Straussler-Scheinker syndrome. Acta Neuropathol (Berl) 83: 333–339

13. Miyozano M, Kitamoto T, Doh-Ura K, Iwaki T, Tatieshi J (1992) Creutzfeldt-Jakob disease with codon 129 polymorphism (Valine): a comparative study of patients with codon 102 point mutation or without mutations. Acta Neuropathol (Berl) 84: 349–354

14. Prusiner SB (1992) Molecular biological studies of prion disorders in humans and animals. In: Roos RP (ed) Molecular Neurovirology. Humana Press Inc, Totawa, pp 473–501

15. Renkawek K, de Jong WW, Merck KB, Frenken CWGM, van Workum EPA, Bosman GJCG (1992) Alpha beta-crystallin is present in reactive glia in Creutzfeldt-Jakon disease. Acta Neuropathol (Berl) 83: 324–327

16. Scott JR, Fraser H (1992) Scrapie in the central nervous system: neuranatomical spread of infection and *Sinc* control of pathogenesis. J Gen Virol 73: 1637–1644

17. Taraboulos A, Raeber AJ, Borchelt DR, Serban D, Prusiner SB (1992) Synthesis and tracfficking of prion proteins in cultures cells. Mol Cell Biol 3: 851–863

18. Taraboulos A, Jendroska K, Serban d, Yang S-L, DeArmond SJ, Prusiner SB (1992) Regional mapping of prion proteins in brain. Proc Natl Acad Sci USA 89: 7620–7624

19. Xi YG, Infrosso L, Ladogana A, Masullo C, Pocchiari M (1992) Amphotericin B treatment dissociates *in vitro* replication of the scrapie agent from PrP accumulation. Nature 356: 598–601

References

1. Adam J, Crow TJ, Duchen LW, Scarvilli F, Spokes E (1982) Familial cerebral amyloidosis and spongiform encephalopathy. J Neurol Neurosurg Psychiatr 45: 37–45
2. Adams DH (1972) Studies on DNA from normal and scrapie-affected mouse brain. J Neurochem 19: 1869–1882
3. Adams DH, Caspary EA, Field EJ (1970) The incorporation of [^{3}H] thymidine and [^{14}C] glucosamine into a DNA-polysaccharide complex in normal and scrapie-affected mouse brain. Arch Ges Virusforsch 30: 224–237
4. Adams DH, Caspary EA, Field EJ (1969) The incorporation of [^{3}H] thymidine, [^{14}C] orotic acid, [^{14}C] uridine-diphosphoglucose and [^{14}C] glucosamine into a post-ribosomal fraction of normal and scrapie-affected mouse brain. J Gen Virol 4: 89–100
5. Adams H, Beck E, Shenkin AM (1974) Creutzfeldt-Jakob disease: further similarities with kuru. J Neurol Neurosurg Psychiatr 37: 195–200
6. Adornato B, Lampert PW (1971) Status spongiosus of the nervous tissue. Electron microscopic studies. Acta Neuropathol (Berl) 19: 271–289
7. Agamanolis DP, Victor M, Harris JW, Hines JD, Chester JM, Kark J (1978) An ultrastructural study of subacute combined degeneration of the spinal cord in vitamin B12 deficient rhesus monkeys. J Neuropathol Exp Neurol 37: 273–299
8. Agostini L, Fiume S (1958) Contributo allo studio della malattia di Jakob-Creutzfeldt o degenerazione cortico-strio-spinale (Wilson) attraverso la descrizione di un caso. Lavoro Neuropsichiatr 23: 213–244
9. Aiken JM, Williamson JL, Borchardt LM, Marsh RF (1990) Presence of mitochondrial D-loop DNA in scrapie-infected brain preparations enriched for the prion protein. J Virol 64: 3265–3268
10. Aiken JM, Williamson JL, Marsh F (1989) Evidence of mitochondrial involvement in scrapie infection. J Virol 63: 1686–1694
11. Ajuriaguerra JD, Hecaen H, Sadouin R (1953) Degeneration cortico-strio-spinale etude anatomo-clinique a propos de la maladie de Creutzfeldt-Jakob. Rev Neurol 89: 81–100
12. Akowitz A, Sklaviadis T, Manuelidis EE, Manuelidis L (1990) Nuclease-resistant polyadenylated RNAs of significant size are detected by PCR in highly purified Creutzfeldt-Jakob disease preparations. Microbial Pathogenesis 9: 33–45
13. Allen LB, Cochram KW (1977) Acceleration of scrapie in mice by target-organ treatment with interferon inducers. Ann New York Acad Sci 284: 676–681
14. Allen IV, Dermott E, Conolly JH, Hurvitz LJ (1971) A study of a patient with the amyotrophic form of Creutzfeldt-Jakob disease. Brain 94: 715–724
15. Alper T (1985) Scrapie agent unlike viruses in size and susceptibility to inactivation by ionizing or ultraviolet radiation. Nature 317: 750

16. Alper T (1987) Radio- and photobiological techniques in the investigations of prions. In: Prusiner SB, McKinley MP (eds) Prions. Novel Infectious Pathogens Causing Scrapie and Creutzfeldt-Jakob Disease. Academic Press, New York, pp 113–148

17. Alper T, Cramp WA, Haig DA, Clarke MC (1967) Does the agent of scrapie replicate without nucleic acid? Nature 214: 764–766

18. Alper T, Haig DA (1968) Protection by anoxia of the scrapie agent and some DNA and RNA viruses irradiated as dry preparations. J Gen Virol 3: 157–166

19. Alper T, Haig DA, Clarke MC (1966) The exceptionally small size of the scrapie agent. Bioch Biophys Res Comm 22: 278–284

20. Alper T, Haig DA, Clarke MC (1978) The scrapie agent: evidence against its dependence for replication on intrinsic nucleic acid. J Gen Virol 41: 503–515

21. Alpers BJ (1931) Diffuse progressive degeneration of the gray matter of the cerebrum. Arch Neurol Psychiatry 25: 469–505

22. Allsop D, Ikeda S, Bruce M, Glenner GG (1988) Cerebrovascular amyloid in scrapie-affected sheep reacts with antibodies to prion protein. Neurosci Letters 92: 234–239

23. Amyot R, Gauthier C (1964) Sur la maladie de Creutzfeldt-Jakob. Deux observations anatomo-cliniques. Conception uniciste. Rev Neurol 110: 473–488

24. Annunziata P, Federico A (1981) Brain glycosidases in Creutzfeldt-Jakob disease. J Neurol Sci: 325–328

25. Anonymous (1874) Maladie tremblante des moutons. Ann Med Vet 23: 357–360

26. Aoki T, Gibbs CJ Jr, Sotelo J, Gajdusek DC (1982) Heterogeneic autoantibody against neurofilament protein in the sera of animals with experimental kuru and Creutzfeldt-Jakob disease and natural scrapie infection. Infect Immun 38: 316–324

27. Arendt T, Bigl V, Arendt A (1984) Neurone loss in the nucleus basalis Meynert in Creutzfeldt-Jakob disease. Acta Neuropathol (Berl) 65: 85–88

28. Arsenio-Nunes ML, Goutieres F, Aicardi J (1981) An ultramicroscopic study of skin and conjunctival biopsies in chronic neurological disorders of childhood. Ann Neurol 9: 163–173

29. Asher DM, Gajdusek DC, Gibbs CJ Jr (1976) Pathogenesis of subacute spongiform encephalopathies. Ann Clin Med 6: 84–103

30. Asher DM, Masters CL, Gajdusek DC, Gibbs CJ Jr (1983) Familial spongiform encephalopathies. In: Ket SS, Rowland LP, Sidman RL (eds) Genetics of Neurologic and Psychiatric Disorders. Raven Press, New York, pp 273–291

31. Azzarelli B, Muller J, Ghetti B, Dyken M, Conneally PM (1985) Cerebellar plaques in familial Alzheimer's disease (Gerstmann-Sträussler-Scheinker variant)? Acta Neuropathol (Berl) 65: 235–246

32. Baas NH, Hess HH, Pope A (1974) Altered cell mebranes in Creutzfeldt-Jakob disease. Arch Neurol 31: 174–182

33. Bahmanyar S, Moreau-Dubois M-C, Brown P, Cathala F, Gajdusek DC (1983) Serum antibodies to neurofilament antigens in patients with neurological and other diseases and in healthy controls. J Neuroimmunol 5: 191–196

34. Bahmanyar S, Liem RKH, Griffin JW, Gajdusek DC (1984) Characterization of antineurofilament autoantibodies in Creutzfeldt-Jakob disease. J Neuropathol Exp Neurol 43: 369–375

35. Bahmanyar S, Williams ES, Johnson FB, Young S, Gajdusek DC (1985) Amyloid plaques in spongiform encephalopathy of mule deer. J Comp Pathol 95: 1–5

36. Baker HF, Duchen LW, Jacobs JM, Ridley RM (1990) Spongiform encephalopathy transmitted experimentally from Creutzfeldt-Jakob and familial Gerstmann-Sträussler-Scheinker diseases. Brain 113: 1891–1909

37. Baker HF, Poulter M, Crow TJ, Frith CD, Lofthouse R, Ridley RM (1991) Aminoacid polymorphism in human prion protein and age in inherited prion disease. Lancet 337: 1286

38. Ball MJ (1980) Features of Creutzfeldt-Jakob disease in brains of patients with familial dementia of Alzheimer type. J Can Sci Neurol 7: 51–57

39. Baringer JR, Bowman KA, Prusiner SB (1983) Replication of the scrapie agent in hamster brain precedes neuronal vacuolation. J Neuropathol Exp Neurol 42: 539–547

40. Baringer JR, Prusiner SB (1978) Experimental scrapie in mice: ultrastructural observations. Ann Neurol 4: 205–210

41. Baringer JR, Prusiner SB, Wong JS (1981) Scrapie-associated particles in postsynaptic processes. Further ultrastructural studies. J Neuropathol Exp Neurol 40: 281–288

42. Baringer JR, Wong J, Klassen T, Prusiner SB (1979) Further observations on the neuropathology of experimental scrapie in mouse and hamster. In: Prusiner SB, Hadlow WJ (eds) Slow Transmissible Diseases of the Nervous System, vol 2. Academic Press, New York, pp 111–121

43. Barlow RM, Middleton DJ (1990) Dietary transmission of bovine spongiform encephalopathy to mice. Vet Rec 126: 111–112

44. Baron H, Baron-Van Evercooren A, Brucher J-M (1988) Antiserum to scrapie-associated fibril protein react with amyloid plaques in familial transmissible dementia. J Neuropathol Exp Neurol 47: 158–165

45. Baron H, Cathala F, Brown P, Chatelain J, Gajdusek DC (1988) Familial Creutzfeldt-Jakob disease in France: an analysis of 38 familial cases. In: Court LA, Dormont D, Brown P, Kingsbury DT (eds) Unconventional Virus Diseases of the Central Nervous System. Commissariat a' l'Energie Atomique, Paris, pp 70–73

46. Barron KD, Dentinger MP, Nelson LR, Micy JE (1975) Ultrastructure of axonal reaction in red nucleus of cat. J Neuropathol Exp Neurol 34: 222–248

47. Barron RD, Means ED, Larsen E (1973) Ultrastructure of retrograde degeneration in thalamus of rat. I. Neuronal somata and dendrites. J Neuropathol Exp Neurol 32: 218–244

48. Barry RA, Kent SBH, McKinley MP, Meyer RK, DeArmond SJ, Hood LE, Prusiner SB (1986) Scrapie and cellular prion proteins share polypeptide epitopes. J Infect Dis 153: 848–854

49. Barry RA, McKinley MP, Bendheim PE, Lewis GK, DeArmond SJ, Prusiner SB (1985) Antibodies to the scrapie protein decorate prion rods. J Immunol 135: 603–613

50. Barry RA, Prusiner SB (1986) Monoclonal antibodies to the cellular and scrapie prion proteins. J Infect Dis 154: 518–521

51. Barry RA, Vincent MT, Kent SBH, Hood LE, Prusiner SB (1988) Characterization of prion proteins with monospecific antisera to synthetic peptides. J Immunol 140: 1188–1193

52. Basler K, Oesch B, Scott M, Westeway D, Walchli M, Groth DF, McKinley MP, Prusiner SB, Weissmann C (1986) Scrapie and cellular PrP isoforms are encoded by the same chromosomal gene. Cell 46: 417–428

53. Bass NH, Hess HH, Pope A (1986) Altered cell membranes in Creutzfeldt-Jakob disease. Arch Neurol 31: 174–182

54. Bassant MH, Baron H, Gumpel M, Cathala F, Court L (1986) Spread of scrapie agent to the central nerbous system: study of a rat model. Brain Res 383: 397–401

55. Bassant MH, Picard M, Olichon D, Cathala F, Court L (1986) Changes in the serotoninergic, noradrenergic and dopaminergic levels in the brain of scrapie-infected rats. Brain Res 367: 360–363

56. Bastian FO (1987) Spiroplasma-like inclusions in Creutzfeldt-Jakob disease. Arch Pathol Lab Med 103: 665–669

57. Bastian FO, Jennings RA, Gardner WA (1987) Antiserum to scrapie-associated fibril protein cross-reacts with Spiroplasma mirum fibril proteins. J Clin Microbiol 25: 2430–2431

58. Bazan JF, Fletterick RJ, McKinley MP, Prusiner SB (1987) Predicted secondary structure and membrane topology of the scrapie prion protein. Protein Engin 1: 125–135

59. Beck E (1988) Lesions akin to transmissible spongiform encephalopathy in the brains of rats inoculated with immature cerebellum. Their significance in the aetiology of these diseases. Acta Neuropathol 76: 295–305

60. Beck E, Bak IJ, Christ JF, Gajdusek DC, Gibbs CJ Jr, Hassler R (1975) Experimental kuru in the Spider monkey. Histopathological and ultrastructural studies of the brain during early stages of incubation. Brain 98: 595–612

61. Beck E, Daniel PM (1979) Kuru and Creutzfeldt-Jakob disease: neuropathological lesions and their significance. In: Prusiner SB, Hadlow WJ (eds) Slow Transmissible Diseases of the Nervous System, vol 1. Academic Press, New York, pp 253–270

62. Beck E, Daniel PM (1965) Kuru and scrapie compared: are they examples of system degeneration? In: Gajdusek DC, Gibbs CJ Jr, Alpers MP (eds) Slow, Latent, and Temperate Virus Infections. U.S. Dept Health, Education, Welfare, Washington, D.C., pp 85–93

63. Beck E, Daniel PM (1987) Neuropathology of transmissible spongiform encephalopathies. In: Prusiner SB, McKinley MP (eds) Prions. Novel Infectious Pathogens causing Scrapie and Creutzfeldt-Jakob Disease. Academic Press, New York, pp 331–386

64. Beck E, Daniel PM, Alpers MP, Gajdusek DC, Gibbs CJ Jr (1969) Neuropathological comparisons of experimental kuru in Chimpanzees with human kuru. With a note on its relation to scrapie and spongiform encephalopathy. In: Burdzy K, Kallos P (eds) Pathogenesis and Etiology of Demyelinating Diseases, Proc. of the Workshop on Contributions to the Pathogenesis and Etiology of Demylinating Conditions, Locarno, Switzerland, May 31–June 3, 1967. Additamentum to Arch Allergy Appl Immunol, vol 36. S. Karger, Basel, pp 553–562

65. Beck E, Daniel PM, Asher DM, Gajdusek DC, Gibbs CJ Jr (1973) Experimental kuru in chimpanzee: a neuropathological study. Brain 96: 441–462

66. Beck E, Daniel PM, Alpers M, Gajdusek DC, Gibbs CJ (1966) Experimental "kuru" in chimpanzees. A pathological report. Lancet ii: 1056–1059

67. Beck E, Daniel PM, Davey AJ, Gajdusek DC, Gibbs CJ Jr (1985) A note on membrane lamellation. Brain 108: 153–154

68. Beck E, Daniel PM, Davey AJ, Gajdusek DC, Gibbs CJ Jr (1982) The pathogenesis of transmissible spongiform encephalopathy. An ultrastructural study. Brain 105: 755–786

69. Beck E, Daniel PM, Matthews WB, Stevens DL, Alpers MP, Gajdusek DC, Gibbs CJ Jr (1969) Creutzfeldt-Jakob disease. The neuropathology of a transmission experiment. Brain 92: 699–716

70. Belinger-Kawahara CG, Cleaver JE, Diener TO, Prusiner SB (1987) Purified scrapie prions resist inactivation by UV irradiation. J Virol 61: 159–166

71. Belinger-Kawahara CG, Diener TO, McKinley MP, Groth D, Smith DR, Prusiner SB (1987) Purified scrapie prions resist inactivation by procedures that hydrozlyze, modify, or sheer nucleic acids. Virology 160: 271–274

72. Belinger-Kawahara CG, Kempner E, Groth D, Gabizon R, Prusiner SB (1988) Scrapie prion liposomes and rods exhibit target sizes of 55 000 Da. Virology 164: 537–541

73. Behrman S, Mandybur T, McMenemey WH (1962) Un cas de maladie de Creutzfeldt-Jacob a la suite d'un traumatisme cerebral. Rev Neurol 107: 453–459

74. Bendheim PE, Barry RA, DeArmond SJ, Stities DP, Prusiner SB (1984) Antibodies to scrapie prion protein. Nature 310: 318–421

75. Bendheim PE, Bolton DC (1991) Against the proposition "The transmissible agent causing scrapie must contain more than protein". Rev Med Virol (in press)

76. Bendheim PE, Brown HR, Rudelli RD, Scala LJ, Goller NL, Wen GY, Kascak RJ, Cashman NR, Bolton DC (1991) Nearly ubiquitous tissue distribution of the scrapie agent precursor protein. Neurology (in press)

77. Bendheim PE, Kascak R, Cashman NR, Bolton DC (1991) Distribution and possible functions of the normal isoform of the scrapie agent protein. Sem Virol (in press)

78. Bendheim PE, Marmorstein AD, Potempska A, Bolton DC (1988) Scrapie agent proteins do not accumulate in grey matter tremor mice. J Gen Virol 69: 961–966

79. Bendheim PE, Potempska A, Kascak R, Bolton DC (1988) Purification and partial characterization of the normal cellular homologue of the scrapie agent protein. J Infect Dis 158: 1198–1208

80. Berciano J, Berciano MT, Polo JM, Figols J, Ciudad J, Lafarga M (1990) Creutzfeldt-Jakob disease with severe involvement of cerebral white matter and cerebellum. Virchows Archiv A Pathol Anatom 417: 533–538

81. Berciano J, Diez C, Polo JM, Pascual J, Figols J (1991) CT appearance of panencephalopathic and ataxic type of Creutzfeldt-Jakob disease. J Comp Assist Tomography 15: 332–334

82. Bertoni JM, Label LS, Sackelleres JC, Hicks SP (1983) Supranuclear gaze palsy in familial Creutzfeldt-Jakob disease. Arch Neurol 40: 618–622

83. Bertrand I, Carre H, Lucam F (1937) La "tremblante" du mouton. Rec Med Vet 113: 586–603

84. Bertrand I, Carre H, Lucam F (1937) La "tremblante du mouton" (recherches histo-pathologiques). Ann Anat Pathol (Paris) 14: 565–586

85. Besnoit C (1899) La tremblante ou nevrite peripherique enzootique du mouton. Rev Vet (Tolouse), 24: 265–277

86. Besnoit M, Morel M Ch (1898) Note sur les lesions nerveuses de la tremblante du mouton. Rev Vet, Toulouse 23: 397–400

87. Bignami A (1974) The ultrastructure of status spongiosus and neuronal vacuolation in Jakob Creutzfeldt disease and other transmissible encephalopathies. In: Subirana A, Espadaler JM, Burrows EH (eds) Neurology. Proc of the Xth Int Cong Neurol. Excerpta Medica, Amsterdam, pp 330–338

88. Bignami A, Beck E, Parry HB (1970) Neurosecretion-like material in the hindbrain of ageing sheep and sheep affected with natural scrapie. Nature 225: 194–196

89. Bignami A, Forno LS (1970) Status spongiosus in Jakob-Creutzfeldt disease. Electron microscopic study of a cortical biopsy. Brain 93: 89–94

90. Bignami A, Parry HB (1971) Aggregations of 35-nanometer particle associated with neuronal cytopathic changes in natural scrapie. Science 171: 389–390

91. Bignami A, Parry HB (1972) Electron microscopic studies of the brain of sheep with natural scrapie. I. The fine structure of neuronal vacuolation. Brain 95: 319–326

92. Bignami A, Parry HB (1972) Electron microscopic studies of the brain of sheep with natural scrapie. Brain 95: 487–494

93. Blakemore WF, Cavanagh JB (1969) "Neuroaxonal dystrophy" occurring in an experimental "dying back" process in the rat. Brain 92: 789–804

94. Blisard KS, Davis LE, Harrington MG, Lovell JK, Kornfeld M, Berger ML (1990) Pre-mortem diagnosis of Creutzfeldt-Jakob disease by detection of abnormal cerebrospinal fluid proteins. J Neurol Sci 99: 75–81

95. Bobowick A, Brody JA, Matthews MR, Roos R, Gajdusek DC (1973) Creutzfeldt-Jakob disease: a case control study. Am J Epidemiol 98: 381–394

96. Bockman JM, Kingsbury DT (1988) Immunological analysis of host and agent effects on Creutzfeldt-Jakob disease and scrapie prion proteins. J Virol 62: 3120–3127

97. Bockman JM, Kingsbury DT, McKinley MP, Bendheim PE, Prusiner SB (1985) Creutzfeldt-Jakob disease prion proteins in human brains. New Engl J Med 312: 73–78

98. Bockman JM, Prusiner SB, Tateishi J, Kingsbury DT (1987) Immunoblotting of Creutzfeldt-Jakob disease prion proteins: host species-specific epitopes. Ann Neurol 21: 589–595

99. Bode L, Pocchiari M, Gelderblom H, Diringer H (1985) Characterization of antisera against scrapie-associated fibrils (SAF) from affected hamster and cross-reactivity with SAF from scrapie-affected mice and from patients with Creuzefeldt-Jakob disease. J Gen Virol 66: 2471–2478

100. Boellaard JW, Kao M, Schlote W, Diringer H (1991) Neuronal autophagy in experimental scrapie Acta Neuropathol (Berl) 82: 225–228

101. Boellaard JW, Schlote W (1980) Subakute spongiforme Encephalopathie mit multiformer Plaquebildung. Eigenartige familiär-hereditäre Kranknheit des Zentralnervensystems [spino-cerebellare Atrophie mit Demenz, Plaques and plaqueähnlichen Ablagerungen im Klein- and Grosshirn (Gerstmann, Sträussler, Scheinker)]. Acta Neuropathol (Berl) 49: 205–212

102. Boellaard JW, Schlote W (1981) Glial plaques: amyloid deposits characteristic of slow transmissible encephalopathies. Virchows Arch [Cell Pathol] 37: 337–341

103. Boellaard JW, Schlote W, Heldt N (1989) The development of amyloid plaques in human transmissible encephalopathy and the role of astrocytes in their formation. In: Court LA, Dormont D, Brown P, Kingsbury DT (eds) Unconventional Virus Diseases of the Central Nervous System. Commissariat a' l'Energie Atomique, Paris, pp 162–171

104. Boellaard JW, Schlote W, Heldt N (1990) Ultrastructure of A4 and transmissible cerebral amyloid types. Abstract IV-C-10 In: Abstracts of XIth International Congress of Neuropathology, Kyoto, September 2–8, pp 267

105. Boellaard JW, Schlote W, Tateishi J (1989) Neuronal autophagy in experimental Creutzfeldt-Jakob disease. Acta Neuropathol (Berl) 78: 410–418

106. Bogaert L van, Dewulf A, Palsson PA (1978) Rida in sheep. Pathological and clinical aspects. Acta Neuropathol (Berl) 41: 201–206

107. Bohnert B, Noetzel H (1974) Beitrag zur familiären spongiosen glio-neuralen dystrophie. Arch Psych Nervenkrank 218: 353–368

108. Bolton DC, Bendheim PE, Marmorstein AD, Potempska A (1987) Isolation and structural studies of the intact scrapie agent protein. Arch Biochem Biophys 258: 579–590

109. Bolton DC, McKinley MP, Prusiner SB (1982) Identification of a protein that purifies with the scrapie prion. Science 218: 1309–1311

110. Bolton DC, McKinley MP, Prusiner SB (1984) Molecular characteristics of the major scrapie prion protein. Biochemistry 23: 5898–5906

111. Bolton DC, Meyer RK, Prusiner SB (1985) Scrapie PrP 27–30 is a sialoglyco-protein. J Virol 53: 596–606

112. Bonduelle M, Escourolle R, Bouygues P, Lormeau G, Ribadeau Dumas JL, Merland JJ (1971) Maladie de Creutzfeldt-Jakob familiale. Observation anatomo-clinique. Rev Neurol 125: 197–209

113. Borchelt DR, Scott M, Taraboulos A, Stahl N, Prusiner SB (1990) Scrapie and cellular prion proteins differ in their kinetics of synthesis and topology in cultured cells. J Cell Biol 110: 743–752

114. Bornstein S, Jervis GA (1955) Presenile dementia of the Jakob type. Arch Neurol Psychiatr 74: 598–610

115. Borras T, Gibbs CJ Jr (1986) Molecular hybridization studies with scrapie brain nucleic acids. I. Search for specific DNA sequences. Arch Virol 88: 67–78

116. Borras TG, Merendino JJ, Gibbs CJ Jr (1986) Molecular hybridization studies with scrapie brain nucleic acids. II. Differential expression in scrapie hamster brain. Arch Virol 88: 79–90

117. Bots GThAM, De Man CH, Verjaal A (1971) Virus-like particles in brain tissue from two patients with Creutzfeldt-Jakob disease. Acta Neuropathol (Berl) 18: 267–270

118. Boudin G, Pepin B, Milhaud M (1965) Maladie de Creutzfeldt-Jakob a symp-tomatologie cerebelleuse dominante. Rev Neurol 113: 73–75

119. Boudouresques J, Toga M, Khail R, Cherif A, Pellissier JF, Gosset A (1976) Degenerescence spino-cerebelleuse tardive avec amyotrophies comportant une atteinte pallido-luysienne severe et des lesions histologiques diffuses de senilite. (Etude anatomo-clinique d'un cas avec discussion nosographique). Rev Neurol 132: 623–637

120. Boudouresques J, Toga M, Roger J, Naguet R, Khalil R, Gosset A, Baurand G, Hassoun J (1968) Encephalopathie presenile d'evolution subaigue type Heiden-hain. Rev Neurol 119: 468–476

121. Bouteille M, Kalifat S, Delarue J (1967) Ultrastructural variations of nuclear bodies in human diseases. J Ultrastruct Res 19: 474–486

122. Braig H, Diringer H (1985) Scrapie: a concept of virus-induced amyloidosis of the brain. EMBO J 4: 2309–2312

123. Brion JP, Fraser H, Flament-Durand J, Dickinson AG (1987) Amyloid scrapie plaques in mice, and Alzheimer senile plaques, share common antigens with tau, a microtubule-associated protein. Neurosci Letters 78: 113–118

124. Brion S (1968) La efermedad de Creutzfeldt-Jakob. Neurol Neurocir Psiq (Mexico) 9: 103–115

125. Brion S, Mikol J, Raverdy P, Isidor P (1969) Etude anatomo clinique d'un cas de maladie de Creutzfeldt-Jakob. Aspects ultra-structuraux. Rev Neurol (Paris) 121: 165–179

126. von Braunmühl A (1954) Uber eine eigenartige hereditaer-familiaere Erkrankung des Zentralnervensystems. Arch Psychiatr Z Neurol 191: 419–449

127. Brosnan SF, Selmaj K, Raine CS (1988) Hypothesis: A role for tumor necrosis factor in immune-mediated demyelination and its relevance to multiple sclerosis. J Neuroimmunol 18: 87–94

128. Brotherston JG, Renwick CC, Stamp JT, Zlotnik I (1968) Spread of scrapie by contact to goats and sheep. J Comp Pathol 78: 9–17

129. Brown G, Buckle D (1939) A case of Jakob's disease. J Mental Sci 85: 562–565

130. Brown HR, Goller NL, Rudelli RD, Merz GS, Wolfe GC, Wisniewski HM, Robakis NK (1990) The mRNA encoding the scrapie agent protein is present in a variety of non-neuronal cells. Acta Neuropathol (Berl) 80: 1–6

131. Brown P (1989) Central nervous system amyloidosis. A comparison of Alzheimer's disease and Creutzfeldt-Jakob disease. Neurology 39: 1103–1105

132. Brown P (1980) An epidemiologic critique of Creutzfeldt-Jakob disease. Epidemiologic Rev 2: 113–135

133. Brown P (1988) Human growth hormone therapy and Creutzfeldt-Jakob disease: a drama in three acts. Pediatrics 81: 85–92

134. Brown P (1989) The relevance of Creutzfeldt-Jakob disease to the study of Alzheimer's Disease. In: Iqbal K, Wisniewski HM, Winblad B (eds) Alzheimer's Disease and Related Disorders. Alan R Liss, pp 535–548

135. Brown P (1990) A therapeutic panorama of the spongiform encephalopathies. Antiviral Chem Chemotherapy 1: 75–83

136. Brown P, Cathala F, Raubertas RF, Gajdusek DC, Castaigne P (1988) Creutzfeldt-Jakob disease in France: epidemiologic data for the 15-year period 1968–1982. In: Court LA, Dormont D, Brown P, Kingsbury DT (eds) Unconventional virus diseases of the central nervous system. Commissariat a' l'Energie Atomique, Paris, pp 94–109

137. Brown P, Cathala F, Raubertas RF, Gajdusek DC, Castaigne P (1987) The epidemiology of Creutzfeldt-Jakob disease: conclusion of a 15-year investigation in France and review of the world literature. Neurology 37: 895–904

138. Brown P, Coker-Vann M, Pomeroy K, Franko M, Asher DM, Gibbs CJ Jr, Gajdusek DC (1986) Diagnosis of Creutzfeldt-Jakob disease by Western blot identification of marker protein in human brain tissue. New Engl J Med 314: 547–551

139. Brown P, Gajdusek DC, Gibbs CJ Jr, Asher DM (1985) Potential epidemic of Cerutzfeldt-Jakob disease from human growth hormone therapy. New Engl J Med 313: 728–731

140. Brown P, Goldfarb LG, Brown WT, Goldgaber D, Rubenstein R, Kascak RJ, Guiroy DC, Piccardo P, Boellaard JW, Gajdusek DC (1991) Clinical and molecular genetic study of a large German kindred with Gerstmann-Sträussler-Scheinker syndrome. Neurology 41: 375–379

141. Brown P, Goldfarb LG, Cathala F, Vbrovska A, Sulima M, Nieto A, Gibbs CJ Jr, Gajdusek DC (1991) The molecular genetics of familial Creutzfeldt-Jakob disease in France. J Neurol Sci 105: 240–246

142. Brown P, Goldfarb LG, Gibbs CJ Jr, Gajdusek DC (1991) The phenotypic expression of different mutations in transmissible familial Creutzfeldt-Jakob disease. Europ J Epidemiol 7: 469–476

143. Brown P, Goldfarb LG, Gajdusek DC (1991) The new biology of spongiform encephalopathy: infectious amyloidoses with a genetic twist. Lancet 337: 1019–1022

144. Brown P, Goldfarb LG, McCombie WR, Nieto A, Trapp S, Squillacote D, Shermata W, Godec M, Gibbs CJ Jr, Gajdusek DC (1991) Atypical Creutzfeldt-Jakob disease in an American family with an insert mutation in the PRNP amyloid precursor gene. Neurology 42 (in press)

145. Brown P, Jannotta F, Gibbs CJ Jr, Guiroy DC, Gajdusek DC (1990) Coexistence of Creutzfeldt-Jakob disease and Alzheimer's disease in the same patients. Neurology 40: 226–228

146. Brown P, Liberski PP, Wolff A, Gajdusek DC (1990) Conservation of infectivity in purified fibrillary extracts of scrapie-infected hamster brain after sequential enzymatic digestion or polyacrylamide gel electrophoresis. Proc Natl Acad Sci USA 87: 7240–7244

147. Brown P, Liberski PP, Wolff A, Gajdusek DC (1990) Resistance of scrapie infectivity to steam autoclaving after formaldehyde fixation and limited survival after ashing at 360°C: practical and theoretical implications. J Infect Dis 161: 467–472

148. Brown P, Rogers-Johnson P, Cathala F, Gibbs CJ Jr, Gajdusek DC (1984) Creutzfeldt-Jakob disease of long duration: clinicopathological characteristics, transmissibility, and differential diagnosis. Ann Neurol 16: 295–304

149. Brown P, Rohwer R, Moreau-Dubois M-C, Green EM, Gajdusek DC (1980) Use of golden syrian hamster in study of scrapie. In: Streilein JW, Hart DA, Stein-Streilein J, Duncan WR, Bilingham RE (eds) Hamster Immune Responses in Infectious and Oncologic diseases. Plenum Press, New York, pp 365–373

150. Brown P, Salazar AM, Gibbs CJ, Gajdusek DC (1982) Alzheimer disease and transmissible virus dementia (Creutzfeldt-Jakob disease). Ann New York Acad Sci 396: 131–143

151. Brownell B, Oppenheimer DR (1965) An ataxic form of subacute presenile polioencephalopathy (Creutzfeldt-Jakob disease). J Neurol Neurosurg Psychiatry 28: 350–361

152. Bruce ME (1985) Agent replication dynamics in a long incubation period model of mouse scrapie. J Gen Virol 66: 2517–2522

153. Bruce ME (1981) Serial studies on the development of cerebral amyloidosis and vacuolar degeneration in murine scrapie. J Comp Pathol 91: 589–597

154. Bruce ME, Dickinson AG (1987) Biological evidence that scrapie agent has an idependent genome. J Gen Virol 68: 79–89

155. Bruce ME, Dickinson AG (1985) Genetic control of amyloid plaque production and incubation period in scrapie-infected mice. J Neuropathol Exp Neurol 44: 285–294

156. Bruce ME, Fraser H (1975) Amyloid plaques in the brains of mice infected with scrapie: morphological variation and staining characteristics. Neuropathol Appl Neurobiol 1: 189–202

157. Bruce ME, Fraser H (1976) Cerebral amyloidosis in scrapie in the mouse: effect of agent strain and mouse genotype. Neuropathol Appl Neurobiol 2: 471–478

158. Bruce ME, Fraser H (1981) Effect of route of infection on the frequency and distribution of cerebral amyloid plaques in scrapie mice. Neuropathol Appl Neurobiol 7: 289–298

159. Bruce ME, Fraser H (1982) Effects of age on cerebral amyloid plaques in murine scrapie. Neuropathol Appl Neurobiol 8: 71–74

160. Bruce ME, Fraser H (1982) Focal and asymetrical vacuolar lesions in the brains of mice infected with certain strains of scrapie. Acta Neuropathol (Berl) 58: 133–143

161. Bruce ME, McBride PA, Farquhar CF (1986) Precise targeting of the pathology of the sialoglycoprotein, PrP, and vacuolar degeneration in mouse scrapie. Neurosci Letters 102: 1–6

162. Bruce ME, McConnell I, Fraser H, Dickinson AG (1991) The disease characteristics of different strains of scrapie in *Sinc* congenic mouse lines: implications for the nature of the agent and host control of pathogenesis. J Gen Virol 72: 595–603

163. Brun A, Gottfries CG, Roos BE (1971) Studies of monoamine metabolism in the central nervous system in Jakob-Creutzfeldt disease. Acta Neurol Scandinav 47: 642–645

164. Bubis JJ, Goldhammer Y, Braham J (1965) Subacute spongiform encephalopathy. Electron microscopic studies. J Neurol Neurosurg Psychiatr 35: 881–887

165. Buening GM, Gustafson DP (1970) Enzymatic changes in astrocytes in mice affected with scrapie. Fed Proc 29: 289

166. Bundza A, Charlton KM (1988) Comparison of spongiform lesions in experimental scrapie and rabies in skunks. Acta Neuropathol 76: 275–280

167. Burger LJ, Rowan AJ, Goldensohn ES (1972) Creutzfeldt-Jakob disease. An electroencephalographic study. Arch Neurol 26: 428–433

168. Carlo J, Willis J, McGarry P, Duncan C (1982) Examination of dentalp pulp to diagnose infantile neuroaxonal dystrophy. Arch Neurol 39: 422–423

169. Carlson GA, Goodman PA, Lovett M, Taylor BA, Marshall ST, Peterson-Torchia M, Westaway D, Prusiner SB (1988) Genetics and polymorphism of the mouse prion gene complex: control of scrapie incubation time. Mol Cell Biol 8: 5528–5540

170. Carlson GA, Hsiao K, Oesch B, Westaway D, Prusiner SB (1991) Genetics of prion infections. Trends Genetics 7: 61–65

171. Carlson GA, Kingsbury DT, Goodman PA, Coleman S, Marshall ST, DeArmond SJ, Westaway D, Prusiner SB (1986) Linkage of prion protein and scrapie incubation time genes. Cell 46: 503–511

172. Carlson GA, Westaway D, DeArmond SJ, Peterson-Torchia M, Prusiner SB (1989) Primary structure of prion protein may modify scrapie isolate properties. Proc Natl Acad Sci USA 86: 7475–7479

173. Carp RI (1982) Transmission of scrapie by oral route: effect of gingival scarification. Lancet i: 170–171

174. Carp RI, Callahan SM (1982) Effect of mouse peritoneal macrophages on scrapie infectivity during extended in vitro incubation. Intervirology 17: 201–207

175. Carp RI, Callahan SM (1985) Effect of prior treatment with thioglycolate on the incubation period of intraperitoneally injected scrapie. Intervirology 24: 170–173

176. Carp RI, Callahan SM (1981) In vitro interaction of scrapie agent and mouse peritoneal macrophages. Intervirology 16: 8–13

177. Carp RI, Merz PA, Kascak RJ, Merz G, Wisniewski HM (1985) Nature of the scrapie agent: current status of facts and hypotheses. J Gen Virol 66: 1357–1368

178. Carp RI, Moretz RC, Natelli M, Dickinson AG (1987) Genetic control of scrapie: incubation period and plaque formation in I mice. J Gen Virol 68: 401–407

179. Carrara E (1946) La malattia di Jakob-Creutzfeldt. Riv Patol Nerv Ment 67: 107–132

180. Carrier H, Joubert L, Lapras M, Gastellu J (1979) Aspects ultrastructuraux des lesions nevraxiques de la tremblante du mouton. Bull Soc Sci Vet Med Comp 75: 101–115

181. Cartier L, Galvez S (1977) Intranuclear corpuscles in the neurons of the substantia nigra. Anatomoclinical report of 11 cases of Jakob-Creutzfeldt disease. Neurocirurgia (Chile) 35: 5–24

182. Cartier L, Galvez S, Gajdusek DC (1985) Familial clustering of the ataxic form of Creutzfeldt-Jakob disease with Hirano bodies. J Neurol Neurosurg Psychiatry 48: 234–238

183. Cartier L, Verdugo R, Vergara C, Galvez S (1989) The nucleus basalis of Meynert in 20 definite cases of Creutzfeldt-Jakob disease. J Neurol Neurosurg Psychiatry 52: 304–309

184. Casaccia P, Fersko R, Kascak RJ, Callahan SM, Carp RI (1990) Regional distribution of PrP following stereotactic injection of different scrapie strains in mice. VIIIth Int Congress Virology, Berlin, p 282

185. Casaccia P, Ladogana A, Xi YG, Pocchiari M (1989) Levels of infectivity in the blood through the incubation period of hamsters peripherally injected with scrapie. Arch Virol 108: 145–149

186. Cashman NR, Loertscher R, Nalbantoglu J, Shaw I, Kascak RJ, Bolton DC, Bendheim PE (1990) Cellular isoform of the scrapie agent protein participates in lymphocyte activation. Cell 61: 185–192

187. Cassirer R (1898) Ueber die Traberkrankheit der Schafe. Pathologisch-anatomische und bakterielle Untersuchungen. Arch Pathol Anat Physiol Klin Med 153: 89–110

188. Castaigne P, Cambier J, Cathala HP, Augustin P (1961) Sur un cas de demence presenile avec cecite corticale et myoclonies. Rev Neurol 104: 431–433

189. Cathala F, Chatelain J, Brown P, Dumas M, Gajdusek DC (1980) Familial Creutzfeldt-Jakob disease. Autosomal dominance in 14 members over 3 generations. J Neurol Sci 47: 343–351

190. Caughey B, Race RE, Chesebro B (1988) Detection of prion protein mRNA in normal and scrapie-infected tissues and cell lines. J Gen Virol 69: 711–716

191. Caughey B, Race RE, Ernst D, Buchmeier MJ, Chesebro B (1989) Prion protein biosynthesis in scrapie-infected and uninfected neuroblastoma cells. J Virol 63: 175–181

192. Caughey B, Neary K, Buller R, Ernst D, Perry LL, Chesebro B, Race R (1990) Normal and scrapie-associated forms of prion protein differ in their sensitivities to phospholipases and proteases in intact neuroblastoma cells. J Virol 64: 1093–1101

193. Cervos-Navarro J, Cruz-Sanchez F, Gimeno-Alava A, Ferszt R (1986) Intranuclear vacuoles in Creutzfeldt-Jakob disease. Neuropathol Appl Neurobiol 12: 477–482

194. Chandler RL (1959) Attempts to demonstrate antibodies in scrapie disease. Vet Rec 71: 58–59

195. Chandler RL (1967) Electron microscopic and other observations on the ependyma and small cerebral blood vessels in mice and rats infected with scrapie. Res Vet Sci 8: 166–169

196. Chandler RL (1962) Encephalopathy in mice. Lancet i: 107–108

197. Chandler RL (1961) Encephalopathy in mice produced by inoculation with scrapie brain material. Lancet i: 1378–1379

198. Chandler RL (1963) Experimental Scrapie in the mouse. Res Vet Sci 4: 276–285

199. Chandler RL (1966) Intranuclear structures in neurons. Nature 209: 1260–1261
200. Chandler RL (1963) Transmissible encephalopathy in mice inoculated with scrapie brain. J Gen Microbiol 31: XIV
201. Chandler RL (1968) Ultrastructural pathology of scrapie in the mouse: an electron microscopic study of spinal cord and cerebellar areas. Br J Exp Pathol 49: 52–59
202. Chandler RL (1968) Ultrastructural observations on scrapie in the gerbil. Res Vet Sci 15: 322–328
203. Chandler RL, Fisher J (1963) Experimental transmission of scrapie to rats. Lancet ii: 1165
204. Charlton KM, Casey GA, Webster WA, Bundza A (1987) Experimental rabies in skunks and foxes. Pathogenesis of the spongiform lesions. Lab Invest 57: 634–645
205. Chateau R, Fau R, Tommasi M, Groslambert R, Perret J, Chatelain R (1966) Amyotrophic distale lente des quatre membres avec evolution dementielle (forme amyotrophique du syndrome de Creutzfeldt-Jakob?). Etude anatomo-clinique. Rev Neurol 115: 955–964
206. Chelle PL (1942) Un cas tremblante chez la chevre. Bull Acad Vet Fr 15: 294–295
207. Check W (1978) Creutzfeldt-Jakob disease: do patients have viremia? JAMA 240: 2037
208. Chesebro B, Race R, Wehrly K, Nishio J, Bloom M, Lechner D, Bergstrom S, Robbins K, Mayer L, Keith JM, Garon C, Haase A (1985) Identification of scrapie prion protein-specific mRNA in scrapie-infected and uninfected brain. Nature 315: 331–333
209. Cho HJ (1980) Requirement of a protein component for scrapie infectivity. Intervirology 14: 213–216
210. Cho HJ (1986) Antibody to scrapie-associated fibril protein identifies a cellular antigen. J Gen Virol 67: 243–253
211. Cho HJ, Greig AS (1975) Isolation of 14-nm virus-like particles from mouse brain infected with scrapie agent. Nature 257: 685–686
212. Cho HJ, Greig AS, Corp CR, Kimberlin RH, Chandler RL, Millson GC (1977) Virus-like particles from both control and scrapie-affected mouse brain. Nature 267: 549–560
213. Chou SM, Martin JD (1971) Kuru-plaques in a case of Creutzfeldt-Jakob disease. Acta Neuropathol (Berl) 17: 150–155
214. Chou SM, Payne WN, Gibbs CJ Jr, Gajdusek DC (1980) Transmission and scanning electron microscopy of spongiform change in Creutzfeldt-Jakob disease. Brain 103: 885–904
215. Chou S-M, Hartmann A (1965) Electron microscopy of focal neuroaxonal lesions produced by beta-beta-iminodipropionitrile (IDPN) in rats. I. The advanced lesions. Acta Neuropathol (Berl) 4: 590–603
216. Clark AW, Manz HJ, White CL, Lehman J, Miller D, Coyle JT (1984) Cortical degeneration with swollen chromatolytic neurons: its relationship to Pick's disease. J Neuropathol Exp Neurol 43: 268–284
217. Clark AW, Parhard IM, Griffin JW, Price DL (1984) Neurofilamentous axonal swellings as a normal finding in the spinal anterior horn of man and other primates. J Neuropathol Exp Neurol 43: 253–262
218. Clarke MC, Haig DA (1971) Multiplication of scrapie agent in mouse spleen. Res Vet Sci 12: 195–197

219. Clarke MC, Haig DA (1967) Presence of the transmissible agent of scrapie in the erum of affected mice and rats. Vet Rec 80: 504

220. Clarke MC, Kimberlin RH (1984) Pathogenesis of mouse scrapie: dustribution of agent in the pulp and stroma of infected spleens. Vet Microbiol 9: 215–225

221. Clarke MC, Millson GC (1976) The membrane location of scrapie infectivity. J Gen Virol 31: 441–445

222. Cochius JI, Burns RJ, Blumbergs PC, Mack K, Alderman CP (1990) Creutzfeldt-Jakob disease in a recipient of human pituitary-derived gonadotrophin. Aust NZ Med 20: 592–593

223. Colinge J, Harding AE, Owen F, Poulter M, Lofthouse R, Boughey AM, Shah T, Crow TJ (1989) Diagnosis of Gerstmann-Sträussler syndrome in familial dementia with prion protein gene analysis. Lancet i: 15–17

224. Collinge J, Owen F, Poulter M, Leach M, Crow TJ, Rossor MN, Hardy J, Mullan MJ, Janota I, Lantos PL (1990) Prion dementia without characteristic pathology. Lancet 336: 7–9

225. Collinge J, Palmer MS, Dryden AJ (1991) Genetic predisposition to iatrogenic Creutzfeldt-Jakob disease. Lancet 337: 1441–1442

226. Collis SC, Kimberlin RH (1983) Further studies on changes in immunoglobulin G in the sera and CSF of Herdwick sheep with natural and experimental scrapie. J Comp Pathol 93: 331–338

227. Collis SC, Kimberlin RH (1985) Long-term persistence of scrapie infection in mouse spleens in the absence of clinical disease. FAEMS Microbiol Lett 29: 111–114

228. Collis SC, Kimberlin RH (1989) Polyclonal increase in certain IgG subclass in mice persistently infected with the 87V strain of scrapie. J Comp Pathol 101: 131–141

229. Collis SC, Kimberlin RH, Millson GC (1979) Immunoglobulin G concentration in the sera of Herdwick sheep with natural scrapie. J Comp Pathol 89: 389–396

230. Cork LC, Tronsco JC, Price DL, Stanley EF, Griffin JW (1983) Canine neuro-axonal dystrophy. J Neuropathol Exp Neurol 42: 286–296

231. Cowen D, Olmstead V (1976) Infantile neuroaxonal dystrophy. J Neuropathol Exp Neurol 22: 175–235

232. Cremieux A, Recordier M, Boudouresques J, Roger J, Toga M, Daniel F, Dubois D (1963) Degenerescence spino-cerebelleuse familiale et maladie d'Alzheimer. Etude anatomo-clinique d'un cas. Rev Neurol 109: 45–54

233. Creutzfeldt HG (1920) Über eine eigenartige herdförmige Erkrankung des Zentralnervensystems. Z Ges Neurol Psychiat 57: 1–18

234. Creutzfeldt HG (1989) On a particular focal disease of the central nervous system (preliminary communication). Alzheimer Dis Assoc Disord 1/2: 15–25

235. Cross AH, McCarron R, McFarlin DE, Raine CS (1987) Adoptively transferred acute and chronic relapsing autoimmune encephalomyelitis in the PL/J mouse and observations on altered pathology by intercurrent virus infection. Lab Invest 57: 499–512

236. Cross AJ, Kimberlin RH, Crow TJ, Johnson JA, Walker CA (1985) Neuro-transmitter metabolism, enzymes and receptors in experimental scrapie. J Neurol Sci 70: 231–241

237. Cruz-Sanchez F, Lafuente J, Gertz H-J, Stoltenburg-Didinger G (1987) Spongi-form encephalopathy with extensive involvement of white matter. J Neurol Sci 82: 81–87

238. Cuille J, Chelle P-L (1938) La tremblante du mouton est bien inoculable. C R Acad Sci 206: 1687–1688

239. Cuille J, Chelle P-L (1938) La tremblante du mouton. Rev Pathol Comp 38: 1358–1372

240. Cuille J, Chelle P-L (1938) La tremblante du mouton est-elle inoculable? C R Acad Sci 203: 1552–1554

241. Cuille J, Chelle P-L (1939) Transmission experimentale de la tremblante a la chevre. C R Acad Sci, Paris 208: 1058–1060

242. Cullen PR, Brownlie J, Kimberlin RH (1984) Sheep lymphocyte antigens and scrapie. J Comp Pathol 94: 405–415

243. Cummings JF, Wood PA, Walkey SU, De Lahunta A, DeForest ME (1985) GM_2 gangliosydosis in a Japanese spaniel. Acta Neuropathol (Berl) 67: 247–253

244. Cunnington PG, Kimberlin RH, Hunter GD, Newsome PM (1976) Absence of antibodies to double-stranded RNA and DNA in the sera of scrapie-infected sheep and mice. IRCS Med Sci 4: 250

245. Czub M, Braig HR, Diringer H (1986) Pathogenesis of scrapie: study of the temporal development of clinical symptoms, of infectivity titres and scrapie-associated fibrils in brains of hamsters infected intraperitoneally. J Gen Virol 67: 2005–2009

246. Czub M, Braig HR, Diringer H (1988) Replication of the scrapie agent in hamsters infected intracerebrally confirms the pathogenesis of an amyloid-inducing virosis. J Gen Virol 69: 1753–1756

247. Daniel PM (1972) Creutzfeldt-Jakob disease. J Clin Pathol 25 (Suppl 6): 97–101

248. Davanipour Z, Alter M, Coslett HB, Sobel E, Kundu S, Hoenig EM (1988) Prolonged progressive dementia with spongiform encephalopathy: a variant of Creutzfeldt-Jakob disease? Neuroepidemiology 7: 56–65

249. David-Ferreira JF, David-Ferreira KL, Gibbs CJ Jr, Morris JA (1968) Scrapie in mice: ultrastructural observations in the cerebral cortex. Proc Soc Exp Biol Med 127: 313–320

250. Davison C (1932) Spastic pseudosclerosis (cortico-pallido-spinal degeneration). Brain 55: 247–264

251. Davison C, Rabiner AM (1940) Spastic pseudosclerosis (disseminated encephalonyelopathy: corticopallidospinal degeneration). Familial and non familial incidence (a clinicopathological study). Arch Neurol Psychiatr 44: 578–598

252. Dawson M, Wells GAH, Parker BNJ (1990) Preliminary evidence of the experimental transmissibility of bovine spongiform encephalopathy to cattle. Vet Rec 126: 112–113

253. Dayan AD (1971) Comparative neuropathology of aging. Studies on the brain of 47 species of vertebrates. Brain 94: 31–34

254. DeArmond SJ, Gonzales M, Mobley WC, Kon AA, Stern A, Prusiner SB (1989) PrPsc in scrapie-infected hamster brain is spatially and temporally related to histopathology and infectivity titer. In: Iqbal K, Wisniewski HM, Winblad B (eds) Alzheimer's Disease and Realted Disorders. Alan R Liss, New York, pp 601–618

255. DeArmond SJ, Kretzschmar HA, McKinley MP, Prusiner SB (1987) Molecular pathology of prion disease. In: Prusiner SB, McKinley MP (eds) Prions. Novel Infectious Pathogens causing Scrapie and Creutzfeldt-Jakob Disease. Academic Press, New York, pp 387–414

256. DeArmond SJ, McKinley MP, Barry RA, Braunfeld MB, McColloch JR, Prusiner SB (1985) Identification of prion amyloid filaments in scrapie-infected brain. Cell 41: 221–235

257. DeArmond SJ, Mobley WC, DeMott DL, Barry RA, Beckstead JH, Prusiner SB (1987) Changes in the localization of brain prion proteins during scrapie infection. Neurology 37: 1721–1780

258. de Courten-Myers G, Mandybur TI (1987) Atypical Gerstmann-Straussler syndrome or familial ataxia and Alzheimer's disease? Neuroloy 37: 269–275

259. Dees C, German TL, Wade WF, Marsh RF (1985) Characterization of lipids in membrane vesicles from scrapie-infected hamster brain. J Gen Virol 66: 861–870

260. Dees C, German TL, Wade WF, Marsh RF (1985) Characterization of proteins in membrane vesicles from scrapie-infected hamster brain. J Gen Virol 66: 851–859

261. Dees C, McMillan BC, Wade WF, German TL, Marsh RF (1985) Characterization of nucleic acids in membrane vesicles from scrapie-infected hamster brain. J Virol 55: 126–132

262. De Iraldi PA, De Robertis E (1968) The neurotubular system of the axon and the origin of granulated and non-granulated vesicles in regenerating nerves. Z Zellforsch 87: 330–344

263. Delez AL, Gustafson DP, Luttrell CN (1957) Some clinical and histological observations on scrapie in sheep. J Am Vet Med Assoc 131: 439–446

264. Dememes D, Fuentes C, Marty R (1974) Cinetique des processus de degenerescence axonique dans le systeme nerveux cenrale a court terme dans la corps calleux chez le rat. Acta Neuropathol (Berl) 29: 311–332

265. Dickinson AG (1974) Natural infection, "spontaneous generation" and scrapie. Nature 252: 179–180

266. Dickinson AG (1976) Scrapie in sheep and goats. In: Kimberlin RH (ed) Slow Virus Diseases of Animals and Man. North-Holland, Amsterdam, pp 209–241

267. Dickinson AG, Bruce ME, Outram GW, Kimberlin RH (1984) Scrapie strain differences: the implications of stability and mutation. In: Tateishi J (ed) Proceedings of Workshop on Slow Virus Transmissible Diseases, Res Comm on Slow Virus Infection. Japanese Ministry of Health and Welfare, Takckashi-Kaikan, pp 105–118

268. Dickinson AG, Fraser H (1979) An assessment of the genetics of scrapie in sheep and mice. In: Prusiner SB, Hadlow WJ (eds) Slow Transmissible Diseases of the Nervous System, vol 1. Academic Press, New York, pp 367–385

269. Dickinson AG, Fraser H (1972) Scrapie: effect of Dh gene on incubation period of extraneurally injected agent. Heredity 29: 91–93

270. Dickinson AG, Fraser H (1977) Scrapie: pathogenesis in inbred mice: an assessment of host control and response involving many strains of agent. In: Ter Meulen V, Katz M (eds) Slow virus infections of the central nervous system. Springer, Berlin Heidelberg New York, pp 3–14

271. Dickinson AG, Fraser H, McConnell, Outram GW (1978) Mitogenic stimulation of the host enhances susceptibility to scrapie. Nature 272: 54–55

272. Dickinson AG, Fraser H, McConnell, Outram GW, Sales DI, Taylor DM (1975) Extraneural competition between different scrapie agent leading to loss of infectivity. Nature 253: 556

273. Dickinson AG, Fraser H, Meikle VMH, Outram GW (1972) Competition between different scrapie agents in mice. Nature 237: 244–245

274. Dickinson AG, Fraser H, Outram GW (1975) Scrapie incubation time can exceed natural life span. Nature 256: 732–733

275. Dickinson AG, Mackay JMK (1964) Genetical control of the incubation period in mice of the neurological disease, scrapie. Heredity 19: 279–288

276. Dickinson AG, Meikle VMH (1969) A comparison of some biological characteristics of the mouse-passaged scrapie agents, 22A and ME7. Genet Res (Camb) 13: 213–225

277. Dickinson AG, Meikle VMH (1971) Host-genotype and agent effects in scrapie incubation: change in allelic interaction with different strains of agent. Mol Gen Genet 112: 73–79

278. Dickinson AG, Meikle VMH, Fraser H (1969) Genetical control of the concentration of ME7 scrapie agent in brain of mice. J Comp Pathol 79: 15–22

279. Dickinson AG, Meikle VMH, Fraser H (1968) Identification of a gene which controls the incubation period of some strains of scrapie agent in mice. J Comp Pathol 78: 293–299

280. Dickinson AG, Stamp JT, Renwick CC, Rennie JC (1977) Some factors controlling the incidence of scrapie in Cheviot sheep injected with a Cheviot-passaged scrapie agent. J Comp Pathol 78: 313–321

281. Dickinson AG, Outram GW (1973) Differences in access into the central nervous system of ME7 scrapie agent from two strains of mice. J Comp Pathol 83: 13–18

282. Dickinson AG, Outram GW (1988) Genetic aspects of unconventional virus infections: the basis of the virino hypothesis. In: Bock G, Marsh J (eds) Novel Infectious Agents and the Central Nervous System. Wiley & Sons, Chichister, pp 63–77

283. Dickinson AG, Outram GW (1979) The scrapie replication-site hypothesis and its implications for pathogenesis. In: Prusiner SB, Hadlow WJ (eds) Slow Transmissible Diseases of the Nervous System, vol 2. Academic Press, New York, pp 13–31

284. Dickinson AG, Outram GW (1983) Operational limitations in the characterisation of the infective units of scrapie. In: Court LA, Cathala F (eds) Virus non Conventionnels at Affections du Systeme Nerveaux Central. Masson, Paris, pp 3–16

285. Dickinson AG, Outram GW, Taylor DM, Foster JD (1988) Further evidence that scrapie agent has an independent genome. In: Court LA, Dormont D, Brown P, Kingsbury DT (eds) Unconventional Virus Diseases of the Central Nervous System. Commissariat à l'Energie Atomique, Paris, pp 446–460

286. Dickson DW, Yen S-H, Suzuki KI, Davies P, Garcia JH, Hirano A (1986) Balooned neurons in select neurodegenerative disease contain phosphorylated neurofilament epitopes. Acta Neuropathol (Berl) 71: 216–223

287. Diener TO (1972) Is the scrapie agent a viroid? Nature, New Biol 235: 218–219

288. Diener TO (1987) PrP and the nature of the scrapie agent. Cell 49: 719–721

289. Diener TO, McKinley MP, Prusienr SB (1982) Viroids and prions. Proc Natl Acad Sci USA 79: 5220–5224

290. Diedrich J, Wietgrafe S, Zupancic M, Staskus K, Retzel E, Haase T, Race R (1987) The molecular pathogenesis in scrapie and Alzheimer's disease. Microb Pathogen 2: 435–442

291. Dimitri V, Aranovich J (1945) Contribucion al conocimiento de la degeneracion corticoestriospinal o "seudoesclerosis espastica" de A. Jakob. Rev Neurol Buenos Aires 10: 225–245

292. Diringer H (1984) Sustained viremia in experimental hamster scrapie. Arch Virol 82: 105–109

293. Diringer H, Blode H, Oberdieck U (1991) Virus-induced amyloidosis in scrapie involves a change in covalent linkages in the preamyloid. Arch Virol 118: 127–131

294. Diringer H, Ehlers B (1991) Chemoprophylaxis of scrapie in mice. Arch Virol 72: 457–460

295. Diringer H, Gerdelblom H, Hilmert H, Ozel M, Edelbluth C, Kimberlin RH (1983) Scrapie infectivity, fibrils and low molecular weight protein. Nature 306: 476–478

296. Diringer H, Hilmert H, Simon D, Werner E, Ehlers B (1983) Towards purification of the scrapie agent. Eur J Biochem 134: 555–560

297. Doh-Ura K, Tateishi J, Kitamoto T, Sasaki H, Sakaki Y (1990) Creutzfeldt-Jakob disease patients with congophilic kuru plaques have the missense variant prion protein common to Gerstmann-Straussler syndrome. Ann Neurol 27: 121–126

298. Dormont D, Delpech B, Delpech A, Courcel M-N, Viret J, Markovits P, Court L (1981) Hyperproduction de proteine gliofibrillaire acide (GFA) au cors de l'evolution de la tremblante experimentale de la Souris. C R Acad Sci, Paris 293: 53–56

299. Dubois-Dalcq M, Rodriguez M, Reese TS, Gibbs CJ Jr, Gajdusek DC (1977) Search for a specific marker in the neural membranes of scrapie mice. Lab Invest 36: 547–553

300. Duguid JR, Dinauer MC (1989) Library subtraction of in vitro cDNA libraries to identify differentially expressed genes in scrapie infection. Nucleic Acid Res 18: 2789–2792

301. Duguid JR, Rohwer R, Seed B (1988) Isolation of cDNA of scrapie-modulated RNAs by subtractive hybridization of cDNA library. Proc Natl Acad Sci USA 85: 5738–5742

302. Economo C, Schilder P (1920) Eine der Pseudosclerose nahestehende Erkrankung im Prasenium. Z Ges Neurol Psychiatr 55: 1–26

303. Ehlers B, Rudolph R, Diringer H (1984) The reticuloendothelial system in scrapie pathogenesis. J Gen Virol 65: 423–428

304. Ehlers B, Diringer H (1984) Dextran sulphate 500 delays and prevents mouse scrapie by impairment of agent replication in spleen. J Gen Virol 65: 1325–1330

305. Eklund CM, Kennedy RC, Hadlow WJ (1965) Pathogenesis of scrapie virus infection in the mouse. In: Gajdusek DC, Gibbs CJ Jr, Alpers MP (eds) Slow, Latent, and Temperate Virus Infections, Monograph no 2. US Dept Health, Education, and Welfare, Washington, pp 207–208

306. Eklund CM, Kennedy RC, Hadlow WJ (1967) Pathogenesis of scrapie virus infection in the mouse. J Infect Dis 117: 15–22

307. Eklund CM, Hadlow WJ (1969) Pathogenesis of slow viral diseases. J Am Vet Med Ass 155: 2094–2099

308. Eklund CM, Hadlow WJ, Kennedy RC (1963) Some properties of the scrapie agent and its behavior in mice. Proc Soc Exp Biol Med 112: 974–979

309. Elleman CJ (1985) ConA induced supressor cells in scrapie-infected mice. Vet Immunol Immunopathol 8: 79–82

310. Eng LF, Vanderhaegen JJ, Bignami A, Gerstl B (1971) An acidic protein isolated from fibrous astrocytes. Brain Res 28: 351–354

311. Eng LE, DeArmond SJ (1983) Immunochemistry of the glial fibrillary acidic protein. In: Zimmerman HM (ed) Progress in Neuropathology, vol 5. Raven Press, New York, pp 19–39

312. Euziere MMJ, Lafon R, Faure JL, Carli G (1950) Maladie de Creutzfeldt-Jakob: pseudosclerose spastique de l'age mur avec desintegration psychique. Montpellier Med 37–38: 375–381

313. Ezrin-Waters C, Resch L, Lang AE (1985) Coexistence of idiopathic Parkinson's disease and Creutzfeldt-Jakob disease. Can J Neurol Sci 12: 272–273
314. Farlow MR, Yee RD, Dlouhy SR, Conneally PM, Azzarelli B, Ghetti B (1989) Gerstmann-Straussler-Scheinker disease. I. Extending the clinical spectrum. Neurology 39: 1446–1452
315. Farquhar CF, Dickinson AG (1986) Prolongation of scrapie incubation period by an injection of dextran sulphate 500 within the month before or after infection. J Gen Virol 67: 463–473
316. Farquhar CF, Somerville RA, Ritchie LA (1989) Post-mortem immunodiagnosis of scrapie and bovine spongiform encephalopathy. J Virol methods 24: 215–222
317. Fattovich G (1960) Considerazioni su un caso di probabile malattia di Jakob-Creutzfeldt. Il Cervello 36: 393–414
318. Fattovich G (1952) La malattia di Jakob-Creutzfeldt. Il Cervello 28: 81–122
319. Fattovich G (1965) Rilievi clinici ed isopatologici sulla malattia di Jakob-Creutzfeldt. Rasseegna Studi Psichiatr 54: 231–253
320. Federico A, Annunziata P, Malentacchi G (1980) Neurochemical changes in Creutzfeldt-Jakob disease. J Neurol 223: 135–146
321. Ferber RA, Wiesenfeld SL, Roos RP, Bobowick AR, Gibbs CJ Jr, Gajdusek DC (1974) Familial Creutzfeldt-Jakob disease: transmission of the familial cases to primates. In: Subirana A, Espadaler JM, Burrows EH (eds) Neurology. Proc of the X Int Congress of Neurology, Barcelona, Spain, September 8–15, 1973, Excerpta Medica, Amsterdam, pp 358–380
322. Ferrer I, Guiionet N, Cruz-Sanchez F, Tunon T (1990) Neuronal alterations in patients with dementia: a Golgi study on biopsy samples. Neurosci Lett 114: 11–16
323. Field EJ (1970) Amyloidosis, Alzheimer's disease, and ageing. Lancet ii: 780–781
324. Field EJ (1967) Invasion of the mouse nervous system by scrapie agent. Br J Exp Pathol 48: 662–664
325. Field EJ (1967) The significance of astroglial hypetrophy in scrapie, kuru, multiple sclerosis and old age together with a note on the possible nature of the scrapie agent. Deutsch Z Nervenheilk 192: 265–274
326. Field EJ (1968) Transmission of kuru to mice. Lancet i: 981–982
327. Field EJ, Caspary EA, Joyce G (1968) Scrapie agent in blood. Vet Rec 83: 109–110
328. Field EJ, Joyce G, Keith A (1971) Viral properties of scrapie. Nature New Biol 230: 56–57
329. Field EJ, Farmer F, Caspary EA, Joyce G (1969) Susceptibility of scrapie agent to ionizing radiation. Nature 222: 90–91
330. Field EJ, Narang HK (1972) An electron-microscopic study of scrapie in the rat: further observations on "inclusion bodies" and virus like particles. J Neurol Sci 17: 347–364
331. Field EJ, Raine CS (1964) An electron microscopic study of scrapie in the mouse. Acta Neuropathol (Berl) 4: 200–211
332. Field EJ, Raine CS, Joyce G (1967) Scrapie in the rat: an electron microscopic study. II. Glial inclusions. Acta Neuropathol 9: 305–315
333. Fields BN (1987) Powerful prions? New Engl J Med 317: 1597–1598
334. Fischer O (1911) Der spongiose Rindenschwund, ein besonderer Destruktions-prozess der Hirnrinde. Z Ges Neurol Psychiatr 7: 1–33

335. Fleischhacker H (1924) Eine familiäre chronisch-progressive Erkrankung des mittleren Lebensalters vom Pseudosklerosetyp. Z Ges Neurol Psychiatr 91: 1–22

336. Foncin J (1967) Etude ultrastructurale de la maladie de Creutzfeldt-Jakob. Acta Neuropathol (Berl) [Suppl] 3: 127–130

337. Foncin J (1971) Neuropathologie ultrastructurale dans les demences dites degeneratives. Neurochir 17 [Suppl 1]: 67–73

338. Foncin J, Gaches J, le Beau J (1964) Encephalopathie spongiforme (apparentee a la maladie de Creutzfeldt-Jakob). Biopsie etude au microscope electronique, confirmation autopsique. Rev Neurol 111: 507–515

339. Fontana A, Fierz W, Wekerle H (1987) Astrocytes present myelin basic protein to encephalitogenic T-cell lines. Nature 307: 273–276

340. Foster JD, Dickinson AG (1988) Genetic control of scrapie in Cheviot and Suffolk sheep. Vet Rec 123: 159

341. Foster JD, Dickinson AG (1988) The unusual properties of CH1641, a sheep-passaged isolate of scrapie. Vet Rec 123: 5–8

342. Fraser H (1982) Neuronal spread of scrapie agent and targeting of lesions within the retino-tectal pathway. Nature 295: 149–150

343. Fraser H (1979) Neuropathology of scrapie: the precision of the lesions and their significance. In: Prusiner SB, Hadlow WJ (eds) Slow Transmissible Diseases of the Nervous System, vol 1. Academic Press, New York, pp 387–406

344. Fraser H (1969) The occurrence of nerve fiber degeneration in brains of mice inoculated with scrapie. Res Vet Sci 10: 338–341

345. Fraser H (1979) The pathogenesis and pathology of scrapie. In: Tyrell DAJ (ed) Aspects of Slow and Persistent Virus Infection. Martinus Nijhoff, The Hague, pp 30–58

346. Fraser H (1979) The pathology of natural and experimental scrapie. In: Kimberlin RH (ed) Slow Virus Diseases of Animals and Man. North-Holland, Amsterdam, pp 267–406

347. Fraser H (1979) Scrapie: a transmissible degenerative cns diseases. In: Behan PO, Rose CF (eds) Progress in Neurological Research. Pitman Medical, London, pp 194–210

348. Fraser H, Bruce ME (1973) Argyrophilic plaques in mice inoculated with scrapie from particular sources. Lancet i: 617–618

349. Fraser H, Bruce ME (1983) Experimental control of cerebral amyloid in scrapie in mice. In: Behan PO, ter Meulen V, Clifford Rose F (eds) Immunology of the Nervous System. Progress in Brain Research, vol 59. Elsevier Science, Amsterdam, pp 281–290

350. Fraser H, Bruce M, Dickinson AG (1974) Quantitative pathology for understanding the nature and pathogenesis of scrapie, using different strains of agent. In: VII International Congress of Neuropathology. Excerpta Medica, Akademiai Kiado, Amsterdam, pp 277–280

351. Fraser H, Dickinson AG (1967) Distribution of experimentally induced scrapie lesions in the brain. Nature 216: 1310–1311

352. Fraser H, Dickinson AG (1970) Pathogenesis of scrapie in the mouse: the role of the spleen. Nature 226: 462–463

353. Fraser H, Dickinson AG (1968) The sequential development of the brain lesions of scrapie in three strains of mice. J Comp Pathol 78: 301–311

354. Fraser H, Dickinson AG (1973) Scrapie in mice. Agent-strain differences in the distribution and intensity of grey matter vacuolation. J Comp Pathol 83: 29–40

355. Fraser H, Dickinson AG (1978) Studies of the lymphoreticular system in the pathogenesis of scrapie: the role of spleen and thymus. J Comp Pathol 88: 563–573

356. Fraser H, Dickinson AG (1985) Trageting of scrapie lesions and spread of agent via the retino-tectal projections. Brain Res 346: 32–41

357. Fraser H, Farquar CF (1987) Ionising radiation has no influence on scrapie incubation period in mice. Vet Microbiol 13: 211–223

358. Fraser H, McBride PA (1985) Parallels and contrast between scrapie and dementia of the Alzheimer type and ageing: strategies and problems for experiments involving life span studies. In: Traber J, Gispen WH (eds) Senile Dementia of the Alzheimer Type. Springer, Berlin, Heidelberg New York Tokyo, pp 251–267

359. Frei K, Fontana A (1989) Immune regulatory functions of astrocytes and microglial cells within the central nervous system. In: Neurimmune Networks: Physiology and Diseases. Alan R Liss, New York, pp 127–136

360. Friede RI, DeJong RN (1964) Neuronal enzymatic failure in Creutzfeldt-Jakob disease. Arch Neurol 10: 181–195

361. Fujisawa K, Shiraki H (1978) Study of axonal dystrophy. I. Pathology of the neuropil of the gracile and cuneate nuclei in aging and old rats. Neuropathol Appl Neurobiol 4: 1–20

362. Fukatsu R, Gibbs CJ, Gajdusek DC (1984) Cerebral amyloid plaques in experimental murine scrapie. In: Tateishi J (ed) Proc of Workshop of slow transmissible diseases. Research Commeetee on slow virus infection, the Japanese Ministry of Health and Welfare. Takekashi-Kaikan, pp 71–84

363. Gabizon R, McKinley MP, Groth D, Kenaga L, Prusiner SB (1988) Properties of scrapie prion protein liposomes. J Biol Chem 263: 4950–4955

364. Gabizon R, McKinley MP, Groth D, Prusiner SB (1988) Immunoaffinity purification and neurtalization of scrapie prion infectivity. Proc Natl Acad Sci USA 85: 6617–6621

365. Gabizon R, McKinley MP, Prusiner SB (1987) Purified prion proteins and scrapie infectivity copartition into liposomes. Proc Natl Acad Sci USA 84: 4017–4021

366. Gachez J, Supino-Viterbo, Foncin JF (1977) Association de maladies d'Alzheimer et de Creutzfeldt-Jakob. Acta Neurol Belg 77: 202–212

367. Gajdusek DC (1990) Annual Report. Laboratory of Central nervous System Studies, October 1, 1989–September 30, 1990. In: National Institute of Neurological Disorders and Stroke. Intramural Research. Annual Report. Fiscal Year 1990. U.S. Dept of Health and Human Services, Public Health Service, National Institutes of Health, Washington, Table 3, pp 1–34

368. Gajdusek DC (1986) Chronic dementia caused by small unconventional viruses apparently containing no nucleic acids. In: Scheibel AB, Wechsler AF, Brazier MA (eds) The Biological Substrate of Alzheimer's Disease. Academic Press, New York, pp 33–54

369. Gajdusek DC (1985) Chronic spongiform virus encephalopathies caused by unconventional viruses. In: Maramorsch K, McKelvey JJ (eds) Subviral Pathogens of Plants and Animals: Viroids and Prions. Academic Press, New York, pp 483–544

370. Gajdusek DC (1985) Hypothesis: interference with axonal transport of neurofilaments as a common pathogenetic mechanism in certain diseases of the central nervous system. New Engl J Med 209: 714–719

371. Gajdusek DC (1985) Interference with axonal transport of neurofilaments as a mechanism of pathogenesis underlying Alzheimer's disease and many other degenerations of the CNS. In: Gottfries CG (ed) Normal Aging, Alzheimer's Disease and Senile Dementia. Aspects of Etiology, Pathogenesis, Diagnosis and Treatment. Edititions de l'Universite Bruxelles, Bruxelles, pp 51–67

372. Gajdusek DC (1984) Interference with axonal transport of neurofilaments as the common etiology and pathogenesis of neurofibrillary tangles, amyotrophic lateral sclerosis, parkinsonism-dementia, and many other degenerations of the cns: a series of hypothesis. In: Chen KM, Yase Y (eds) Amyotrophic Lateral Sclerosis in Asia and Oceania. Proceedings of the Sixth Asian and Oceanian Congress of Neurology. Amyotrophic Lateral Sclerosis Workshop, Taipei, Taiwan, Republic of China, Nov. 14, 1983. Shyan-Fu Chou, National Taiwan University, pp 423–436

373. Gajdusek DC (1987) A newly recognized mechanism of pathogenesis in Alzheimer's disease, amyotrophic lateral sclerosis, and other degenerative neurological diseases: the beta-fibrilloses of brain. In: Jariwalla RJ, Schwoebel SL (eds) Nitrition, Health and Peace. Proc Int Symp Honor of Linus Pauling. Linus Pauling Inst, Palo Alto, pp 21–55

374. Gajdusek DC (1979) Observations on the early history of kuru investigation. In: Prusiner SB, Hadlow WJ (eds) Slow Transmissible Diseases of the Nervous System, vol 1. Academic Press, New York, pp 7–36

375. Gajdusek DC (1970) Physiological and psychological characteristics of stone age man. Engin Sci 33: 26–62

376. Gajdusek DC (1978) Slow infections with unconventional viruses. In: The Harvey Lectures, Series 72. Academic Press, New York, pp 283–353

377. Gajdusek DC (1990) Subacute spongiform encephalopathies: transmissible cerebral amyloidoses caused by unconventional viruses. In: Fields BN, Knipe DM (eds) Fields Virology, 2nd edn, vol 2. Raven Press, New York, pp 2289–2324

378. Gajdusek DC (1991) Transmissible amyloidoses: genetical control of spontaneous generation of infectious amyloid proteins by nucleation of configurational change in host precursors: kuru-CJD-GSS-scrapie-BSE. Europ J Epidemiol 7: 567–578

379. Gajdusek DC (1988) Transmissible and nontransmissible dementias: distinction between primary cause and pathogenetic mechanisms in Alzheimer's disease and aging. Mount Sinai J Med 55: 3–5

380. Gajdusek DC (1984) Unconventional viruses. In: Notkins AL, Oldstone MBA (eds) Concepts in Viral Pathogenesis. Springer, Berlin Heidelberg New York Tokyo, pp 350–357

381. Gajdusek DC (1977) Unconventional viruses and the origin and disappearance of kuru. In: Nobel Fdn (ed) Les Prix Nobel en 1976 PA Norstedt & Soner, Stockholm, pp 167–216

382. Gajdusek DC (1977) Unconventional viruses and the origin and disappearance of kuru. Science 197: 943–960

383. Gajdusek DC, Gibbs CJ Jr (1972) Transmission of kuru from man to rhesus monkey (Macaca mulatta) 8½ years after inoculation. Nature 240: 351

384. Gajdusek DC, Gibbs CJ Jr (1971) Transmission of two subacute spongiform encephalopathies of man (kuru and Creutzfeldt-Jakob disease) to New World monkeys. Nature 230: 588–591

385. Gajdusek DC, Gibbs CJ Jr (1990) Brain amyloidoses. Precursor proteins and the amyloids of transmissible and non transmissible dementias: scrapie-kuruCJD viruses as infectious polypeptides or amyloid engancing factors. In: Goldstein AL (ed) Biomedical Advances in Aging. Plenum Publ, pp 3–24

386. Gajdusek DC, Gibbs CJ (1982) Slow infection of the CNS with unconventional viruses and attempts to demonstrate slow virus etiology in amyotrophic lateral sclerosis and parkinsonism dementia. Proc Int Conf Peripheral Neuropathies. Excerpta Medica, Amsterdam, pp 93–109

387. Gajdusek DC, Gibbs CJ (1978) Unconventional viruses causing the spongiform virus encephalopathies. A fruitless search for the coat and core. In: Kurstak E, Maramorsch K (eds) Viruses and Environment. Academic Press, New York, pp 79–98

388. Gajdusek DC, Gibbs CJ, Alpers MP (1966) Experimental transmission of a kuru-like syndrome to chimpanzees, Nature 209: 794–796

389. Gajdusek DC, Zigas V (1957) Degenerative disease of the central nervous system in New Guinea. The endemic occurrence of "kuru" in the native population. New Engl J Med 257: 974–978

390. Gajdusek DC, Zigas V (1959) Kuru. Clinical, pathological and epidemiological study of an acute progressive gegenerative disease of the central nervous system among natives of the Eastern Highlans of New Papua. Am J Med 26: 442–469

391. Galvez S, Cartier L (1979) A new familial clustering of Creutzfeldt-Jakob disease in Chile. Neurocirurgia 37: 58–65

392. Galvez S, Ferrer S, Cartier L, Palma A (1980) Subacute spongiform encephalopathy (Creutzfeldt-Jakob disease) associated with normal-pressure hydrocephalus. Anatomoclinical report of one case. Acta Neurochir 51: 227–232

393. Galvez S, Masters CL, Gajdusek DC (1980) Descriptive epidemiology of Creutzfeldt-Jakob disease in Chile. Arch Neurol 37: 11–14

394. Gambarelli D, Vuillon-Cacciuttolo G (1983) Experimental kuru in Macaca Nemestrina: new anatomical data. Acta Neuropathol 61: 300–304

395. Garcin R, Bertrand I, Bogaert von L, Gruner J, Brion S (1950) Sur un type nosologique special de syndrome extra-pyramidal avec mouvements involontaires particuliers. Composante psychique variable d'evolution rapidement mortelle. Etude anatomo-clinique. Rev Neurol 83: 161–179.

396. Garcin R, Brion S (1966) La maladie de Creutzfeldt-Jakob et les degenerescences cortico-strird du presenium. Evol Psychiatr 31: 273–285

397. Garcin R, Brion S, Khochneviss A-A (1963) Le syndrome de Creutzfeldt-Jakob et les syndromes cortico-stries du presenium (a l'occasion de 5 observations anatomo-cliniques). Rev Neurol 109: 419–441

398. Gardiner AC (1965) Gel diffusion reaction of tissues and sera from scrapie-affected animals. Res Vet Sci 7: 190–195

399. Gardiner AC, Marucci AA (1969) Immunological responsiveness of scrapie infected mice. J Comp Pathol 79: 233–235

400. Garfin DE, Stities DP, Perlman JD, Cochran SP, Prusiner SB (1978) Mitogen stimulation of splenocytes from mice infected with scrapie agent. J Infect Dis 138: 396–400

401. Garfin DE, Stities DP, Zitnik LA, Prusiner SB (1978) Suppression of polyclonal B cell activation in scrapie-infected C3H/HeJ mice. J Immunol 120: 1986–1990

402. Garzully F, Jellinger K, Pilz P (1971) Subakute spongiöse Encephalopathie (Jakob-Creutzfeldt-Syndrom). Klinisch-morphologische Analyse von 9 Fällen. Arch Psychiatr Nervenk 214: 207–227

403. Ghadially F (1988) Heterolysosomes and autolysosomes. In: Ultrastructural pathology of the cell and matrix, vol 2. Butterworths, London, pp 594–599

404. Gaytan-Garcia S, Gilbert JJ, Deck JHN, Kaufmann JCE (1988) Jakob-Creutzfeldt disease associated with Wernicke encephalopathy. Can J Neurol Sci 15: 156–160

405. German TL, McMillan BC, Castle BE, Dees C, Wade WF, Marsh RF (1985) Comparison of RNA from healthy and scrapie-infected hamster brain. J Gen Virol 66: 839–844

406. Georgsson G, Martin JR, Klein J, Palsson PA, Nathanson N, Petursson G (1982) Primary demyelination in visna. An ultrastructural study of icelandic sheep with clinical signs following experimental infection. Acta Neuropathol (Berl) 57: 171–178

407. Gerstmann J, Sträussler E, Scheinker I (1936) Über eine eigenartige hereditär-familiäre Erkrankung des Zentralnervensystems. Zugleich ein Beitrag zur Frage des vorzeitigen lokalen Alterns. Z Ges Neurol Psychiatr 154: 736–762

408. Gertz HJ, Henkes H, Cervos-Navarro J (1988) Creutzfeldt-Jakob disease: correlation of MRI and neuropathologic findings. Neurology 38: 1481–1482

409. Ghetti B, Tagliavini F, Masters CL, Beyreuther K, Giaccone G, Verga L, Farlow MR, Conneally PM, Dlouhy SR, Azzarelli B, Bugiani O (1989) Gerstmann-Sträussler-Scheinker disease. II. Neurofibrillary tangles and plaques coexist in an affected family. Neurology 39: 1453–1461

410. Ghezzi A, Zaffaroni M, Marforio S, Cazzullo CL, Allegranza A (1989) Two familial cases of Creutzfeldt-Jakob disease in Italy. Ital J Neurol Sci 10: 199–202

411. Giaccone G, Tagliavini F, Verga L, Frangione B, Farlow MR, Bugiani O, Ghetti B (1991) Indiana kindred of Gerstmann-Sträussler-Scheinker disease: neuro-fibrillary tangles and neurites of plaques with PrP amyloid share antigenic determinants with those of Alzheimer's disease. In: Iqbal K, MacLachlan DRC, Winblad B, Wisniewski HM (eds) Alzheimer's Disease: Basic Mechanisms, Diagnosis and Therapeutic Strategies. Wiley & Sons, pp 207–211

412. Gibbs CJ Jr (1980) Virus-induced subacute degnerative diseases of the central nervous system. Ophtalmol 87: 1208–1218

413. Gibbs CJ Jr, Gajdusek DC (1978) Atypical viruses as the cause of sporadic, epidemic, and familial chronic diseases in man: slow viruses and human diseases. In: Pollard M (ed) Perspectives in Virology, vol 10. Raven Press, New York, pp 161–198

414. Gibbs CJ Jr, Gajdusek DC (1978) Subacute spongiform virus encephalopathies: the transmissible virus dementias. In: Katzman R, Terry RD, Bick KL (eds) Alzheimer's Disease: Senile Dementia and Related Disorders. Raven Press, New York, pp 559–574

415. Gibbs CJ Jr, Gajdusek DC (1971) Transmission and characterization of the agents of spongiform virus encephalopathies: kuru, Creutzfeldt-Jakob disease, scrapie and mink encephalopathy. In: Rowland LP (ed) Immunological Disorders of the Nervous System, Association for Research in Nervous & Mental Diseases, Res Publ, vol 49. Williams and Wilkins, Baltimore, pp 383–410

416. Gibbs CJ Jr, Gajdusek DC (1978) Virus-induced subacute slow infections of the brain associated with a cerebellar-type ataxia. In: Kark RAP, Rosenberg RN, Schut LJ (eds) Advances in Neurology, vol 21. Raven Press, New York, pp 359–372

417. Gibbs CJ Jr, Gajdusek DC, Amyx H (1979) Strain variation in the viruses of Creutzfeldt-Jakob disease and kuru. In: Prusiner SB, Hadlow WJ (eds) Slow

Transmissible Diseases of the Nervous System, vol 2. Academic Press, New York, pp 87–110

418. Gibbs CJ Jr, Gajdusek DC, Asher DM, Alpers MP, Beck E, Daniel PM, Matthews WB (1968) Creutzfeldt-Jakob disease (spongiform encephalopathy): transmission to chimpanzee. Science 161: 388–389

419. Gibbs CJ Jr, Gajdusek DC (1973) Experimental subacute spongiform virus encephalopathies in primates and other laboratory animals. Science 182: 67–68

420. Gibbs CJ Jr, Gajdusek DC, Latarjet R (1978) Unusual resistance to ionizing radiation of the viruses of kuru, Creutzfeldt-Jakob disease, and scrapie. Proc Natl Acad Sci USA 75: 6268–6270

421. Gibbs CJ, Joy A, Heffner R, Franko M, Miyazaki M, Asher DM, Parisi JE, Brown P, Gajdusek DC (1985) Clinical and pathological features and laboratory confirmation of Creutzfeldt-Jakob disease in a recipient of pituitary-derived human growth hormone. New Engl J Med 313: 734–738

422. Gibbs CJ, Masters CL, Gajdusek DC (1983) Bibliography of Creutzfeldt-Jakob disease. US Dept Health and Human Services, Bethesda, pp 169

423. Gibbons RA, Hunter GD (1967) Nature of the scrapie agent. Nature 215: 1041–1043

424. Gibson PH (1986) Distribution of amyloid plaques in four regions of the brains of mice infected with scrapie by intracerebral and intraperitoneal routes of injection. Acta Neuropathol (Berl) 69: 322–325

425. Gibson PH (1991) Mitochondrial inclusions and Creutzfeldt-Jakob disease. Lancet 337: 923

426. Gibson PH (1985) Relationship between numbers of cortical argentophilic and congophilic senile plaques in the brains of elderly people with and without senile dementia of Alzheimer type. Gerontology 31: 321–324

427. Gibson PH (1987) Ultrastructural abnormalities in the cerebral neocortex and hippocampus associated with Alzheimer's disease and aging. Acta Neuropathol (Berl) 73: 86–91

428. Gibson PH, Doughty LA (1989) An electron microscopic study of inclusion bodies in synaptic terminals of scrapie-infected animals. Acta Neuropathol (Berl) 77: 420–425

429. Gibson PH, Liberski PP (1987) An electron and light microscopic study of the numbers of dystrophic neurites and vacuoles in the hippocampus of mice infected intracerebrally with scrapie. Acta Neuropathol (Berl) 73: 379–382

430. Gibson PH, Somerville RA, Fraser H, Foster JD, Kimberlin RH (1987) Scrapie associated fibrils in the diagnosis of scrapie in sheep. Vet Rec 120: 125–127

431. Gibson PH, Tomlinson BE (1979) Vacuolation in the human cerebral cortex and its relationship to the interval between death and autopsy and to synapse numbers: an electronmicroscopic study. Neuropathol Appl Neurobiol 5: 1–7

432. Gilmour JS, Bruce ME, MacKellar A (1985) Cerebrovascular amyloidosis in scrapie-affected sheep Neuropathol Appl Neurobiol 11: 173–183

433. Goban Y, Saida T, Saida K, Hishitani H, Kameyama (1986) Ultrastructural study of central nervous system demyelination in galactocerebroside sensitized rabbits. Lab Invest 86: 86–90

434. Goldberg H, Alter M, Kahana E (1979) The Libyan Jewish focus of Creutzfeldt-Jakob disease: a search for the mode of natural transmission. In: Prusiner SB, Hadlow WJ (eds) Slow Transmissible Diseases of the Nervous System, vol 1. Academic Press, New York, pp 195–211

435. Goldfarb L, Brown P, Goldgaber D, Garruto RM, Yanagihara R, Asher DM, Gajdusek DC (1990) Identical mutation in unrelated patients with Creutzfeldt-Jakob disease. Lancet 336: 174–175

436. Goldfarb L, Brown P, Mitrova E, Cervenakova L, Goldin L, Korczyn AD, Chapman J, Galvez S, Cartier L, Rubenstein R, Gajdusek DC (1991) Creutzfeldt-Jakob disease associated with PRNP codon 200^{Lys} mutation: an analysis of 45 families. Europ J Epidemiol 7: 477–486

437. Goldfarb L, Brown P, McCombie WR, Goldgaber D, Swergold GD, Wills PR, Cervenakova L, Baron H, Gibbs CJ Jr, Gajdusek DC (1991) Familial Creutzfeldt-Jakob disease associated with extra actapeptide coding repeats in the PRNP gene: analysis of 3 families with 10, 12 and 13 repeat elements. Proc Natl Acad Sci USA (in press)

438. Goldfarb LG, Haltia M, Brown P, Nieto A, Kovanen J, McCombie WR, Trapp S, Gajdusek DC (1991) New mutation in scrapie amyloid precursor gene (at codon 178) in Finish Creutzfeldt-Jakob disease. Lancet 337: 425

439. Goldfarb LG, Mitrova E, Brown P, Toh BH, Gajdusek DC (1990) Mutation in codon 200 of scrapie amyloid protein gene in two clusters of Creutzfeldt-Jakob disease in Slovakia. Lancet 336: 514

440. Goldfarb LG, Korczyn AD, Brown P, Chapman J, Gajdusek DC (1990) Mutation on codon 200 of scrapie amyloid protein gene linked to Creutzfeldt-Jakob disease in Sephardic Jews of Libyan origin and non-Libyan origin. Lancet 336: 637

441. Goldgaber D, Goldfarb L, Brown P, Asher D, Brown T, Lin S, Teener JW, Feinstone SM, Rubenstein R, Kascak R, Bollaard JW, Gajdusek DC (1989) Mutations in familial Creutzfeldt-Jakob disease and Gerstmann-Straussler-Scheinker syndrome. Exp Neurol 106: 204–206

442. Goldmann W, Hunter N, Foster JD, Salbaum JM, Beyreuther K, Hope J (1990) Two alleles of a neural protein gene linked to scrapie in sheep. Proc Natl Acad Sci USA 87: 2476–2480

443. Goldmann W, Hunter N, Martin T, Dawson M, Hope J (1991) Different forms of bovine PrP gene have five or six copies of a short G-C rich element within protein-codin exon. J Gen Virol 72: 201–204

443a. Gomori AJ, Partnow MJ, Horoupian DS, Hirano A (1973) The ataxic form of Creutzfeldt-Jakob disease. Arch Neurol 29: 318–323

444. Gonatas NK (1967) Axonic and synaptic lesions in human neuropsychiatric disorders. Nature 214: 653

445. Gonatas NK, Anderson KW, Evangelista I (1967) The contribution of altered synapses in the senile plaque: amd electron microscopic study in Alzheimer's dementia. J Neuropathol Exp Neurol 26: 25–39

446. Gonatas NK, Baird HW, Evangelista I (1968) The fine structure of neocortical synapses in infantile amaurotic idiocy. J Neuropathol Exp Neurol 27: 39–49

447. Gonatas NK, Moss A (1968) Pathological axons and synapses in human neuro-psychiatric disorders. Hum Pathol 6: 571–582

448. Gonatas NK, Terry RD, Weiss M (1965) Electron microscopic study in two cases of Jakob-Creutzfeldt disease. J Neuropathol Exp Neurol 24: 575–598

449. Gordon WS (1946) Louping ill, tickborne fever and scrapie. Vet Rec 47: 516–520

450. Goudsmit J, Rohwer RG, Silbergeld E, Gajdusek DC (1981) Hypersensitivity to central serotonin receptor activation in scrapie-infected hamsters and the effect of serotoninergic drugs on scrapie symptoms. Brain Res 220: 372–377

451. Grabov JD, Campbell RJ, Okazaki H, Schut L, Zollman PE, Kurland LT (1976) A transmissible subacute spongiform encephalopathy in a visitor to the Eastern Highlands of New Guinea. Brain 99: 637–658

452. Gray A, Francis RJ, Scheltz CL (1980) Spiroplasma and Creutzfeldt-Jakob disease. Lancet 2: 152

453. Gray EG (1985) Membrane lamellation in brain unrelated to spongiform encephalopathy. Brain 108: 139–152

454. Gray EG (1986) Spongiform encephalopathy: a neurocytologist's viewpoint with a note of Alzheimer's disease. Neuropathol Appl Neurobiol 12: 149–172

455. Gresser I, Maury C, Chandler RL (1983) Failure to modify scrapie in mice by administration of interferon or anti-interferon globulin. J Gen Virol 64: 1387–1389

456. Griffin JW, Hoffman PN, Price DL (1982) Axonal transport in beta-beta-iminodipropionitrile neuropathy. In: Weiss DG, Gorio A (eds) Axoplasmic Transport in Physiology and Pathology. Springer, Berlin Heidelberg New York, pp 109–118

457. Griffith JS (1968) Self-replication and scrapie. Nature 215: 1043–1044

458. Grimaldi LME, Martino GV, Franciotta DM, Brustia R, Castagna A, Pristera R, Lazzarin A (1990) Elevated alpha-tumor necrosis factor levels in spinal fluid from HIV-1-infected patients with central nervous system involvement. Ann Neurol 29: 21–25

459. Grunnet ML (1975) Nuclear bodies in Creutzfeldt-Jakob and Alzheimer's diseases. Neurology 25: 1091–1093

460. Guiroy DC (1990) Amyloid plaques in chronic wasting disease, Masters Thesis, Hood College, Frederick, Maryland

461. Guiroy DC, Gajdusek DC (1988) Fibril-derived amyloid engancing factors as nucleating agents in Alzheimer's disease and transmissible virus dementia. Discussions in Neurosci 5: 69–73

462. Guiroy DC, Gajdusek DC (1990) Modification of host precursor proteins to amyloid fibrils in Alzheimer's disease. In: Harrison DE (ed) Genetic Effects on Aging. Telford Press, Caldwell, pp 543–556

463. Guiroy DC, Liberski PP, Papierz W, Gajdusek DC (1991) Amyloid beta-protein present in brain amyloid of a 32-year old with progressive dementia. Acta Neuropathol (Berl) 82: 523–526

464. Guiroy DC, Marsh R, Gajdusek DC (1991) Immunolocalization of scrapie amyloid in non-congophilic and non-birefringent deposits in golden Syrian hamsters with transmissible mink encephalopathy. Neurosci Lett (in press)

465. Guiroy DC, Shankar SK, Gibbs CJ Jr, Messenheimer JA, Das S, Gajdusek DC (1989) Neuronal degeneration and neurfilament accumulation in the trigeminal ganglia in Creutzfeldt-Jakob disease. Ann Neurol 25: 102–106

466. Guiroy DC, Williams ES, Liberski PP, Gajdusek DC (1991) Ultrastructure of chronic wasting disease in mule deer and elk. Acta Neuropathol (in press)

467. Guiroy DC, Williams E, Liberski P, Yanagihara R (1990) Amyloid plaques of chronic wasting disease in captive mule deer, hybrids of mule deer and white-tailed deer, Rocky Mountain Elk and experimental scrapie in hamsters sre immunoreactive to scrapie-associated fibril/protease-resistant protein and contain sulphated glycosaminoglycans. Ann Neurol 28: 290–291

468. Guiroy DC, Williams ES, Yanagihara R, Gajdusek DC (1991) Topographic distribution of scrapie amyloid-immunoreactive plaques in chronic wasting disease in captive mule deer (Odocoileus heminonus heminonus). Acta Neuropathol (Berl) 81: 475–478

469. Guiroy DC, Williams ES, Yanagihara R, Gajdusek DC (1991) Immunolocalization of scrapie amyloid (PrP27–30) in chronic wasting disease of Rocky Mountain elk and hybrids of captive mule deer and white-tailed deer. Neurosci Lett
126: 195–198

470. Guiroy DC, Yanagihara R, Gajdusek DC (1991) Localization of amyloidogenic
proteins and sulfated glycosaminoglycans in nontransmissible and transmissible
cerebral amyloidoses. Acta Neuropathol (Berl) 82: 87–92

471. Guo Y-P, Yang Y-C, Zhang X-G, Huang H-F, Feng Y-K (1986) Virus-induced
subacute spongiform encephalopathy (Creutzfeldt-Jakob disease). A report of
two Chinese cases with clinico-pathologic studies. J Neurol Sci 73: 137–144

472. Hadlow WJ (1961) The pathology of experimental scrapie in dairy goats. Res
Vet Sci 2: 289–315

473. Hadlow WJ, Eklund CM, Kennedy RC, Jackson TA, Whitford HW, Boyle C
(1974) Course of experimental scrapie virus infection in the goat. J Infect Dis
129: 559–567

474. Hadlow WJ, Kennedy RC, Race RE, Eklund CM (1980) Virologic and histo-
logic findings in dairy goats affected with natural scrapie. Vet Pathol 17: 187–199

475. Hadlow WJ, Kennedy RC, Race RE (1982) Natural infection of Suffolk sheep
with scrapie virus. J Infect Dis 146: 657–664

476. Hadlow WJ, Race RE, Kennedy RC, Eklund CM (1979) Natural infection
of sheep with scrapie virus. In: Prusiner SB, Hadlow WJ (eds) Slow Transmissible Diseases of the Nervous System, vol 2. Academic Press, New York,
pp 3–12

477. Haig DA, Clarke MC, Blum E, Alper T (1969) Further studies on the inactivation of the scrapie agent by ultraviolet light. J Gen Virol 5: 455–457

478. Haltia M, Kovanen J, Golfrab LG, Brown P, Gajdusek DC (1991) Familial
Creutzfeldt-Jakob disease in Finland: epidemiological, clinical, pathological and
molecular genetic studies. Europ J Epidemiol 7: 494–500

479. Haltia M, Kovanen J, Van Crevel H, Bots GThAM, Stefanko S (1979) Familial
Creutzfeldt-Jakob disease. J Neurol Sci 42: 381–389

480. Haraguchi T, Fisher S, Olofsson S, Endo T, Groth D, Tarentino A, Borchelt
DR, Teplow D, Hood L, Burlingame A, Lycke E, Kobata A, Prusiner SB
(1989) Asparagine-linked glycosylation of the scrapie and cellular prion proteins.
Arch Biochem Biophys 274: 1–13

481. Hardy J (1989) Slow virus dementias: prion gene holds the key. Trends Neurosci
12: 168–169

482. Harris DA, Falls DL, Johnson FA, Fischbach GD (1991) A prion-like protein
from chicken brain copurifies with an acetylcholine receptor-inducing activity.
Proc Natl Acad Sci USA 88: 7664–7668

483. Hart J, Gordon B (1990) Early-onset dementia and extrapyramidal disease:
clinicopathological variant of Gerstmann-Sträussler-Scheinker or Alzheimer's
disease. J Neurol Neurosurg Psychiatr 53: 932–934

484. Haseltine WA, Patarca R (1986) AIDS virus and scrapie agent share protein.
Nature 323: 115–116

485. Hauser SL, Doolittle TH, Lincoln R, Brown RH, Dinarello CA (1990) Cytokine
accumulations in CNS of multiple sclerosis patients: frequent detection of
interleukin-1 and tumor necrosis factor but not interleukin-6. Ann Neurol 40:
1735–1739

486. Hay B, Barry RA, Lieberburg I, Prusiner SB, Lingappa VR (1987) Biogenesis
and transmembrane orientation of the cellular isoform of the scrapie prion
protein. Mol Cell Biol 7: 914–920

487. Hay B, Prusiner SB, Lingappa VR (1987) Evidence for a secretory form of the cellular protein. Biochemistry 26: 8110–8115

488. Hayek J, Ulrich J (1975) Kuru-plaques in Creutzfeldt-Jakob disease. Europ Neurol 13: 251–257

489. Hayrech SMS, McDonnell DE, Aschenbrener CA (1979) Meningioma with Creutzfeldt-Jakob disease. Arch Neurol 36: 179–180

490. Heidenhain A (1929) Klinische und Anatomische Untersuchungen über eine eigenartige organische Erkrankung des Zentralnervensystems im Prasenium. Z Ges Neurol Psychiatr 118: 49–114

491. Heldt N, Floquet J, Warter JM, Steinmetz G, Weber M, Rohmer F (1983) Syndrome de Gerstmann-Sträussler-Scheinker: neuropathologie de trois cas dans une familie alsacienne. In: Court LA, Cathala F (eds) Virus Nob Conventionnels et Affections du Systeme Nerveux Central. Masson, Paris, pp 290–295

492. Hemm DM, Carlton WW (1971) Ultrastructural changes of cuprizone encephalopathy in mice. Toxicol Appl Pharmacol 18: 869–882

493. Hemphill RE, Stengel ES (1941) Subacute combined degeneration of unknown origin with extensive involvement of the brain. J Mental Sci 87: 77–87

494. Hevinson RG, Lowings JP, Dawson MD, Woodward MJ (1991) Anti-prions and other agents. Nature 352: 291

495. Hikita K, Tateishi J, Nagara H (1985) Morphogenesis of amyloid plaques in mice with Creutzfeldt-Jakob disease. Acta Neuropathol (Berl) 68: 138–144

496. Hilmert H, Diringer H (1984) A rapid and efficient method to enrich SAF-protein from scrapie brains of hamsters. Biosc Rep 4: 165–170

497. Hirano A (1991) Neurons, astrocytes and ependyma. In: Davies RL, Roberts DM (eds) Textbook of Neuropathology. 2nd Ed. Williams & Wilkins, pp 1–94

498. Hirano A, Ghatak NR, Johnson AB, Partnov MJ, Gomori AJ (1972) Argentophilic plaques in Creutzfeldt-Jakob disease. Arch Neurol 26: 530–542

499. Hirano T, Tsuchiyama H, Kawai, Mori KK (1977) An autopsy case of Creutzfeldt-Jakob disease with kuru-like neuropathological changes. Acta Pathol Jap 27: 231–238

500. Hoare M, Davies DC, Pattison IH (1977) Experimental production of scrapie-resistant Swaledale sheep. Vet Rec 101: 482–484

501. Hogan RN, Baringer JR, Prusiner SB (1987) Scrapie infection diminishes spines and increases varicosities of dendrites in hamsters: a quantitative Golgi analysis. J Neuropathol Exp Neurol 46: 461–473

502. Holthoff VA, Sandmann J, Pawlik G, Shroder R, Heiss W-D (1990) Positron emission tomography in Creutzfeldt-Jakob disease. Arch Neurol 47: 1035–1038

503. Hope J, Morton LJD, Farquhar CF, Multhaup G, Beyreuther K, Kimberlin RH (1986) The major polypeptide of scrapie-associated fibrils (SAF) has the same size, charge distribution and N-terminal protein sequence as predicted for the normal protein (PrP) (1986) EMBO J 5: 2591–2597

504. Hope J, Multhaup G, Reekie LJ, Kimberlin R, Beyreuther K (1988a) Molecular pathology of scrapie-associated fibril protein (PrP) in mouse brain affected by the ME7 strain of scrapie. Eur J Biochem 172: 271–277

505. Hope J, Reekie LJ, Hunter N, Multhaup G, Beyreuther K, White K, Scott AC, Stack J, Dawson M, Wells GEH (1988b) Fibrils from brains of cows with new cattle disease contain scrapie-associated protein. Nature 336: 390–392

506. Hope J, Ritchie L, Farquhar C, Hunter N (1989) Bovine spongiform encephalopathy: a scrapie like disease of British cattle. In: Iqbal K, Wisniewski HM, Winblad B (eds) Alzheimer Disease and Related Disorders. Alan R. Liss, pp 659–667

507. Hopf HC, Althaus HH, Sabuneu S (1974) "Typ Heidenhain" der subakuten spongiosen Encephalopathie (SSE) Creutzfeldt-Jakob. Z Neurol 206: 149–156.

508. Hornabrook RW (1979) Kuru and clinical neurology. In: Prusiner SB, Hadlow WJ (eds) Slow Transmissible Diseases of the Nervous System, Academic Press, New York, pp 37–66

509. Hornabrook RW (1978) Slow virus infections of the central nervous system. In: Vinken PJ, Bruyn GW (eds) Handbook of Clinical Neurology, vol 34. North-Holland, Amsterdam, pp 275–290

510. Horoupian DS, Powers JM, Schaumburg HH (1972) Kuru-like neuropathological changes in a North American. Arch Neurol 27: 555–561

511. Hsiao K, Baker HF, Crow TJ, Poulter M, Owen F, Terwilliger JD, Westaway D, Ott J, Prusiner SB (1989) Linkage of a prion protein missense variant to Gerstmann-Straussler syndrome. Nature 338: 342–345

512. Hsiao K, Doh-Ura K, Kitamoto T, Tateishi J, Prusiner SB (1989) A prion protein amino acid substitution in ataxic Gerstmann-Straussler syndrome. Ann Neurol 26: 106

513. Hsiao KK, Groth D, Scott M, Yang S-L, Serban A, Rapp D, Foster D, Torchia M, DeArmond SJ, Prusiner SB (1991) Neurologic disease of transgenic mice which express GSS mutant prion proteins is transmissible to inoculated recipient animals. In: Abstracts of the Prion Diseases in Humans and Animals, September 2–4, 1991

514. Hsiao K, Meiner Z, Kahana E, Cass C, Kahana I, Avraham D, Scarlato G, Abramsky O, Prusiner SB, Gabizon R (1991) Mutation of the prion protein in Libyan Jews with Creutzfeldt-Jakob disease. New Engl J Med 324: 1091–1097

515. Hsiao K, Scott M, Foster D, Groth DF, DeArmond SJ, Prusiner SB (1990) Spontaneous neurodegeneration in transgenic mice with mutant prion protein. Science 250: 1587–1590

516. Hudson AJ, Farrell MA, Kalnins R, Kaufmann CE (1983) Gerstmann-Sträussler-Scheineker disease with coincidental familial onset. Ann Neurol 14: 670–678

517. Hunter GD, Collis SC, Milson GC, Kimberlin RH (1976) Search for scrapie-specific RNA and attempts to detect an infectious DNA or RNA. J Gen Virol 32: 157–162

518. Hunter GD, Kimberlin RH, Gibbons RA (1968) Scrapie: a modified membrane hypothesis. J Theoret Biol 20: 355–357

519. Hunter GD, Millson GC (1966) Distribution and activation of lysosomal enzyme activities in subcellular components of normal and scrapie-affected mouse brain. J Neuroch 13: 375–383

520. Hunter N, Foster JD, Benson G, Hope J (1991) Restriction fragment length polymorphisms of scrapie-associated fibril protein (PrP) gene and their association with susceptibility to natural scrapie in British sheep. J Gen Virol 72: 1287–1292

521. Hunter N, Foster JD, Dickinson AG, Hope J (1989) Linkage of the gene for the scrapie-associated fibril protein (PrP) to the SIP gene in Cheviot sheep. Vet Rec 124: 364–366

522. Hunter N, Hope J, McConnell I, Dickinson AG (1987) Linkage of the scrapie-associated fibril protein (PrP) gene and Sinc using congenic mice and restriction fragment length polymorphism analysis. J Gen Virol 68: 2711–2716

523. Huxtable CR, Dorling PR, Walkey SU (1982) Onset and regression of neuro-axonal lesions in sheep with mannosidosis induced experimentally with swainsonine. Acta Neuropathol (Berl) 58: 27–33

524. Iqbal K, Somerville RA, Thompson CH, Wisniewski HM (1985) Brain glutamate decarboxylase and cholinergic enzyme activities in scrapie. J Neurol Sci 67: 345–350

525. Isomura H, Shinagawa M, Ikegami Y, Sasaki K, Ishiguro N (1990) Morphological and biochemical evidence that scrapie-associated fibrils are derived from aggregated amyloid-like fibrils. Virus Res 18: 191–202

526. Ito M, Shinagawa M, Doi S, Sasaki S, Isomura H, Takahashi K, Goto H, Sato G (1988) Effects of the antiserum against the fraction enriched in scrapie-associated fibrils on the scrapie incubation period in mice. Microbiol Immunol 32: 749–753

527. Jacob H, Pyrkosch W, Strube H (1950) Die erbliche Form der Creutzfeldt-Jakobschen Krankheit (Familie Backer). Arch Psych Nervenkr Z Ges Neurol Psychiatr 184: 653–674

528. Jakob A (1989) Concerning a disorder of the central nervous system clinically resembling multiple sclerosis with remarkable anatomic findings (spastic pseudosclerosis): report of 5 cases. Alzheimer Diseases and Associated Disorders 1/2: 37–45

529. Jakob A (1921) Über eigenartige Erkrankungen des Zentralnervensystems mit bemerkenswertem anatomischem Befunde (spastische Pseudosclerose-Encephalopathie mit disseminierten Degenerationsherden). Deutsch Z Nervenheilk 70: 132–146

530. Jakob A (1921) Uber eine der multiplen Sklerose klinisch nahestehende Erkrankung des Centralnervensystems (spastische Pseudosklerose) mit bemerkenswertem anatomischen Befunde. Med Klin 13: 372–376

531. Jansen J, Monrad-Krohn GH (1938) Uber die Creutzfeldt-Jakobsche Krankheit. Z Ges Neurol Psychiatr 163: 670–704

532. Jeffrey M, Scott JR, Fraser H (1991) Scrapie inoculation of mice: light and electron microscopy of the superior colliculus. Acta Neuropathol (Berl) 81: 562–571

533. Jellinger K (1971) Nuclear inclusions in subacute spongiform encephalopathy. Acta Neuropathol (Berl) 17: 283–286

534. Jellinger K (1973) Neuroaxonal dystrophy. Its natural history and related disorders. In: Zimmerman HM (ed) Progress in Neuropathology, vol 2. Grune & Stratton, New York, pp 129–180

535. Jellinger K, Heiss WD, Deisenhammer E (1974) The ataxic (cerebellar) form of Creutzfeldt-Jakob disease. J Neurol 207: 289–305

536. Jellinger K, Jirasek A (1971) Neuroaxonal dystrophy in man: character and natural histotry. Acta Neuropathol (Berl) [Suppl] 5: 3–16

537. Jellinger K, Seitelberger F, Heiss ED, Holczabek W (1972) Konjugale Form der subakuten spongiosen Enzephalopathie. Wien Klin Wochenschr 84: 245–249

538. Johnson JE, Miguel J (1974) Fine structure changes in the lateral vestibular nucleus of ageing rats. Mech Age Develop 3: 203–224

539. Jones MZ, Cunningham JG, Dade AW, Alessi DM, Mostosky UV, Vorro JR, Benitez JT, Lovell KL (1983) Caprine beta-mannosidosis: clinical and pathological features. J Neuropathol Exp Neurol 42: 268–285

540. Josephy H (1936) Jakob-Creutzfeldt Krankheit (Spastische Pseudosklerose Jakob). In: Bumke O, Foerster O (eds) Handbuch der Neurologie. Springer, Berlin, pp 882–886

541. Kao CC, Chang LW, Bloodworth MB (1977) Electron microscopic observations on mechanism of terminal club formation in transsected spinal cord axons. J Neuropathol Exp Neurol 36: 140–156

542. Katzman R, Kagan EH, Zimmerman HM (1961) A case of Jakob-Creutzfeldt disease. 1. Clinicopathological analysis. J Neuropathol Exp Neurol 20: 78–94

543. Kascak RJ, Rubenstein R, Merz PA, Carp RI, Robakis NK, Wisniewski HM, Diringer H (1986) Immunological comparison of scrapie-associated fibrils isolated from animals infected with four different scrapie strains. J Virol 59: 676–683

544. Kascak RJ, Rubenstein R, Merz PA, Carp RI, Wisniewski HM, Diringer H (1985) Biochemical differences among scrapie-associated fibrils support the biological diversity of scrapie agents. J Gen Virol 66: 1715–1722

545. Kasper KC, Stities DP, Bowman KA, Panitch H, Prusiner SB (1982) Immunological studies of scrapie infection. J Neuroimmunol 3: 187–201

546. Keohane C, Petfield R, Duchen LW (1985) Subacute spongiform encephalopathy (Creutzfeldt-Jakob disease) with amyloid angiopathy. J Neurol Neurosurg Psychiatr 48: 1175–1178

547. Khurana RK, Garcia JH (1981) Autonomic dysfunction in subacute spongiform encephalopathy. Arch Neurol 38: 114–117

548. Kidd M (1967) Some electron microscopical observations on status spongiosus. Acta Neuropathol (Berl) [Suppl 3]: 137–144

549. Kim JK, Lach B, Manuelidis EE (1988) Creutzfeldt-Jakob disease with intranuclear vacuolar inclusions: a biopsy case of negative light microscopic fidings and successful animal transmission. Acta Neuropathol (Berl) 76: 422–426

550. Kim JH, Manuelidis EE (1989) Neuronal alterations in experimental Creutzfeldt-Jakob disease: a Golgi study. J Neurol Sci 89: 93–101

551. Kim JH, Manuelidis EE (1983) Pathology of human and experimental Creutzfeldt-Jakob disease. Pathol Ann 18: 359–373

552. Kim JH, Manuelidis EE (1986) Serial ultrastructural study of experimental Creutzfeldt-Jakob disease in guinea pigs. Acta Neuropathol (Berl) 69: 81–90

553. Kim JH, Manuelidis EE (1983) Ultrastructural findings in experimental Creutzfeldt-Jakob disease in guinea pigs. J Neuropathol Exp Neurol 42: 29–43

554. Kim SU, Rizutto N (1975) Neuroaxonal degeneration induced by diethylthiocarbamate in cultures of central nervous tissue. J Neuropathol Exp Neurol 34: 531–541

555. Kim YS, Carp RI, Callahan S, Natelli BS, Wisniewski HM (1990) Vacuolization, incubation period and survival time in three mouse genotypes injected stereotactically in three brain regions with the 22L scrapie strain. J Neuropathol Exp Neurol 49: 106–113

556. Kim YS, Carp RI, Callahan S, Wisniewski HM (1990) Incubation periods and histopathological changes in mice injected stereotactically in different brain areas with 87V scrapie strain. Acta Neuropathol (Berl) 80: 388–392

557. Kim YS, Carp RI, Callahan S, Wisniewski HM (1990) Pathogenesis and pathology of scrapie after stereotactic injection of strain 22L in intact and bisected cerebella. J Neuropathol Exp Neurol 49: 114–121

558. Kimberlin RH (1979) An assessment of genetical methods in the control of scrapie. Livestock Prod Sci 6: 233–242

559. Kimberlin RH (1979) Early events in the pathogenesis of scrapie in mice: biological and biochemical studies. In: Prusiner SB, Hadlow WJ (eds) Slow Transmissible Diseases of the Nervous System, vol 2. Academic Press, New York, pp 33–54

560. Kimberlin RH (1972) The nature of the increased rate of DNA synthesis in scrapie-affected mouse brain. J Neurochem 19: 2767–2778

561. Kimberlin RH (1982) Reflections on the nature of scrapie agent. Trends Bioch Sci 7: 392–394

562. Kimberlin RH (1980) Scrapie. Br Vet J 137: 105–112

563. Kimberlin RH (1982) Scrapie agent: prions or virinos? Nature 297: 107–108

564. Kimberlin RH (1990) Scrapie and possible relationships with viroids. Sem Virol 1: 153–162

565. Kimberlin RH (1984) Scrapie: the disease and the infectious agent. Trends Neurosci 7: 312–316

566. Kimberlin RH (1986) Scrapie: how much do we really understand? Neuropathol Appl Neurobiol 12: 131–147

567. Kimberlin RH, Cole S, Walker CA (1987) Temporary and permanent modifications to a single strain of mouse scrapie on transmission to rats and hamsters. J Gen Virol 68: 1875–1881

568. Kimberlin RH, Collis SC, Walker CA (1976) Profiles of brain glycosidase activity in cuprizone-fed syrian hamsters and in scrapie-affected mice, rats, chinese hamsters and syrian hamsters. J Comp Pathol 86: 135–141

569. Kimberlin RH, Cunnington PG (1978) Reduction of scrapie incubation time in mice and hamsters by a single injection of methanol extraction residue of BCG. FEBS Microbiol Lett 3: 169–172

570. Kimberlin RH, Field HJ, Walker CA (1983) Pathogenesis of mouse scrapie: evidence for spread of infection from central to peripheral nervous system. J Gen Virol 64: 713–716

571. Kimberlin RH, Hall SM, Walker CA (1983) Pathogenesis of mouse scrapie. Evidence for direct neural spread of infection to the CNS after injection of sciatic nerve. J Neurol Sci 61: 315–325

572. Kimberlin RH, Hope J (1987) Genes and genomes in scrapie. Trends Genetics 3: 117–118

573. Kimberlin RH, Hunter GD (1967) DNA synthesis in scrapie-affected mouse brain. J Gen Virol 1: 115–124

574. Kimberlin RH, Marsh RF (1975) Comparison of scrapie and transmissible mink encephalopathy in hamsters. I. Bichemical studies of brain during development of disease. J Infect Dis 131: 97–103

575. Kimberlin RH, Millson GC (1976) The effect of cuprizone toxicity on the incubation period of scrapie in mice. J Comp Pathol 86: 489–496

576. Kimberlin RH, Millson GC, Bountiff L, Collis SC (1974) A comparison of the biochemical changes induced in mouse brain by cuprizone toxicity and by scrapie infection. J Comp Pathol 84: 263–270

577. Kimberlin RH, Shirt DB, Collis SC (1974) The turnover of isotopically labelled DNA *in vivo* in developing, adult and scrapie-affected mouse brain. J Neurochem 23: 241–248

578. Kimberlin RH, Walker CA (1979) Antiviral compound effective against experimental scrapie. Lancet ii: 591–592

579. Kimberlin RH, Walker CA (1983) The antiviral compound HPA-23 can prevent scrapie when administered at the time of infection. Arch Virol 78: 9–18

580. Kimberlin RH, Walker CA (1977) Characteristics of a short incubation model of scrapie in the Golden hamsters. J Gen Virol 34: 295–304

581. Kimberlin RH, Walker CA (1984) Competition between strains of scrapie depends on the blocing agent being infectious. Intervirology 23: 74–81

582. Kimberlin RH, Walker CA (1978) Evidence that the transmission of one source of scrapie agent to hamsters involve separation of agent strains from a mixture. J Gen Virol 39: 487–496

583. Kimberlin RH, Walker CA (1988) Incubation periods in six models of intraperitoneally injected scrapie depend mainly on the dynamics of agent replication within the nervous system and not the reticuloendothelial system. J Gen Virol 69: 2953–2960

584. Kimberlin RH, Walker CA (1990) Intraperitoneal infection with scrapie is established within minutes of injection and is non-specifically enhanced by a variety of different drugs. Arch Virol 112: 103–114

585. Kimberlin RH, Walker CA (1983) Invasion of the CNS by scrapie agent and its spread to different parts of the brain. In: Court LA, Cathala F (eds) Virus Non Cenventionnels et Affecions du Systeme Nerveaux Central. Masson, Paris, pp 17–33

586. Kimberlin RH, Walker CA (1979) Pathogenesis of scrapie: agent multiplication in brain at the first and second passage of hamster scrapie in mice. J Gen Virol 42: 107–117

587. Kimberlin RH, Walker CA (1979) Pathogenesis of scrapie: dynamics of agent replication in spleen, spinal cord and brain after infection with different routes. J Comp Pathol 89: 551–562

588. Kimberlin RH, Walker CA (1980) Pathogenesis of mouse scrapie: evidence for neural spread of infection to the CNS. J Gen Virol 51: 183–187

589. Kimberlin RH, Walker CA (1982) Pathogenesis of mouse scrapie: patterns of agent replication in different parts of CNS following intraperitoneal infection. J Royal Soc Med 75: 618–624

590. Kimberlin RH, Walker CA (1989) Pathogenesis of scrapie in mice after intragstric infection. Virus Res 12: 213–220

591. Kimberlin RH, Walker CA (1987) Pathogenesis of scrapie is faster when infection is intraspinal instead of intracerebral. Microbial Pathogenesis 2: 404–415

592. Kimberlin RH, Walker CA (1986) Pathogenesis of scrapie (strain 263K) in hamsters infected intracerebrally, intraperitoneally or intraocularly. J Gen Virol 67: 255–263

593. Kimberlin RH, Walker CA (1989) The role of the spleen in the neuoinvasion of scrapie in mice. Virus Res 12: 201–212

594. Kimberlin RH, Walker CA (1986) Supression of scrapie infection in mice by heteropolyanion 23, dextran sulfate, and some other polyanions. Antimicrobial Agents Chemoth 30: 409–413

595. Kingsbury DT (1990) Genetics of response to slow virus (prion) infection. Annu Rev Genet 24: 115–132

596. Kingsbury DT, Carlson GA, Prusiner SB (1987) Genetic control of prion replication. In: Prusiner SB, McKinley MP (eds) Prions. Novel Infectious Pathogens causing Scrapie and Creutzfeldt-Jakob Disease. Academic Press, New York, pp 316–350

597. Kingsbury DT, Kasper KC, Stities DP, Watson JD, Hogan RN, Prusiner SB (1983) Genetic control of scrapie and Creutzfeldt-Jakob disease. J Immunol 131: 491–496

598. Kingsbury DT, Smeltzer DA, Gibbs CJ Jr, Gajdusek DC (1981) Evidence for normal cell-mediated immunity in scrapie-infected mice. Infect Immun 32: 1176–1180

599. Kirschbaum WR (1924) Zwei eigenartige Erkrankungen des Zentralnerven-systems nach Art der spastischen Pseudosklerose (Jakob). Z Ges Neurol Psychiatr 92: 175–220

600. Kirschbaum WR (1968) Jakob-Creutzfeldt Disease (spastic pseudosclerosis A. Jakob; Heidenheim syndrome; subacute spongiform encephalopathy). American Elsevier, New York, pp 251

601. Kirschbaum WR (1971) Jakob-Creutzfeldt disease (Spastic Pseudosclerosis, cortico-striato-spinal degeneration, Heidenheim's syndrome, subacute spongiform encephalopathy). In: Mincler J (ed) Pathology of the Nervous System, vol 2. McGraw-Hill Comp, New York, pp 1410–1419

602. Kitagawa Y, Gotoh F, Koto A, Okayasu H, Ishii T, Matsuyama H (1983) Creutzfeldt-Jakob disease: a case with extensive white matter degeneration and optic atrophy. J Neurol 229: 97–101

603. Kitamoto T, Mohri S, Tateishi J (1989) Organ distribution of protease-resistant prion protein in humans and mice with Creutzfeldt-Jakob disease. J Gen Virol 70: 3371–3379

604. Kitamoto T, Ogomori K, Tateishi J, Prusiner SB (1987) Formic acid pretreat-ment enhances immunostaining of cerebral and systemic amyloids. Lab Invest 57: 230–236

605. Kitamoto T, Tateishi J (1988) Immunohistochemical confirmation of Creutzfeldt-Jakob disease with a long clinical course with amyloid plaque core antibodies. Am J Pathol 131: 435–443

606. Kitamoto T, Tateishi J, Sato Y (1988) Immunohistochemical verification of senile and kuru plaques in Creutzfeldt-Jakob disease and the allied disorders. Ann Neurol 24: 537–542

607. Kitamoto T, Tateishi J, Sawa H, Doh-Ura K (1989) Positive transmission of Creutzfeldt-Jakob disease verified by murine kuru plaques. Lab Invest 60: 507–512

608. Kitamoto T, Tateishi J, Tashima T, Takeshita J, Barry RA, DeArmond SJ, Prusiner SB (1986) Amyloid plaques in Creutzfeldt-Jakob disease stain with prion protein antibodies. Ann Neurol 20: 204–208

609. Kitamoto T, Yamaguchi K, Doh-Ura K, Tateishi J (1991) A prion protein missense variant is integrated in kuru plaque cores in patients with Gerstmann-Straussler syndrome. Ann Neurol 41: 306–310

610. Kitamoto T, Yi R, Mohri S, Tateishi Y (1990) Cerebral amyloid in mice with Creutzfeldt-Jakob disease is influenced by the strain of the infectious agent. Brain Res 508: 165–167

611. Kitagawa Y, Gotoh F, Koto A, Ebihara S, Okayashu H, Ishii T, Matsuyama H (1983) Creutzfeldt-Jakob disease: a case with extensive white matter degener-ation and optic atrophy. J Neurol 229: 97–101

612. Klatzo I, Gajusek DC (1959) Pathology of kuru. Lab Invest 8: 799–847

613. Klatzo I, Gajusek DC, Zigas V (1959) Evaluation of pathological findings in twelve cases of kuru. In: Van Boagert L, Radermecker J, Hozay J, Lowenthal A (eds) Encephalitides. Elsevier Publ, Amsterdam, pp 172–190

614. Knox CA, Yates RD, Chen I-L (1980) Brain ageing in normotensive and hypetrtensive strains of rats. II. Ultrastructural changes in neurons and glia. Acta Neuropathol (Berl) 52: 7–15

615. Koch TK, Berg BO, DeArmond SJ, Gravina RF (1985) Creutzfeldt-Jakob disease in a young adult with idiopathic hypopituitarism. New Engl J Med 313: 731–733

616. Korey SR, Katzman R, Orloff J (1961) A case of Jakob-Creutzfeldt Disease. 2. Analysis of some constituents of the brain of a patient with Jakob-Creutzfeldt disease. J Neuropathol Exp Neurol 20: 95–104

617. Kosaka K, Arai H, Ikeda K (1985) Familial presenile dementia with CJD-like lesions: preliminary results. Clin Neuropathol 4: 149–155

618. Kovanen J, Erkinjuntti T, Ivanainen M, Ketonen L, Haltia M, Sulkava R, Sipponen JT (1985) Cerebral MR and CT imaging in Creutzfeldt-Jakob disease. J Comp Ass Tomogr 9: 125–128

619. Kovanen J, Haltia M (1988) Descriptive epidemiology of Creutzfeldt-Jakob disease in Finland. Acta Neurol Scand 77: 474–480

620. Kovanen J, Tilikainen A, Haltia M (1980) Histocompatibility antigens in familial Creutzfeldt-Jakob disease. J Neurol Sci 45: 317–321

621. Kreindler A, Hornet T, Petresco A (1964) Encephalopathie spongioeuse subaugue. Acta Neurol Psychiatr Belgica 64, 223: 223–243

622. Kretzschmer HA, Honold G, Seitelberger F, Feucht M, Wessely P, Mehraein P, Budka H (1991) Prion protein mutation in family first reported by Gerstmann, Straussler, and Scheinker. Lancet 337: 1160

623. Kretzschmer HA, Kitamoto T, Doerr-Schott J, Mehraein P, Tateishi J (1991) Diffuse deposition of immunohistochemically labeled prion protein in the granular layer of the cerebellum in a patient with Creutzfeldt-Jakob disease. Acta Neuropathol (Berl) 82: 536–540

624. Kretzschmer HA, Stowring LE, Westaway D, Stubblebine WH, Prusiner SB, DeArmond SJ (1986) Molecular cloning of human prion protein cDNA. DNA 4: 315–324

625. Krigman MR, Feldman RG, Bensch K (1965) Alzheimer's presenile dementia. A histochemical and electron microscopic study. Lab Invest 14: 381–396

626. Krucke W, Beck E, Vitzthum H (1973) Creutzfeldt-Jakob disease. Some unusual morphological features reminiscent of kuru. Z Neurol 206: 1–24

627. Kruger H (1990) Panencephalopathic type of Creutzfeldt-Jakob disease with primary involvement of white matter. Eur Neurol 30: 115–119

628. Kuroda Y, Gibbs CJ Jr, Amyx HL, Gajdusek DC (1983) Creutzfeldt-Jakob disease in mice: persistent viremia and preferential replication of virus in low-density lymphocytes. Infect Immun 41: 154–161

629. Kusaka H, Hirano A, Bornstein MB, Raine CS (1985) Fine structure of astrocytic processes during serum-induced demyelination in vitro. J Neurol Sci 69: 255–267

630. Kuzuhara S, Kanazawa I, Sasaki H, Nakanishi T, Shimamura K Gerstmann-Sträussler-Scheinker's disease. Ann Neurol 14: 216–225

631. Lafarga M, Berciano MT, Suarez I, Viadero CF, Andres MA, Berciano J (1991) Cytology and organization of reactive astroglia in human cerebellar cortex with severe loss of granule cells: a study on the ataxic form of Creutzfeldt-Jakob disease. Neuroscience 40: 337–352

632. Lafon R, Labauge R, van Boagert L, Castan Ph (1965) Sur l'unite histopathologique des encephalopathies subaigues (types Creutzfeldt-Jakob, Heidenhain et Nevin). Rev Neurol 112: 201–227

633. Lamar CH, Gustafson DP, Krasovich M, Hinsman EJ (1974) Ultrastructural studies of spleens, brains, and brain cell cultures of mice with scrapie. Vet Pathol 11: 13–19

634. Lampert PW (1967) A comparative electron microscopic study of reactive, degenerating regenerating, and dystrophic axons. J Neuropathol Exp Neurol 26: 345–368

635. Lampert P (1967) Electron microscopic studies and hyperacute experimental allergic encephalomyelitis. J Neuropathol Exp Neurol 9: 99–126

636. Lampert PW (1971) Fine structural changes of neurites in Alzheimer's disease. Acta Neuropathol (Berl) [Suppl 5]: 49–53

637. Lampert PW, Blumberg JM, Pentschew A (1964) An electron microscopic study of dystrophic axons in the gracile nucleus and cuneate nuclei of vitamin deficient rats. J Neuropathol Exp Neurol 23: 60–77

638. Lampert P, Carpenter S (1967) Electron microscopic studies on the vascular permeability and the mechanism of demyelination in experimental allergic encephalomyelitis. J Neuropathol Exp Neurol 24: 371–384

639. Lampert PW, Cressman MR (1966) Fine structural changes of myelin sheaths after axonal degeneration in the spinal cord of rats. Am J Pathol 49: 1139–1155

640. Lampert PW, Earle KM, Gibbs CJ Jr, Gajdusek DC (1970) Electron microscopic studies on experimental spongiform encephalopathies (kuru and Creutzfeldt-Jakob disease) in chimpanzees. In: Proc VIth Congres International de Neuropathologie. Masson, Paris, pp 916–930

641. Lampert PW, Earle KM, Gibbs CJ Jr, Gajdusek DC (1969) Experimental kuru encephalopathy in chimpanzees and spider monkeys. Electron microscopic studies. J Neuropathol Exp Neurol 28: 353–370

642. Lampert PW, Gajdusek DC, Gibbs CJ (1971) Experimental spongiform encephalopathy (Creutzfeldt-Jakob disease) in chimpanzees. Electron microscopic studies. J Neuropathol Exp Neurol 30: 20–32

643. Lampert PW, Gajdusek DC, Gibbs CJ Jr (1972) Subacute spongiform virus encephalopathies. Scrapie, kuru and Creutzfeldt-Jakob disease: a review. Am J Pathol 68: 626–652

644. Lampert PW, Hooks J, Gibbs CJ, Gajdusek DC (1971) Altered plasma membranes in experimental scrapie. Acta Neuropathol (Berl) 19: 81–93

645. Landis MD, Williams RS, Masters CL (1981) Golgi and electron microscopic studies of spongiform encephalopathy. Neurology 31: 538–549

646. Laplanche J-L, Chatelain J, Launay J-M, Gazengel G, Vidaud M (1990) Deletion in prion protein gene in a Moroccan family. Nucleic Acids Res 18: 6745

647. Latarjet R, Muel B, Haig DA, Clarke MC, Alper T (1970) Inactivation of the scrapie agent by near monochromatic ultraviolet light. Nature 227: 1341–1343

648. Latarjet R (1979) Inactivation of the agents of scrapie, Creutzfeldt-Jakob disease, and kuru by irradiations. In: Prusiner SB, Hadlow WJ (eds) Slow Transmissible Diseases of the Nervous System, vol 2. Academic Press, New York, pp 387–407

649. Lax AJ, Millson GC, Manning EJ (1983) Involvement of protein in scrapie agent infectivity. Res Vet Sci 34: 155–158

650. Lavi E, Suzumura A, Murasko DM, Murray EM, Silberberg DH, Weiss SR (1988) Tumor necrosis factor induces expression of MHC class I antigens on mouse astrocytes. J Neuroimmunol 18: 245–253

651. Lavelle GC (1973) Multiplicity of scrapie virus in infected mouse spleen cells in vivo. Infect Immun 7: 918–921

652. Lavelle GC, Sturman L, Hadloe WJ (1972) Isolation from mouse spleen of cell populations with high specific infectivity for scrapie virus. Infect Immun 5: 319–323

653. Layton DD (1961) Jakob-Creutzfeldt disease: report of case with necropsy findings. Arch Neurol 4: 207

654. Leach RH, Matthews WB, Will R (1983) Creutzfeldt-Jakob disease. Failure to detect spiroplasmas by cultivation and serological tests. J Neurol Sci 59: 349–353

655. Lesser RL, Albert DM, Bobowick AR, O'Brien FH (1979) Creutzfeldt-Jakob disease and optic atrophy. Am J Ophtalmol 87: 317–321

656. Levin PK, Edwards V (1991) Mitochondrial inclusions in neurons of Creutzfeldt-Jakob disease. Lancet 337: 236–237

657. Levine P (1972) Scrapie: an infective polypeptide? Lancet i: 748

658. Liao Y-C, Tokes Z, Lim E, Lackey A, Woo CH, Button JD, Clawson GA (1987) Cloning of rat "prion-related protein" cDNA. Lab Invest 57: 370–374

659. Liberski PP (1987) Astrocytic reaction in experimental scrapie. J Comp Pathol 97: 73–78

660. Liberski PP (1987) The brain fine structure in experimental scrapie. The 263K strain in golden syrian hamsters. Neuropatol Polska 25: 35–51

661. Liberski PP (1986) Degenerative neurites in experimental scrapie. Neuropatol Polska 24: 79–88

662. Liberski PP (1987) Electron microscopic observations on dystrophic neurites in hamster brains infected with the 263K strain of scrapie. J Comp Pathol 97: 35–39

663. Liberski PP (1986) Gliocytosis in experimental scrapie (263K strain of scrapie) in golden syrian hamsters. Neuropatol Polska 24: 221–230

664. Liberski PP (1987) The nature of spiroplasma-like inclusions in experimental scrapie. Neuropatol Polska 25: 53–57

665. Liberski PP (1988) The occurrence of cytoplasmic lamellar bodies in scrapie infected and normal hamster brains. Neuropatol Polska 26: 79–85

666. Liberski PP (1990) Ultrastructural neuropathologic features of bovine spongiform encephalopathies. J Am Vet Med Ass 196: 1682

667. Liberski PP, Alwasiak J (1983) Neuropathology of experimental transmissible spongiform encephalopathy. The 263K strain of scrapie in Golden syrian hamsters. Neuropatol Polska 21: 378–392

668. Liberski PP, Alwasiak J, Papierz W (1985) Brain fine structure in Creutzfeldt-Jakob disease with plaques and tangles. II. Neurofibrillary tangles composed of paired helical filaments. Neuropat Polska 23: 521–529

669. Liberski PP, Asher DM, Yanagihara R, Gibbs CJ Jr, Gajdusek DC (1989) Serial ultrastructural studies of scrapie in hamsters. J Comp Pathol 101: 429–442

670. Liberski PP, Budka H, Kitamoto T, Tateishi J, Linke RP, Kascak R, Brown P (1991) PrP plaques in human transmissible spongiform encephalopathies: low frequency in European Creutzfeldt-Jakob disease. Abstract In: Second Congress of the Paneurropean Society of Neurology, Vienna, December 7–12

671. Liberski PP, Budka H, Sluga E, Barcikowska M, Kwiecinski H (1991) Tubulovesicular structures in human and experimental Creutzfeldt-Jakob disease. Europ J Epidemiol 7: 551–555

672. Liberski PP, Gibson PH (1987) Cerebellar lamellar bodies in two strains of murine scrapie. J Comp Pathol 97: 491–493

673. Liberski PP, Kwiecinski H, Barcikowska M, Mirecka B, Kulczycki J, Kida E, Brown P, Gajdusek DC (1991) PrP amyloid plaques in Creutzfeldt-Jakob disease of short duration: immunohistochemical studies of 5 cases from Poland. Europ J Epidemiol 7: 505–510

674. Liberski PP, Papierz W, Alwasiak J (1985) Brain fine structure in Creutzfeldt-Jakob disease with plaques and tangles. I. Neuritic plaques. Neuropat Polska 23: 508–519

675. Liberski PP, Papierz W, Alwasiak J (1987) Creutzfeldt-Jakob disease with plaques and paired helical filaments. Acta Neurol Scandinav 76: 428–432

676. Liberski PP, Papierz W, Alwasiak J, Szulc-Kuberska J, Rozniecki J (1988) Diagnostic difficulties in a case of subacute sclerosing encephalitis. Light and electron microscopic studies. Neuropatol Polska 26: 97–110

677. Liberski PP, Plucienniczak A, Hrabec E, Bogucki A (1989) Isolation and purification of scrapie associated fibrils and prion protein from scrapie-infected hamster brain. J Comp Pathol 100: 178–185

678. Liberski PP, Shankar SK, Gibbs CJ Jr, Gajdusek DC (1987) Electron microscopic and immunohistochemical evidence for impairment of axonal transport in Creutzfeldt-Jakob disease. Ann Neurol 22: 157

679. Liberski PP, Yanagihara R, Asher DM, Gibbs CJ Jr, Gajdusek DC (1990) Reevaluation of the ultrastructural pathology of experimental Creutzfeldt-Jakob disease. Brain 113: 121–137

680. Liberski PP, Yanagihara R, Gibbs CJ Jr, Gajdusek DC (1990) Appearance of tubulovesicular structures in experimental Creutzfeldt-Jakob disease and scrapie preceeds the onset of clinical disease. Acta Neuropathol 79: 349–354

681. Liberski PP, Yanagihara R, Gibbs CJ Jr, Gajdusek DC (1991) Mechanisms of the damage to myelinated axons in experimental Creutzfeldt-Jakob disease in mice: an ultrastructural study. Europ J Epidemiol 7: 545–550

682. Liberski PP, Yanagihara R, Gibbs CJ Jr, Gajdusek DC (1989) Neuroaxonal dystrophy: an ultrastructural link between subacute spongiform virus encephalopathies and Alzheimer's disease. In: Iqbal K, Wisniewski HM, Winblad B (eds) Alzheimer's Disease and Related Disorders. Alan R Liss, pp 549–557

683. Liberski PP, Yanagihara R, Gibbs CJ Jr, Gajdusek DC (1989) Scrapie as a model for neuroaxonal dystrophy: ultrastructural studies. Exp Neurol 106: 133–144

684. Liberski PP, Yanagihara R, Gibbs CJ Jr, Gajdusek DC (1990) Spread of Creutzfeldt-Jakob disease virus along visual pathways after intraocular inoculation. Arch Virol 111: 141–147

685. Liberski PP, Yanagihara R, Gibbs CJ Jr, Gajdusek DC (1988) Tubulovesicular structures in experimental Creutzfeldt-Jakob disease and scrapie. Intervirology 29: 115–119

686. Liberski PP, Yanagihara R, Gibbs CJ Jr, Gajdusek DC (1989) White matter ultrastructural pathology of experimental Creutzfeldt-Jakob disease in mice. Acta Neuropathol 79: 1–9

687. Lieberman AP, Pitha PM, Shin HS, Shin ML (1989) Production of tumor necrosis factor and other cytokines by astrocytes stimulated with lipopoly-saccharide or a neurotropic virus. Proc Natl Acad Sci USA 86: 6348–6352

688. Locht C, Chesebro B, Race R, Keith JM (1986) Molecular cloning and complete sequence of prion protein cDNA from mouse brain infected with the scrapie agent. Proc Natl Acad Sci USA 83: 6372–6376

689. Loesch A, Belai A, Lincoln J, Burnstock G (1986) Enterick nerves in diabetic rats: electron microscopic evidence for neuropathy of vasoactive intestinal polypeptide-containing fibers. Acta Neuropathol (Berl) 70: 161–168

690. Lopez CD, Yost CS, Prusiner SB, Myers RM, Lingappa VR (1990) Unusual topogenic sequence directs prion protein biogenesis. Science 248: 226–229

691. Love S, Duchen LW (1982) Familial cerebellar ataxia with cerebrovascular amyloid. J Neurol Neurosurg Psychiatr 45: 271–273

692. Lovell KL, Jones MZ (1985) Axonal and myelin lesions in beta-mannosidosis: ultrastructural characteristics. Acta Neuropathol (Berl) 65: 293–299

693. Lowe J, McDermott H, Kenward N, Landon M, Mayer RJ, Bruce M, McBride P, Somerville RA, Hope J (1990) Ubiquitin conjugate immunoreactivity in the brains of scrapie infected mice. J Pathol 162: 61–66

694. Lowenstein DH, Butler DA, Westaway D, McKinley MP, DeArmond SJ, Prusiner SB (1990) Three hamster species with different scrapie incubation times and neuropathological features encode distinct prion proteins. Mol Cell Biol 10: 1153–1163

695. Lugaresi E, Gambetti P, Dazzi P, Castan Ph (1965) Sur une observation d'encephalopathie spongieuse presenile revetant siccessivement un aspect amaurotique, myoclonique et dyskinetique. Psychiat Neurol Neurochir 68: 242–258

696. Luse S, Smith KR (1964) The ultrastructure of senile plaques. Am J Pathol 44: 553–563

697. Machado-Salas JP (1986) Dendritic and axonal spherules in the neocortex of a patient with Creutzfeldt-Jakob disease (CJD): Golgi and electron-microscopical observation – neurobiological significance. Clin Neuropathol 5: 176–184

698. Macchi G, Abbamondi AL, Di Trapani G, Sbriccoli A (1984) On the white matter lesions of the Creutzfeldt-Jakob disease. A new subentity be recognized in man? J Neurol Sci 63: 197–206

699. Mackenzie A (1983) Immunohistochemical demonstration of glial fibrillary acidic protein in scrapie. J Comp Pathol 93: 251–259

700. Majtenyi C (1988) Study on the cases of Creutzfeldt-Jakob disease in Hungary. In: Court LA, Dormont D, Brown P, Kingsbury DT (eds) Unconventional virus diseases of the central nervous system. Commissariat a' l'Energie Atomique, Paris, pp 30–36

701. Malamud N (1979) Creutzfeldt-Jakob disease: a clinicopatghological study. In: Prusiner SB, Halow WJ (eds) Slow Transmissible Diseases of the Nervous System, vol 1. Academic Press, New York, pp 271 285

702. Malmström-Groth AG, Kristensson K (1982) Neuroaxonal dystrophy in childhood. Report of two second cousins with Hallervorden-Spatz disease, and a case with Seitelberger's disease. Acta Paediatr Scandinav 71: 1045–1049

703. Malone TG, Marsh RF, Hanson RP, Semancik JS (1979) Evidence for the low molecular weight nature of scrapie agent. Nature 278: 575–576

704. Malone TG, Marsh RF, Hanson RP, Semancik JS (1978) Membrane-free scrapie infectivity. J Virol 25: 933–935

705. Mancardi GL, Mandybur TI, Liwnicz BH (1982) Spongiform-like changes in Alzheimer's disease. Acta Neuropathol (Berl) 56: 146–150

706. Manetto V, Sternberger NH, Perry G, Sternberger LA, Gambetti P (1988) Phosphorylation of neurofilaments is altered in amyotrophic lateral sclerosis. J Neuropathol Exp Neurol 47: 642–653

707. Manning EJ, Millson GC (1983) Infectivity of liposomally encapsulated nucleic acids isolated from EMC virus and scrapie-infected mouse brain. Intervirology 20: 164–168

708. Manolidis LS, Balojannis SJ (1983) Ultrastructural alterations of the vestibular nuclei in Jakob-Creutzfeldt disease. Acta Otolarungol 95: 508–521

709. Manuelidis EE (1985) Creutzfeldt-Jakob disease. J Neuropathol Exp Neurol 44: 1–17

710. Manuelidis EE, Gorgacz EJ, Manuelidis L (1978) Viremia in experimental Creutzfeldt-Jakob disease. Science 200: 1069–10771

711. Manuelidis EE, Kim JH, Manuelidis L (1986) Destructive white matter lesions in experimental Creutzfeldt-Jakob disease. In: Court LA, Dormont D, Brown P, Kingsbury DT (eds) Unconventional Virus Diseases of the Central Nervous System. Commissariat à l'Energie Atomique, Paris, pp 221–230

712. Manuelidis EE, Manuelidis L (1979) Clinical and morphological aspects of transmissible Creutzfeldt-Jakob disease. In: Zimmerman HM (ed) Progress in Neuropathology, vol 4. Raven Press, New York, pp 1–26

713. Manuelidis EE, Manuelidis L (1979) Observations on Creutzfeldt-Jakob disease propoagated in small rodents. In: Prusiner SB, HaAdlow WJ (eds) Slow Transmissible Diseases of the Nervous System, vol 2. Academic Press, New York, pp 147–172

714. Manuelidis L, Manuelidis EE (1986) Recent developments in scrapie and Creutzfeldt-Jakob disease. Prog Med Virol 33: 78–98

715. Manuelidis L, Manuelidis EE (1981) Search for specific DNA in Creutzfeldt-Jakob infectious brain fractions using "nick translation". Virology 109: 435–443

716. Manuelidis L, Sklaviadis T, Manuelidis EE (1987) Evidence suggesting that PrP is not the infectious agent in Creutzfeldt-Jakob disease. EMBO J 6: 341–347

717. Manuelidis L, Tesin D, Sklaviadis T, Manuelidis EE (1987) Astrocyte gene expression in Creutzfeldt-Jakob disease. Proc Natl Acad Sci USA 84: 5937–5941

718. Manuelidis L, Valley S, Manuelidis EE (1985) Specific proteins associated with Creutzfeldt-Jakob disease and scrapie share antigenic and carbohydrate determinants. Proc Natl Acad Sci USA 82: 4263–4267

719. Marin O, Vial JD (1964) Neuropathological and ultrastructural findings in two cases of subacute spongiform encephalopathy. Acta Neuropathol (Berl) 4: 218–229

720. Marsh RF (1987) The scrapie agent. In: Semancik JS (ed) Viroids and Viroid-like Pathogens. CRC Press, Boca Raton, pp 161–168

721. Marsh RF, Hanson RP (1969) Physical and chemical properties of the transmissible mink encephalopathy agent. J Virol 3: 176–180

722. Marsh RF, Kimberlin RH (1975) Comparison of scrapie and transmissible mink encephalopathy in hamsters. II. Clinical signs, pathology and pathogenesis. J Infect Dis 131: 104–110

723. Marsh RF, Malone TG, Lancaster WD, Hanson RP, Semancik JS (1978) Scrapie: virus, viroid, or voodoo? In: Persistent Viruses. Academic Press, New York, pp 581–590

724. Marsh RF, Malone TG, Semancik J, Lancaster WD, Hanson RP (1978) Evidence for an essential DNA component in the scrapie agent. Nature 275: 146–147

725. Marsh RF, Semancik JS, Medappa KC, Hanson RP, Rueckert RR (1974) Scrapie and transmissible mink enecphalopathy: serach for infectious nucleic acid. J Virol 13: 993–996

726. Marsh RF, Sipe JC, Morse SS, Hanson RP (1976) Transmissible mink encephalopathy. Reduced spongiform degeneration in aged mink of the Chediak-Higashi genotype. Lab Inv 34: 381–386

727. Massa PT, Schimpl A, Wecker E, ter Meulen V (1987) Tumor necrosis factor amplifies measles virus-mediated Ia function on astrocytes. Proc Natl Acad Sci USA 84: 7242–7245

728. Masters CL, Gajdusek DC (1982) The spectrum of Creutzfeldt-Jakob disease and the virus induced subacute spongiform encephalopathies. In: Smith TJ, Cavanagh JB (eds) Recent Advances in Neuropathology. Churchill Livingstone, Edinburgh, pp 139–163

729. Masters CL, Gajdusek DC, Gibbs CJ Jr (1981) Creutzfeldt-Jakob disease virus isolations from the Gerstmann-Straussler syndrome. With an analysis of the various forms of amyloid plaque deposition in the virus induced spongiform encephalopathies. Brain 104: 559–588

730. Masters CL, Gajdusek DC, Gibbs CJ Jr (1981) The familial occurrence of Creutzfeldt-Jakob disease and Alzheimer's disease. Brain 104: 535–558

731. Masters CL, Gajdusek DC, Gibbs CJ Jr, Bernoulli C, Asher DM (1979) Familial Creutzfeldt-Jakob disease and other familial dementias. In: Prusiner SB, Hadlow WJ (eds) Slow Transmissible Diseases of the Nervous System, vol 1. Academic Press, New York, pp 143–194

732. Masters CL, Harris JO, Gajdusek DC, Gibbs CJ Jr, Bernoulli C, Asher DM (1979) Creutzfeldt-Jakob disease: patterns of worldwide occurrence. In: Prusiner SB, Hadlow WJ (eds) Slow Transmissible Diseases of the Nervous System, vol 1. Academic Press, New York, pp 113–142

733. Masters CL, Kakulas BA, Alpers MP, Gajdusek DC, Gibbs CJ Jr (1976) Preclinical lesions and their progression in the experimental spongiform encephalopathies (kuru and Creutzfeldt-Jakob disease) in primates. J Neuropathol Exp Neurol 35: 593–605

734. Masters CL, Richardson EP (1978) Subacute spongiform encephalopathy (Creutzfeldt-Jakob disease). The nature and progression of spongiform change. Brain 101: 333–344

735. Masters CL, Rohwer RG, Franko MC, Brown P, Gajdusek DC (1984) The sequential development of spongiform change and gliosis of scrapie in the golden syrian hamsters. J Neuropathol Exp Neurol 43: 242–252

736. Masullo C, Pocchiari M, Gibbs CJ Jr, Gajdusek DC (1984) Choline acetyl-transferase activity and [^{3}H]quinuclidinylbenzilate binding in brains of scrapie-infected hamsters. Neurosci Lett 51: 87–92

737. Masurovsky EB, Bunge RP (1971) Patterns of myelin degeneration following the rapid death of cells in cultures of peripheral nervous tissue. J Neuropathol Exp Neurol 30: 311–324

738. Matthews WB (1985) Creutzfeldt-Jakob disease. In: Fredericks (ed) Handbook of Clinical neurology. (Neurobehavioural Disorders, vol 46.) Elsevier, New York, pp 289–299

739. Matthews WB, Will RG (1981) Creutzfeldt-Jakob disease in a lifelong vegetarian. Lancet ii: 937

740. Maier H, Budka H, Lassmann H, Pohl P (1989) Vacuolar myelopathy with multinucleated giant cells in the aquired immune deficiency syndrome (AIDS). Light and electron microscopic distribution of human immunodeficiency virus (HIV) antigens. Acta Neuropathol (Berl) 78: 497–503

741. May WG (1968) Creutzfeldt-Jakob disease. I. Survey of the literature and clinical diagnosis. Acta Neurol Scandinav 32: 1–32

742. McBride PA, Bruce ME, Fraser H (1988) Immunostaining of scrapie cerebral amyloid plaques with antisera raised to scrapie-associated fibrils (SAF). Neuropathol Appl Neurobiol 14: 325–336

743. McComb JG, Davis RL (1985) Blood-brain and Blood-CSF barriers. In: Davis RL, Robertson DM (eds) Textbook of Neuropathology. Williams & Wilkins, Baltimore, pp 147–175

744. McDermott JR, Fraser H, Dickinso AG (1978) Reduced choline-acetyl-transferase activity in scrapie mouse brain. Lancet ii: 318–319

745. McFarlin DE, Raff MC, Simpson E, Nehlsen SH (1971) Scrapie in immunologically deficient mice. Nature 233: 336

746. McFaydean J (1918) Sarcosporidia as the cause of scrapie. J Comp Pathol 31: 290–300

747. McKinley MP, Barry RA, Prusiner SB, Dearmond SJ (1986a) Ultrastructural and immunological investigations of scrapie prions. In: Scheibel AB, Wechsler AF, Brazier MA (eds) The Biological Substrate of Alzheimer Disease. Academic Press, New York, pp 145–160

748. McKinley MP, Bolton DC, Prusiner SB (1984) A protease resistant protein is a structural component of the scrapie prion. Cell 35: 57–62

749. McKinley MP, Braunfeldt MB, Bellinger CG, Prusiner SB (1986b) Molecular characteristics of prion rods purified from scrapie-infected hamster brains. J Infect Dis 143: 110–120

750. McKinley MP, Braunfeldt MB, Prusiner SB (1987) Scrapie prion ultrastructure. In: Prusiner SB, McKinley MP (eds) Prions. Novel Infectious Pathogens Causing Scrapie and Creutzfeldt-Jakob Disease. Academic Press, New York, pp 197–237

751. McKinley MP, DeArmond SJ, Torchia M, Mobley WC, Prusiner SB (1989) Acceleration of scrapie in neonatal Syrian hamsters. Neurology 39: 1319–1324

752. McKinley MP, Hay B, Lingappa VR, Liberburg I, Prusiner SB (1987) Developmental expression of prion protein gene in brain. Develop Biol 121: 105–110

753. McKinley MP, Masiarz FR, Isaacks ST, Hearts JE, Prusiner SB (1983) Resistance of the scrapie agent to inactivation by psoralens. Photochem Photobiol 37: 539–545

754. McKinley MP, Masiarz FR, Prusiner SB (1981) Reversible chemical modification of the scrapie agent. Science 214: 1529–1261

755. McKinley MP, Meyer RK, Kenaga L, Rahbar F, Cotter R, Serban A, Prusiner SB (1991) Scrapie prion rod formation *in vitro* requires both detergent extraction and limited proteolysis. J Virol 65: 1340–1351

756. McMenemey WH, Pollak E (1941) Presenile dementia of the nervous system. Report of unusual case. Arch Neurol Psychiatr 45: 683–697

757. Meggendorfer F (1930) Klinische und genealogische Beobachtungen bei einem Fall von spastischer pseudosklerose Jakobs. Z Ges Neurol Psychiatr 128: 337–341

758. Meier C (1980) Occurrence of lymphocytes in the cortical neuropil in a case of Creutzfeldt-Jakob disease. Acta Neuropathol (Berl) 52: 69–72

759. Merz PA (1988) Final general discussion. In: Bock G, Marsh J (eds) Novel Infectious Agents and the Nervous System, Ciba Foundation Symposium 135. John Wiley & Sons, Chichister, pp 261–266

760. Merz PA, Kascak RJ, Rubenstein R, Carp RJ, Wisniewski HM (1987) Antisera to scrapie-associated fibril protein and prion protein decorate scrapie-associated fibrils. J Virol 61: 42–49

761. Merz PA, Rohwer RG, Kascak R, Wisniewski HM, Somerville RA, Gibbs CJ Jr, Gajdusek DC (1984) Infection specific particle from the unconventional slow virus diseases. Science 225: 437–476

762. Merz PA, Somerville RA, Wisniewski HM, Manuelidis L, Manuelidis EE (1983a) Scrapie-associated fibrils in Creutzfeldt-Jakob disease. Nature 306: 474–476

763. Merz PA, Somerville RA, Wisniewski HM (1983b) Abnormal fibrils in scrapie and senile dementia of Alzheimer type. In: Court LA, Cathala F (eds) Virus Unconventionnels et Affections du Systeme Nerveaux Central. Masson Paris, pp 259–281

764. Merz PA, Somerville RA, Wisniewski HM, Iqbal J (1981) Abnormal fibrils from scrapie-infected brain. Acta Neuropathol (Berl) 60: 63–74

765. Merz PA, Wisniewski HM, Somerville RA, Bobin SA, Masters CL, Iqbal K (1983c) Ultrastructural morphology of amyloid fibrils from neuritic and amyloid plaques. Acta Neuropathol (Berl) 60: 113–124

766. Meyer RK, McKinley MP, Bowman KA, Braunfeld MB, Barry RA, Prusiner SB (1986) Separation and properties of cellular and scrapie prion proteins. Proc Natl Acad Sci USA 83: 2310–2314

767. M'Fayden J (1918) Scrapie. J Comp Pathol 31: 102–129

768. Mighelli A, Attanasio A, Claudia M, Schiffer D (1991) Dystrophic neurites around amyloid plaques of human patients with Gerstmann-Sträussler-Scheinker disease contain ubiquitinated inclusions. Neurosci Letters 121: 55–58

769. Millot P (1978) The major histocompatibility complex of sheep (OLA) and two minor loci. Anim Blood Groups Biochem Genet 9: 115–121

770. Millot P, Chatelain J, Cathala F (1985) Sheep major histocompatibility complex OLA: gene frequencies in two French breeds with scrapie. Immunogenetics 21: 117–123

771. Millot P, Chatelain J, Dautheville C, Salmon D, Cathala F (1988) Sheep major histocompatibility (OLA) complex: linkage between a scrapie susceptibility/ resistance locus and the OLA complex in Ile-de-France sheep progenies. Immunogenetics 27: 1–11

772. Millson GC (1965) Lysosomal enzymes in normal and scrapie mouse brain. J Neurochem 12: 461–468

773. Millson GC, Bountiff L (1973) Glycosidases in normal and scrapie mouse brain. J Neurochem 20: 541–546

774. Millson GC, Hunter GD, Kimberlin RH (1976) The physico-chemical nature of the scrapie agent. In: Kimberlin RH (ed) Slow Virus Diseases of Animals and Man. North-Holland, Amsterdam, pp 243–266

775. Millson GC, Kimberlin RH, Manning EJ, Collis SC (1979) Early distribution of radioactive liposomes and scrapie infectivity in mouse tissues following administration by different routes. Vet Microbiol 4: 89–99

776. Mitrova E (1988) A case of Creutzfeldt-Jakob disease related to familial retinitis pigmentosa patients. Eur J Epidemiol 4: 55–59

777. Mizusawa H, Hirano A, Llena JF (1987) Involvement of hippocampus in Creutzfeldt-Jakob disease. J Neurol Sci 82: 13–26

778. Mizusawa H, Ohkoshi N, Sasaki H, Kanazawa I, Nakanishi T (1988) Degeneration of the thalamus and inferior olives associated with spongiform encephalopathy of the cerebral cortex. Clin Neuropathol 7: 81–86

779. Mizutani T (1981) Neuropathology of Creutzfeldt-Jakob disease in Japan. With special reference to the panencephalopathic type. Acta Pathol JPN 31: 903–922

780. Mizutani T (1985) Chronic spongiform encephalopathy with plaques characterized by antecedent and longlasting cerebellar ataxia. In: Mizutami T, Shiraki H (eds) Clinicopathological Aspects of Creutzfeldt-Jakob Disease. Elsevier, Nishimura, Amsterdam, Niigata, pp 163–175

781. Mizutani T (1985) Panencephalopathic type of Creutzfeldt-Jakob disease. In: Mizutami T, Shiraki H (eds) Clinicopathological Aspectss of Creutzfeldt-Jakob Disease. Elsevier, Nishimura, Amsterdam, Niigata, pp 123–162

782. Mizutani T, Okumara A, Oda M, Shiraki H (1981) Panencephalopathic type of Creutzfeldt-Jakob disease: primary involvement of the cerebral white matter. J Neurol Neurosurg Psychiatr 44: 103–115

783. Mizusawa H, Hirano A, Llena JF (1987) Involvement of hippocampus in Creutzfeldt-Jakob disease. J Neurol Sci 82: 13–26

784. Miyakawa T, Katsuragi S, Koga Y, Moriyama S (1986) Status spongiosus in Creutzfeldt-Jakob disease. Clin Neuropathol 5: 146–152

785. Mobley WC, Neve RL, Prusiner SB, McKinley MP (1988) Nerve growth factor increases mRNA levels for the prion protein and the beta-amyloid precursor protein in developing hamster brain. Proc Natl Acad Sci USA 85: 9811–9815

786. Mohri S, Tateishi J (1989) Host genetic control of incubation periods of creutzfeldt-Jakob diseases in mice. J Gen Virol 70: 1391–1400

787. Motgomery DL, Storts RW (1984) Hereditary striato-nigral and cerebello-olivary degneration of Kerry blue terrier. II. Ultrastructural lesions in the caudate nucleus and cerebellar cortex. J Neuropathol Exp Neurol 43: 263–275

788. Moore GRW, McCarron RM, McFarlin DE, Raine CS (1987) Chronic relapsing necrotizing encephalomyelitis produced by basic myelin protein in mice. Lab Invest 57: 157–167

789. Morris J, Gajdusek DC (1963) Encephalopathy in mice following inoculation of scrapie sheep brain. Nature 197: 1084–1086

790. Mould DL, Dawson A McL, Rennie JC (1970) Very early replication of scrapie in lymphocytic tissue. Nature 228: 779–780

791. Mukoyama M, Yamazaki K, Kikuchi T, Tomita T (1989) Neuropathology of gracile axonal dystrophy (GAD) mouse. An animal model of central distal axonopathy in primary sensory neurons, Acta Neuropathol (Berl) 79: 294–299

792. Multhaup G, Diringer H, Hilmert H, Prinz H, Heukeshoven J, Beyreuther K (1985) The protein component of scrapie-associated fibrils is a glycosylated low molecular weight protein. EMBO J 4: 1495–1501

793. Museteanu C, Diringer H (1981) Perivascular infiltrates of leukocytes in brains of scrapie-infected mice. Nature 294: 360–361

794. Myrianthopoulos NC, Smith KJ (1962) Amyotrophic lateral sclerosis with progressive dementia. Arch Neurol 12: 603–610

795. Nagashima K, Suzuki S, Ichikawa E, Uchida S, Honma T, Kuroume T, Hirato J, Ogawa J, Ishida Y (1985) Infantile neuroaxonal dystrophy: perinatal onset with symptoms of diencephalic syndrome. Neurology 35: 735–736

796. Nakazato Y, Hirato Y, Ishida Y, Hoshi S, Hasegawa M, Fukuda T (1990) Swollen cortical neurons in Creutzfeldt-Jakob disease contain a phosphorylated neurofilament epitope. J Neuropathol Exp Neurol 49: 197–205

797. Narang HK (1988) A chronological study of experimental scrapie in mice. Virus Res 9: 293–306

798. Narang HK (1974) An electron microscopic study of the scrapie mouse and rat: further observations on virus-like particles with rutenium red and lanthanum nitrate as a possible trace and negative stain. Neurobiology 4: 349–363

799. Narang HK (1974) An electron microscopic study of natural scrapie sheep brain: further observations on virus-like particles and paramyxovirus-like tubules. Acta Neuropathol (Berl) 28: 317–329

800. Narang HK (1974) Ruthenium red and lanthanum nitrate a possible tracer and negative stain for scrapie "particles"? Acta Neuropathol (Berl) 29: 37–43

801. Narang HK (1987) Scrapie, an unconventional virus: the current views. Proc Soc Exp Biol Med 184: 375–388

802. Narang HK (1975) Virus-like particles in Creutzfeldt-Jakob biopsy material. Acta Neuropathol (Berl) 32: 163–168

803. Narang HK (1973) Virus-like particles in natural scrapie of the sheep. Res Vet Sci 14: 108–110

804. Narang HK, Asher DM, Pomeroy KL, Gajdusek DC (1987) Abnormal tubulovesicular particles in brains of hamsters with scrapie. Proc Soc Exp Biol Med 184: 504–509

805. Narang HK, Asher DM, Gajdusek DC (1988) Evidence that DNA is present in abnormal tubulofilamentous structures found in scrapie. Proc Natl Acad Sci USA 85: 3375–3579

806. Narang HK, Asher DM, Gajdusek DC (1987) Tubulofilaments in negatively stained scrapie-infecetd brains: relationships to scrapie-associated fibrils. Proc Natl Acad Sci USA 84: 7730–7734

807. Narang HK, Chandler RL, Anger HS (1980) Further observations on particulate structures in scrapie affected brain. Neuropathol Appl Neurobiol 6: 23–28

808. Neugut RH, Neugut AI, Kahana E, Stein Z, Alter M (1979) Creutzfeldt-Jakob disease: familial clustering among Libyan-born Israelis. Neurology 29: 225–231

809. Neuman MA, Gajdusek DC, Zigas V (1964) Neuropathologic findings in exotic neurologic disorder among natives of the Highlands of New Guinea. J Neuropathol Exp Neurol 23: 486–507

810. Nevin S, McMenemey WH (1959) A form of subacute encephalopathy of uncertain etiology. Proc R Soc Med 31: 533–539

811. Nevin S, McMenemey WH, Behrman S, Jones DP (1960) Subacute spongiform encephalopathy – a subacute form of encephalopathy attributable to vascular dysfunction (spongiform cerebral atrophy). Brain 83: 519–564

812. Nieto A, Goldfarb LP, Brown P, McCombie WR, Trapp S, Asher DM, Gajdusek DC (1991) Codon 178 mutation in ethically diverse Creutzfeldt-Jakob disease. Lancet 337: 662–663

813. Nisipeanu P, El Ad B, Korczyn AD (1990) Spongiform encephalopathy in an Israeli born to immigrants from Libya. Lancet 336: 686

814. Nochlin D, Sumi SM, Bird TD, Snow AD, Leventhal CM, Beyreuther K, Masters CL (1989) Familial dementia with PrP positive amyloid palques: a variant of Gerstmann-Straussler syndrome. Neurology 39: 910–918

815. Nowak JZ, Bogucki A, Liberski PP (1986) Brain histamine metabolism in transmissible spongiform encephalopathy (scrapie). J Neural Transm 65: 187–192

816. Nussbaum RE, Henderson WM, Pattison IH, Elcock NV, Davies DC (1975) The establishment of sheep flocks of predictable susceptibility to experimental scrapie. Res Vet Sci 18: 49–58

817. Nyberg P, Almay BGL, Carlsson A, Masters CL, Winblad B (1982) Brain monoamine abnormalities in the two types of Creutzfeldt-Jakob disease. Acta Neurol Scand 66: 16–24

818. Oesch B, Groth DF, Prusiner SB, Weissmann C (1988) Search for scrapie-specific nucleic acid: a progress report. In: Bock G, Marsh J (eds) Novel Infectious Agents and the Central Nervous System. John Wiley & Sons, Chichister, pp 209–223

819. Oesch B, Teplow DB, Stahl N, Serban D, Hood LE, Prusiner SB (1990) Identification of cellular proteins binding to the scrapie prion protein. Biochemistry 29: 5848–5855

820. Oesch B, Westaway D, Wälchlii M, McKinley MP, Kent SBH, Aebersold R, Barry RA, Tempst P, Teplow DB, Hood LE, Prusiner SB, Weissmann C (1985) A cellular gene encodes scrapie PrP 27–30 protein. Cell 40: 735–746

821. Osetowska E (1980) The Neuropathology of Viral and Allergic Encephalitides. Foreign Scientific Publications Department of the National Center for Scientific, Technical and Economic Information, Warsaw, pp 1–317

822. Outram GW (1976) The pathogenesis of scrapie in mice. In: Kimberlin RH (ed) Slow Virus Diseases of Animals and MAn, North-Holland, Amsterdam, pp 326–357

823. Outram GW, Dickinson AG, Fraser H (1973) Developmental maturation of susceptibility to scrapie in mice. Nature 241: 536–537

824. Outram GW, Dickinson AG, Fraser H (1974) Reduced susceptibility to scrapie in mice after steroid administration. Nature 249: 855–856

825. Outram GW, Dickinson AG, Fraser H (1975) Slow encephalopathies, inflammatory responses, and arachis oil. Lancet ii: 198–200

826. Outram GW, Fraser H, Wilson DT (1973) Scrapie in mice. Some effects on the brain lesion profile of ME7 agent due to genotype of donor, route of injection and genotype of recipient. J Comp Pathol 83: 19–28

827. Owen F, Poulter M, Collinge J, Crow T (1989) A codon 129 polymorphism in the PrP gene. Nucl Acid Res 18: 3103

828. Owen F, Poulter M, Collinge J, Crow T (1990) Codon 129 changes in the prion protein gene in caucasians. Am J Hum Genet 46: 1215–1216

829. Owen F, Poulter M, Collinge J, Leach M, Shah T, Lofthouse R, Chen Y, Crow TJ, Harding AE, Hardy J, Rossor MN (1991) Insertions in the prion protein gene in atypical dementia. Exp Neurol 112: 240–242

830. Owen F, Poulter M, Lofthouse R, Collinge J, Crow TJ, Risby D, Baker HF, Ridley RM, Hsiao K, Prusiner SB (1989) Insertion in prion protein gene in familial Creutzfeldt-Jakob disease. Lancet i: 51–52

831. Owen F, Poulter M, Shah T, Collinge J, Lofthouse R, Baker H, Ridley R, McVey J, Crow TJ (1991) An in-frame insertion in the prion protein gene in familial Creutzfeldt-Jakob disease. Molec Brain Res 7: 273–276

832. Palmer AC (1960) Distribution of vacuolated neurons in brains of scrapie affected sheep. J Neuropathol Exp Neurol 19: 102–110

833. Palmer AC (1957) Vacuolated neurones in sheep affected with scrapie. Nature 179: 480–481

834. Palmer AC (1968) Wallerian type degeneration in sheep scrapie. Vet Rec 729–731

835. Palmer MS, Dryden AJ, Hughes JT, Collinge J (1991) Homozygous prion protein genotype predisposes to sporadic Creutzfeldt-Jakob disease. Nature 352: 340–342

836. Park TS, Kleinman GM, Richardson EP (1980) Creutzfeldt-Jakob disease with extensive degeneration of white matter. Acta Neuropathol (Berl) 52: 239–242

837. Parry HB (1962) Scrapie: a transmissible and hereditary disease of sheep. Heredity 17: 75–105

838. Pattison IH (1966) The relative susceptibility of sheep, goats, and mice to two types of the goat scrapie agent. Res Vet Sci 7: 207–212

839. Pattison IH (1974) Scrapie in sheep selectively bred for high susceptibility. Nature 248: 594–595

840. Pattison IH, Jebbett JN (1971) Histopathological similarities between scrapie and cuprizone toxicity in mice. Nature 230: 115–117

841. Pattison IH, Jones KM (1967) The astrocytic reaction in experimental scrapie in the rat. Res Vet Sci 8: 160–165

842. Pattison IH, Jones KM (1968) Modification of a strain of mouse-adapted scrapie by passage through rats. Res Vet Sci 9: 408–410

843. Pattison IH, Millson GC (1962) Distribution of the scrapie agent in the tissues of experimentally inoculated goats. J Comp Pathol 72: 233–244

844. Pattison IH, Millson GC (1960) Further observations on the experimental production of scrapie in goats and sheep. J Comp Pathol 70: 182–193

845. Pattison IH, Smith K (1963) Histological observations on experimental scrapie in the mouse. Res Vet Sci 4: 269–274

846. Payne CM, Sibley WA (1975) Intranuclear inclusions in a case of Creutzfeldt-Jakob disease. Acta Neuropathol (Berl) 31: 353–361

847. Pearl GS, Anderson RE (1989) Creutzfeldt-Jakob disease: high caudate signal on magnetic resonance imaging. Southern Med J 82: 1177–1180

848. Pearlman RL, Towfighi J, Pezeshkpour GH, Tenser RB, Turel AP (1988) Clinical significance of types of cerebellar amyloid plaques in human spongiform encephalopathies. Neurology 38: 1249–1254

849. Peat A, Field EJ (1970) An unusual structure in Kuru brain. Acta Neuropathol (Berl) 15: 288–292

850. Peiffer J (1982) Gerstmann-Straussler's disease, atypical multiple sclerosis and carcinomas in a family of sheepbreeders. Acta Neuropathol (Berl) 56: 87–92

851. Pentschew A, Schwartz K (1962) Systemic axonal dystrophy in vitamin E deficient adult rats. Acta Neuropathol (Berl) 31: 303–334

852. Piccardo P, Safar J, Ceroni M, Gajdusek DC, Gibbs CJ Jr (1990) Immuno-histochemical localization of prion protein in spongiform encephalopathies and normal brain tissue. Neurology 40: 518–522

853. Pleasure DE, Mishler KC, Engel WK (1987) Axonal transport proteins in experimental neuropathies. J Neuropathol Exp Neurol 36: 214–227

854. Pocchiari M, Casaccia P, Ladogana A (1989) Amphotericin B: a novel class of antiscrapie agent. J Infect Dis 160: 795–802

855. Pocchiari M, Masullo C, Lust WD, Gibbs CJ Jr, Gajdusek DC (1985) Isonicotinic hydrazide causes seizures in scrapie-infected hamsters with shorter latency than in control animals: a possible GABAergic defect. Brain Res 326: 117–123

856. Pocchiari M, Munson PJ, Costa T, Gajdusek DC, Gibbs CJ Jr (1985) Serotononergic system in scrapie-infected hamsters. J Neurochem 44: 862–868

857. Pocchiari M, Schmittinger S, Masullo C (1987) Amphotericin B delays the incubation period of scrapie in intracerebrally inoculated hamsters. J Gen Virol 68: 219–223

858. Prineas JW (1985) The neuropathology of multiple sclerosis. In: Koetsier JC (ed) Handbook of Clinical Neurology, vol 3 (47). Elsevier Sience, Amsterdam, pp 213–257

859. Prineas J, Raine CS, Wisniewski HM (1969) An ultrastructural study of experimental demyelination and remyelination. III. Chronic experimental allergic encephalomyelitis in the central nervous system. Lab Invest 21: 472–483

860. Pro JD, Smith CH, Sumi M (1980) Presenile Alzheimer disease: amyloid plaques in the cerebellum. Neurology 30: 820–825

861. Prusiner SB (1989) Creutzfeldt-Jakob disease and scrapie prions. Alzheimer Dis Assoc Disord 3: 52–78

862. Prusiner SB (1991) Molecular biology of prion diseases. Science 252: 1515–1522

863. Prusiner SB (1982) Novel proteinaceous infection particles cause scrapie. Science 216: 136–144

864. Prusiner SB (1988) Molecular structure, biology, and genetics of prions. Adv Virus Res 35: 83–136

865. Prusiner SB (1989) Scrapie prions. Ann Rev Microbiol 43: 345–374

866. Prusiner SB (1984) Some speculations about prions, amyloid, and Alzheimer's disease. New Engl J Med 310: 661–663

867. Prusiner SB (1987) Terminology. In: Prusiner SB, McKinley MP (eds) Prions. Novel Infectious Pathogens Causing Scrapie and Creutzfeldt-Jakob Disease. Academic Press, New York, pp 38–53

868. Prusiner SB, Bolton DC, Groth DF, Bowman KA, Cochran SP, McKinley MP (1982) Further purification and characterization of scrapie prions. Biochemistry 21: 6942–6950

869. Prusiner SB, Bowman KA, Groth DF (1987) Purification of scrapie prions. In: Prusiner SB, McKinley MP (eds) Prions. Novel Infectious Pathogens causing Scrapie and Creutzfeldt-Jakob Disease. Academic Press, New York, pp 149–171

870. Prusiner SB, Cochran SP, Groth DF, Downey DE, Bowman KA, Martinez HM (1982) Measurement of the scrapie agent using an incubation time interval assay. Ann Neurol 11: 353–358

871. Prusiner SB, Cochran SP, Groth DF, Hadley D, Martinez HM, Hadlow WJ (1980) Aging and the nervous system and prolonged incubation periods of the spongiform encephalopathies. In: Amaducci L (ed) Aging of the Brains and Dementia. Aging, vol 13. Raven Press, New York, pp 205–216

872. Prusiner SB, DeArmond SJ (1987) Prions causing nervous system degeneration. Lab Invest 56: 349–362

873. Prusiner SB, Gabizon R, McKinley MP (1987) On the biology of prions. Acta Neuropathol 72: 299–314

874. Prusiner SB, Garfin DE, Baringer JR, Cochran SP (1978) Evidence for multiple molecular forms of the scrapie agent. In: Stevens J, Todaro G, Fox CF (eds) Persistent Viruses. Academic Press, New York, pp 591–613

875. Prusiner SB, Garfin DE, Cochran SP, Baringer JR, Hadlow WJ, Eklund CM, Race RE (1978) Evidence for hydrophobic domains on the surface of the scrapie agent. Trans Am Neurol Associat 103: 1–4

876. Prusiner SB, Garfin DE, Cochran SP, McKinley MP, Groth DF, Hadlow WJ, Race RE, Eklund CM (1980) Experimental scrapie in the mouse: electrophoretic and sedimentation properties of the partially purified agent. J Neurochem 35: 574–582

877. Prusiner SB, Groth DF, Bildstein C, Masiarz FR, McKinley MP, Cochran SP (1980) Electrophoretic properties of the scrapie agent in agarose gels. Proc Natl Acad Sci USA 77: 2984–2988

878. Prusiner SB, Groth DF, Bolton DC, Kent SB, Hood LE (1984) Purification and structural studies of a major scrapie prion protein. Cell 38: 127–134

879. Prusiner SB, Groth DF, Cochran SP, McKinley MP, Masiarz FR (1980) Gel electrophoresis and glass permeation chromatography of the hamster scrapie agent after enzymatic digestion and detergent extraction. Biochemistry 19: 4892–4898

880. Prusiner SB, Groth DF, Cochran SP, Masiarz FR, McKinley MP, Martinez H (1980) Molecular properties, partial purification, and assay by incubation period measurements of the hamster scrapie agent. Biochemistry 21: 4883–4891

881. Prusiner SB, Groth DF, McKinley MP, Cochran SP, Bowman KA, Kasper KC (1981) Thiocyanate and hydroxyl ions inactivate the scrapie agent. Proc Natl Acad Sci USA 78: 4606–4610

882. Prusiner SB, Hadlow WJ, Eklund CA, Race RE (1977) Sedimentation properties of the scrapie agent. Proc Natl Acad Sci USA 74: 4656–4660

883. Prusiner SB, Hadlow WJ, Eklund CA, Race RE, Cochran SP (1978) Sedimentation characteristics of the scrapie agent from murine spleen and brain. Biochemistry 17: 4987–4992

884. Prusiner SB, Hadlow WJ, Garfin DE, Cochran SP, Baringer JR, Race ER, Eklund CM (1978) Partial purification and evidence for multiple molecular forms of the scrapie agent. Biochemistry 17: 4993–4999

885. Prusiner SB, Hsiao KK, Bredesen DE, DeArmond SJ (1989) Prion disease. In: Vinken PJ, Bruyn GW, Klawans HL (eds) Handbook of Clinical Neurology, vol 12. Elsevier Sci, Amsterdam, pp 543–580

886. Prusiner SB, McKinley MP, Bowman KA, Bolton DC, Bendheim PE, Groth DC, Glenner GG (1983) Scrapie prions aggregate to form amyloid-like birefringent rods. Cell 35: 349–358

887. Prusiner SB, McKinley MP, Groth DF, Bowman KA, Mock NI, Cochran SP, Masiarz FR (1981) Scrapie agent contains a hydrophobic protein. Proc Natl Acad Sci USA 78: 6675–6679

888. Race RE, Caughey B, Graham K, Ernst D, Chesebro B (1988) Analysis of frequency of infection, specific infectivity, and prion protein biosynthesis in scrapie-infected neuroblastoma cell clones. J Virol 62: 2845–2849

889. Race RE, Graham K, Ernst D, Caughey B, Chesebro B (1990) Analysis of linkage between scrapie incubation period and the prion protein gene in mice. J Gen Virol 71: 493–497

890. Race RE, Fadness LH, Chesebro B (1987) Characterization of scrapie infection in mouse neuroblastoma cells. J Gen Virol 68: 1391–1399

891. Rafalowska J, Strugalska H (1971) Przypadek o klinicznym przebiegu stwardnienia zanikowego bocznego z rozleglymi zmianami w istocie szarej (amiotroficzna postac choroby Creutzfeldta-Jakoba) [Case of lateral amyotrophic sclerosis with extensive changes in the gray matter]. Neuropathol Polska 9: 13–21

892. Raine CS (1984) Biology of disease. Analysis of autoimmune demyelination: its impact upon multiple sclerosis. Lab Invest 50: 608–635

893. Raine CS (1985) Experimental allergic encephalomyelitis and experimental allergic neuritis. In: Koetsier JC (ed) Handbook of Clinical Neurology, vol 3 (47). Elsevier Science, New York, pp 429–466

894. Raine CS (1983) Multiple sclerosis and chronic relapsing EAE: comparative ultrastructural neuropathology. In: Hallpike J, Adams CWM, Tourtellotte WW (eds) Multiple Sclerosis – Pathology, Diagnosis and Its Management. Chapman & Hall, London, 413–460

895. Raine CS (1984) The neuropathology of myelin diseases. In: Morell P (ed) Myelin. Plenum Press, New York, pp 259–310

896. Raine CS (1978) Pathology of demyelination. In: Waxman SG (ed) Physiology and Pathobiology of Axons. Raven Press, New York, pp 283–310

897. Raine CS, Barnett LB, Brown A, Behar T, McFarlin DE (1980) Neuropathology of experimental allergic encephalomyelitis in inbred strains of mice. Lab Invest 43: 150–157

898. Raine CS, Bornstein MB (1970) Experimental allergic encephalomyelitis: an ultrastructural study of experimental demyelination in vitro. J Neuropathol Exp Neurol 29: 177–179

899. Raine CS, Cross H (1989) Axonal dystrophy as a consequence of long term demyelination. Lab Invest 60: 714–725

900. Rees SA (1975) A quantitative electron-microscopic study of atypical structures in normal human cortex. Anat Embryol 148: 303–331

901. Rees SA (1976) A quantitative electron-microscopic study of the aging human cerebral cortex. Acta Neuropathol (Berl) 36: 347–362

902. Reyes JM, Hoenig EM (1981) Intracellular spiral inclusions in cerebral processes in Creutzfeldt-Jakob disease. Exp Neurol 40: 1–8

903. Ribadeau-Dumas JL, Escourolle R (1974) The Creutzfeldt-Jakob syndrome. A neuropathological and electron-microscopic study. In: Subirana A, Espadaler JM, Burrows EH (eds) Neurology. Proc of the Xth Int Cong Neurol. Excerpta Medica, Amsterdam, pp 315–329

904. Ribadeau-Dumas JL, Escourolle R, Castaigne P (1969) Syndrome de Creutzfeldt-Jakob: etude ultrastructurale de trois observations. Rev Neurology (Paris) 121: 405–422

905. Rice GPA, Paty DW, Ball MJ, Tatham R, Kertesz A (1980) Spongiform encephalopathy of long duration. A family study. J Can Sci Neurol 7: 171–176

906. Robakis NK, Sawh PR, Wolfe GC, Rubenstein R, Carp RI, Innis MA (1986) Isolation of cDNA clone encoding the leader peptide of prion protein and expression of the homologous gene in various tissues. Proc Natl Acad Sci USA 83: 6377–6381

907. Roberts GW, Lofthouse R, Allsop D, Landon M, Kidd M, Prusiner SB, Crow TJ (1990) CNS amyloid proteins in neurodegenerative diseases. Neurology 38: 1534–1540

908. Robertson HD, Branch AD, Dahlberg JE (1985) Focusing on the nature of the scrapie agent. Cell 40: 725–727

909. Robbins DS, Shirazi Y, Drysdale B-E, Liberman A, Shin HS, Shin ML (1987) Production of cytotoxic factor for oligodendrocytes by stimulated astrocytes. J Immunol 139: 2593–2597

910. Robinson N (1969) Creutzfeldt-Jakob disease: a histochemical study. Brain 92: 581–588

911. Rohwer R (1986) Estimation of scrapie nucleic acid MW from standard curves for virus sensitivity to ionizing radiation. Nature 320: 381

912. Rohwer R (1984) Scrapie-associated fibrils. Lancet ii: 36

913. Rohwer RG, Brown P, Gajdusek DC (1979) The use of sedimentation to equilibrium as a step in the purification of the scrapie agent. In: Prusiner SB, Hadlow WJ (eds) Slow Transmissible Diseases of the Nervous System, vol 2. Academic Press, New York, pp 465–478

914. Rohwer R, Gajdusek DC (1980) Scrapie – virus or viroid. The case for a virus. In: Boese A (ed) Search for the Cause of Multiple Sclerosis and Other Chronic Diseases of the Central nervous System, Proc. of the First International Symposium of the Hertie Foundation, Frankfurt/Main, Sept. 1979. Verlag Chemie, Weinheim, pp 333–355

915. Roos R, Gajdusek DC, Gibbs CJ, Jr (1973) The clinical characteristics of transmissible Creutzfeldt-Jakob disease. Brain 96: 1–20

916. Rosenthal P, Keesey J, Crandall B, Brown J (1976) Familial neurological disease associated with spongiform encephalopathy. Arch Neurol 33: 252–259

917. Roy S, Gupta PC, Sethi U (1972) Creutzfeldt-Jakob disease: an electron-microscopic study with demonstration of virus-like particles. Neurology India 20: 226–230

918. Royden-Jones H, Hedley-Whyte T, Freidberg SR, Baker RA (1985) Ataxic Creutzfeldt-Jakob disease: diagnostic techniques and neuropathologic observations in early disease. Neurology 35: 254–257

919. Rubenstein R, Deng H, Scalici CL, Papini MC (1991) Alterations in neuro-transmitter-related enzyme activity in scrapie-infected PC12 cells. J Gen Virol 72: 1279–1285

920. Rubenstein R, Kascak RI, Merz PA, Papini MC, Carp RI, Robakis NI, Wisniewski HM (1986) Detection of scrapie-associated fibrils (SAF) proteins using anti-SAF antibody in non-purified tissue preparations. J Gen Virol 67: 671–681

921. Rubenstein R, Merz PA, Kascak RJ, Carp RI, Scalici CL, Fama CL, Wisniewski HM (1987) Detection of scrapie-associated fibrils (SAF) and SAF proteins from scrapie-affected sheep. J Infect Dis 156: 36–42

922. Rubenstein R, Merz PA, Kascak RJ, Scalici CL, Papini MC, Carp RI, Kimberlin RH (1991) Scrapie-infected spleens: analysis of infectivity, scrapie-associated fibrils, and protease-resistant protein. J Infect Dis 164: 29–35

923. Rutter G, Asher DM, Rohwer RG, Gibbs CJ Jr, Gajdusek DC (1981) Increased concanavalin A capping in cells from brains of scrapie-infected hamsters. Arch Virol 68: 129–133

924. Sadeh M, Chagnac Y, Goldhammer Y (1990) Creutzfeldt-Jakob disease associated with peripheral neuropathy. Isr J Med Sci 26: 220–222

925. Safar J, Wang W, Padgett MP, Ceroni M, Piccardo P, Zopf D, Gajdusek DC, Gibbs CJ Jr (1990) Molecular mass, biochemical composition, and physico-chemical behavior of the infectious form of the scrapie precursor protein monomer. Proc Natl Acad Sci USA 87: 6373–6377

926. Saito K, Matsumoto S, Yokoyoma T, Okaniwa M, Kamoshita S (1982) Pathology of chronic vitamin E deficiency in fatal familial intrahepatic choleostasis (Byler disease). Virch Arch (Pathol Anatom Histol) 396: 319–330

927. Salazar AM, Brown P, Gajdusek DC, Gibbs CJ Jr (1983) Relation to Creutzfeldt-Jakob disease and other unconventional virus diseases. In: Reisberg B (ed) Alzheimer's disease. Free Press, New York, pp 311–318

928. Salazar AM, Masters CL, Gajdusek DC, Gibbs CJ Jr (1983) Syndromes of amyotrophic lateral sclerosis and dementia: relation to transmissible Creutzfeldt-Jakob disease. Ann Neurol 14: 17–26

929. Sandbank U, Lerman P, Geifman M (1970) Infantile neuroaxonal dystrophy: cortical axonic and presynaptic changes. Acta Neuropathol (Berl) 16: 342–352

930. Sasaki S, Mizoi S, Akashima A, Shinagawa M (1986) Spongiform encephalopathy in sheep scrapie: electron microscopic observations. Jpn J Vet Sci 4: 791–796

931. Sato Y, Ohta M, Tateishi J (1980) Experimental transmission of human subacute spongiform encephalopathy to small rodents. II. Ultrastructural study of spongy state in gray and white matter. Acta Neuropathol (Berl) 51: 135–140

932. Schlapfer WW (1987) Neurofilaments: structure, metabolism and implications in disease. J Neuropathol Exp Neurol 46: 117–129

933. Schlenska GK, Walter GF (1989) Serial computed tomography findings in Creutzfeldt-Jakob disease. Neuroradiol 31: 303–306

934. Schlote W (1970) Subakute prasenile spongiforme Encephalopathie mit occipitalem Schwerpunkt und Rindenblindheit (Heidenhain-Syndrom). Arch Psychiatr Nervenkr 213: 345–369

935. Schlote W (1971) How long can degenerating axons in central nervous system produce reactive changes. Acta Neuropathol (Berl) [Suppl] 5: 40–48

936. Schlote W, Boellaard JW, Schumm F, Stohr M (1980) Gerstmann-Straussler-Scheinker's disease. Electron microscopic observations on a brain biopsy. Acta Neuropathol (Berl) 52: 203–211

937. Schmidt RE, Plurad SB (1985) Ultrastructural appearance of intentionally frustrated axonal regeneration in rat sciatic nerve. J Neuropathol Exp Neurol 44: 130–146

938. Schmidt RE, Santiago BP, Modert CW (1983) Neuroaxonal dystrophy in the autonomic ganglia of aged rats. J Neuropathol Exp Neurol 42: 376–390

939. Schochet SS (1971) Mitochondrial changes in axonal dystrophy produced by vitamin E deficiency. Acta Neuropathol (Berl) [Suppl] 5: 54–60

940. Schoene WC, Masters CL, Gibbs CJ Jr, Gajdusek DC, Tyler HR, Moore FD, Dammin GJ (1981) Transmissible spongiform encephalopathy (Creutzfeldt-Jakob disease). Atypical clinical and pathological findings. Arch Neurol 38: 473–477

941. Schumm F, Boellaard JW, Schlote W, Stohr M (1981) Morbus Gerstmann-Sträussler-Scheinker. Familie Sch. – Ein Bericht über drei Kranke. Arch Psychiatr Nervenkr 230: 179–196

942. Scott AC, Done SH, Venables C, Dawson M (1987) Detection of scrapie-associated fibrils as an aid to the diagnosis of natural sheep scrapie. Vet Rec 120: 280–281

943. Scott AC, Wells GAH, Stack MJ, White H, Dawson M (1990) Bovine spongiform encephalopathy: detection and quantitation of fibrils, fibril protein (PrP) and vacuolation in brain. Vet Microbiol 23: 295–304

944. Scott JR, Fraser H (1989) Enucleation after intraocular scrapie injection delays the spread of infection. Brain Res 504: 301–305

945. Scott JR, Fraser H (1984) Degenerative hippocampal pathology in mice infected with scrapie. Acta Neuropathol (Berl) 65: 62–68

946. Scott MRD, Butler DA, Bredesen DE, Wälchli M, Hsiao KK, Prusiner SB (1988) Prion protein gene expression in cultured cells. Protein Engin 2: 69–76

947. Scott M, Foster D, Mirenda C, Serban D, Coufal F, Wälchli M, DeArmond SJ, Westaway D, Prusiner SB (1989) Transgenic mice expressing hamster protein produce species-specific scrapie infectivity and amyloid plaques. Cell 59: 847–857

948. Scrimgeour EM, Masters CL, Alpers MP, Kaven J, Gajdusek DC (1983) A clinico-pathological study of case of kuru. J Neurol Sci 59: 265–275

949. Seitelberger F (1962) Eigenartige familiär-hereditäre Krankheit des Zetralnervensystems in einer niederösterreichischen Sippe. Wien Klin Wochenschr 74: 687–691

950. Seitelberger F (1986) Neuroaxonal dystrophy: its relation to aging and neurological diseases. In: Vinken PJ, Bruyn GW, Klawans HL (eds) Handbook of Clinical Neurology, vol 5 (49). Elsevier Sci, Amsterdam, pp 391–415

951. Seitelberger F (1971) Neuropathological conditions related to neuroaxonal dystrophy. Acta Neuropathol (Berl) [Suppl] 7: 17–29

952. Selmaj K, Cannella B, Brosnan CF, Raine CS (1990) Characterization of TCR gamma-delta lymphocytes in patients with multiple sclerosis, Abstract no VI-B-6, In: Abstracts of the XIth International Congress of Neuropathology, Kyoto, Japan, September 2–8, pp 278

953. Selmaj KW, Raine CS (1988) Tumor necrosis factor mediates myelin and oligodendrocyte damage in vitro. Ann Neurol 23: 339–347

954. Shehan BJ, Barret PN, Atkins GJ (1981) Demyelination in mice resulting from infection with a mutant of Semliki forest virus. Acta Neuropathol (Berl) 53: 129–136

955. Shibayama Y, Sakaguchi Y, Nakata K, Goto T, Nakai M, Takai T, Shirakata S (1982) Creutzfeldt-Jakob disease with demonstration of virus-like particles. Acta Pathol Jpn 32: 695–702

956. Shimono M, Ohta M, Asada M, Kuroiwa Y (1986) Infantile neuroaxonal dystrophy. Ultrastructural study of peripheral nerve. Acta Neuropathol (Berl) 36: 71–76

957. Shinagawa M, Munekata E, Doi S, Takanashi K, Goto H, Sato G (1986) Immunoreactivity of synthetic pentadecapeptide corresponding to the N-terminal region of the scrapie prion protein. J Gen Virol 67: 1745–1750

958. Shiraki H (1985) Kuru in New Guinea. In: Mizutami T, Shiraki H (eds) Clinicopathological Aspectss of Creutzfeldt-Jakob Disease. Elsevier, Nishimura, Amsterdam, Niigata, pp 176–186

959. Shiraki H, Mizutami T (1985) Ataxic form of Creutzfeldt-Jakob disease. In: Mizutami T, Shiraki H (eds) Clinicopathological Aspectss of Creutzfeldt-Jakob Disease. Elsevier, Nishimura, Amsterdam, Niigata, pp 77–108

960. Siakatos AN, Gajdusek DC, Gibbs CJ Jr, Traub RD, Bucana C (1976) Partial purification of the scrapie agent from mouse brain by pressure disruption and zonal centrifugation in cucrose-sodium chloride gradients. Virology 70: 230–237

961. Siakatos AN, Raveed D, Longa G (1979) The discovery of a particle unique to brain and spleen subcellular fractions from scrapie-infected mice. J Gen Virol 43: 417–422

962. Siedler H, Malamud N (1963) Creutzfeldt-Jakob disease. Clinicopathological report of 15 cases and review of the literature (with special reference to a related disorder designated as subacute spongiform encephalopathy). J Neuropathol Exp Neurol 22: 381–402

963. Sigvald J, Chomette G, Raverdy Ph, Vedrenne C, Bouttier D, Auriol M (1967) Forme de Heidenhain de syndrome de Creutzfeldt-Jakob, (observation anatomo-clinique). Rev Neurol 115: 258–265

964. Sigurdson B (1954) Rida, a chronic encephalitis of sheep. With general remarks on infections which develop slowly and some of their special characteristics. Br Vet J 110: 341–359

965. Silberman J, Cravioto H, Feigin I (1961) Cortico-striatal degeneration of the Creutzfeldt-Jakob disease. J Neuropathol Exp Neurol 20: 105–118

966. Skarpheoinsson S, Johannsdottir R, Sigurdarsson S, Georgsson G (1990) Detection of scrapie associated fibrils (SAF) and protease resistant proteins (PrP) in preclinical scrapie of sheep. Abstract no P28-007. In: Abstracts of the VIIIth International Congress of Virology of the Virology Division of the International Union of Microbiological Societies, Berlin, August, 24–26, 1990, pp 283

967. Sklaviadis T, Akowitz A, Manuelidis L, Manuelidis EE (1990) Nuclease treatment resuls in high specific purification of Creutzfeldt-Jakob disease infectivity with a density characteristic of nucleic acid-protein complexes. Arch Virol 112: 215–229

968. Sklaviadis T, Manuelidis L, Manuelidis EE (1986) Characterization of major peptides in Creutzfeldt-Jakob disease and scrapie. Proc Natl Acad Sci USA 83: 6146–6150

969. Sluga E (1969) Beitrag zur Feinstruktur der Läsionen bei der multiplen Sklerose des Menschen. Z Nervenheilkd Grenzgeb [Suppl 2]: 373–377

970. Sluga E, Seitelberger F (1967) Beitrag zur spongiosen Encephalopathie. Acta Neuropathol (Berl) [Suppl] 3: 60–72

971. Smith MC (1955) Argyrophilic bodies in the human spinal cord. J Neurol Neurosurg Psychiatr 18: 13–16

972. Somerville RA (1985) Ultrastructural links between scrapie and Alzheimer's disease. Lancet i: 504–506

973. Somerville RA, Merz PA, Carp RI (1986) Partial copurification of scrapie-associated fibrils and scrapie infectivity. Intervirology 25: 48–55

974. Somerville RA, Ritchie LA (1989) Are scrapie associated fibrils a pathological product of infection? In: Court LA, Dormont D, Brown P, Kingsbury DT (eds) Unconventional Virus Diseases of the Central Nervous System. Commisariat à l'Energie Atomique, Paris, pp 521–535

975. Somerville RA, Ritchie LA (1990) Differential glycosylation of the protein (PrP) forming scrapie-associated fibrils. J Gen Virol 71: 833–839

976. Somerville RA, Ritchie LA, Gibson PE (1989) Structural and biochemical evidence that scrapie-associated fibrils assemble *in vitro*. J Gen Virol 70: 25–35

977. Sotelo J, Gibbs CJ Jr, Gajdusek DC (1980) Autoantibodies against axonal neurofilaments in patients with kuru and Creutzfeldt-Jakob disease. Science 210: 190–193

978. Sotelo J, Gibbs CJ Jr, Gajdusek DC, Toh BH, Wurth M (1980) Method for preparing cultures of central neurons: cytochemical and immunochemical studies. Proc Natl Acad Sci USA 77: 653–657

979. Sotelo C, Palay SL (1971) Altered axons and axon terminals in the lateral vestibular nucleus of the rat. Possible example of axonal remodelling. Lab Invest 25: 653–671

980. Spencer PS, Griffin JW (1982) Disruption of axoplasmic Transport by neurotoxic agents: the 2,5-hexanedione model. In: Weiss DG, Gorio A (eds) Axoplasmic transport in Physiology and Pathology. Springer, Berlin Heidelberg New York, pp 92–103

981. Stahl N, Baldwin MA, Burlingame AL, Prusiner SB (1990) Identification of glycoinositol phospholipid linked and truncated forms of the scrapie prion protein. Biochemistry 29: 8879–8884

982. Stahl N, Borchelt DR, Hsiao K, Prusiner SB (1987) Scrapie prion protein contains a phosphatydylinositol glycolipid. Cell 51: 229–240

983. Stahl N, Borchelt DR, Prusiner SB (1990) Differential release of cellular and scrapie prion proteins from cellular membranes by phophatidylinositol-specific phospholipase C. Biochemistry 29: 5405–5412

984. Stamp JT, Brotherson JG, Zlotnik I, Mackay JMK, Smith W (1959) Further studies on scrapie. J Comp Pathol 69: 268–280

985. Stender A (1930) Weitere Beitrage zum Kapitel "Spastische Pseudosklerose Jakobs". Z Ges Neurol Psychiatr 128: 528–543

986. Stities DP, Garfin DE, Prusiner SB (1979) The immunology of scrapie. In: Prusiner SB, Hadlow WJ (eds) Slow Transmissible Diseases of the Nervous System, vol 2. Academic Press, New York, pp 211–221

987. Stockman S (1918) Scrapie: an obscure disease of sheep. J Comp Pathol 26: 317–326

988. Strain GM, Barta O, Olcott BM, Braun WF (1984) Serum and cerebrospinal fluid concentrations of immunoglobulin G in Suffolk sheep with scrapie. J Vet Res 45: 1812–1813

989. Suzuki K, Andrews JM, Waltz JM, Terry RD (1969) Ultrastructural studies of multiple sclerosis. Lab Invest 20: 444–454

990. Suzuki K, Plaff LD (1973) Acrylamide neuropathy in rats. An electron microscopic study of regeneration and degeneration. Acta Neuropathol (Berl) 24: 197–213

991. Szumanska G, Vorbrodt AW, Wisniewski HM (1986) Lectin histochemistry of scrapie amyloid plaques. Acta Neuropathol (Berl) 69: 205–212

992. Tagliavini F, Prelli F, Ghiso J, Bugiani O, Serban D, Prusiner SB, Farlow MR, Ghett B, Frangione B (1991) Amyloid protein of Gerstmann-Sträussler-Scheinker (Indiana kindred) is an 11 kd fragment of prion protein with an N-terminal glycine at codon 58. EMBO J 10: 513–519

993. Takahashi K, Shinagawa M, Doi S, Sasaki S, Goto S, Sato G (1986) Purification of scrapie agent from infected animal brains and raising of antibodies to the purified fraction. Microbiol Immunol 30: 123–131

994. Tamai Y, Koijima H, Ikuta F, Kumanishi T (1978) Alterations in the composition of brain lipids in patients with Creutzfeldt-Jakob disease. J Neurol Sci 35: 59–76

995. Tamai Y, Ohtani Y, Miura S, Narita Y, Iwata T, Kaiya H, Namba M (1979) Creutzfeldt-Jakob disease – alterations in ganglioside sphingosine in the brain of a patient. Neurosci Letters 11: 81–86

996. Taraboulos A, Rogers M, Borchelt DR, McKinley MP, Scott M, Serban D, Prusiner SB (1990) Acquisition of protease resistance by prion proteins in scrapie-infected cells does not require asparagine-linked glycosylation. Proc Natl Acad Sci USA 87: 8262–8266

997. Taraboulos A, Serban D, Prusiner SB (1990) Scrapie prion proteins accumulate in the cytoplasm of persistently infected cultured cells. J Cell Biol 110: 2117–2131

998. Taratuto AL, Piccardo P, Leiguarda R, Granillo R, Monti A, Scarlatti A, Leits A, Morasso C, Vigo CM, Vila J, Gutierrez A (1989) Creutzfeldt-Jakob disease. Report of 10 neuropathologically-verified cases in Argentina. Medicina (Buenos Aires) 49: 293–303

999. Tateishi J, Doi H, Sato Y, Suetsugu M, Ishii K, Kuroiwa Y (1981) Experimental transmission of human subacute spongiform encephalopathy to small rodents. Acta Neuropathol (Berl) 53: 161–163

1000. Tateishi J, Hikita K, Kitamoto T, Nagara H (1987) Experimental Creutzfeldt-Jakob disease: induction of amyloid plaques in rodents. In: Prusiner SB, McKinley MP (eds) Prions. Novel Infectious Pathogens causing Scrapie and Creutzfeldt-Jakob Disease. Academic Press, New York, pp 415–426

1001. Tateishi J, Kitamoto T, Hashigushi H, Shii (1988) Gerstmann-Sträussler-Scheinker disease: immunological and experimental studies. Ann Neurol 24: 35–40

1002. Tateishi J, Kitamoto T, Tashima T, Shii H, Hashiguschi H, Yoshimura T (1989) Gerstmann-Sträussler-Scheinker's disease: dissimilarities in pathology and transmission. In: Court LA, Dormont D, Brown P, Kingsbury DT (eds) Unconventional Virus Diseases of the Central Nervous System, Commisariat à l'Energie Atomique, Paris, pp 151–157

1003. Tateishi J, Nagara H, Hikita K, Sato Y (1984) Amyloid plaques in brains of mice with Creutzfeldt-Jakob disease. Ann Neurol 15: 278–280

1004. Tateishi J, Sato J, Koga M, Ohta M, Kuroiwa Y (1979) A transmissible variant of Creutzfeldt-Jakob disease with kuru plaques. In: Prusiner SB, Hadlow WJ (eds) Slow Transmissible Diseases of the Nervous System, vol 2. Academic Press, New York, pp 175–185

1005. Tateishi J, Sato Y, Nagara H, Boellaard JW (1984) Experimental transmission of human subacute spongiform encephalopathy to small rodents. IV. Positive transmission from a typical case of Gerstmann-Sträussler-Scheinker's disease. Acta Neuropathol 64: 85–88

1006. Tateishi J, Sato J, Ohta M (1983) Creutzfeldt-Jakob disease in humans and laboratory animals. In: Zimmerman HM (ed) Progress in Neuropathology. Raven Press, pp 195–221

1007. Tellez-Nagel I, Korthalas JK, Vlassara HV, Cerami A (1977) An ultrastructural study of chronic sodium cya nate induced neuropathy. J Neuropathol Exp Neurol 36: 351–363

1008. Terry RD, Gonatas NK, Weiss M (1964) Ultrastructural studies in Alzheimer's presenile dementia. Am J Pathol 44: 269–297

1009. Thibault J (1972) Neuroaxonal dystrophy. A case of nonpigmented type and protracted course. Acta Neuropathol (Berl) 21: 232–238

1010. Tietjen GE, Drury I (1990) Familial Creutzfeldt-Jakob disease without periodic EEG activity. Ann Neurol 28: 585–588

1011. Tiller-Borcich JK, Urich H (1986) Abnormal arborisations of Purkinje cell dendrites in Creutzfeldt-Jakob disease: a manifestation of neuronal plasticity? J Neurol Neurosurg Psychiatr 49: 581–584

1012. Tinter R, Brown P, Hedley-Whyte T, Rappaport EB, Piccardo CP, Gajdusek DC (1986) Neuropathologic verification of Creutzfeldt-Jakob disease in the exhumed American recipient of human pituitary growth hormone: epidemiologic and pathogenetic implications. Neurology 36: 932–936

1013. Torack RM (1966) Ultrastructure and histochemical studies in a case of progressive dementia and its relationship to protein metabolism. Am J Pathol 49: 77–97

1014. Torack RM (1969) Ultrastructural and histochemical studies of cortical biosies in subacute dementia. Acta Neuropathol (Berl) 13: 43–55

1015. Trabbatoni G, Lechi A, Bettoni L, Macchi G, Brown P (1990) Considerations on a group of 13 patients with Creutzfeldt-Jakob disease in the region of Parma (Italy). Europ J Epidemiol 6: 239–243

1016. Tracey KJ, Cerami A (1990) The biology of cachectin/tumor necrosis factor. In: Habenicht A (ed) Growth Factors, Differentiation Factors, and Cytokines. Springer, Berlin Heidelberg New York Tokyo, pp 356–365

1017. Tracey KJ, Cerami A (1989) The role of cachectin/tumor necrosis factor in AIDS. Cancer Cells 1: 62–63

1018. Tracey KJ, Vlassara H, Cerami A (1989) Cachectin/tumor necrosis factor. Lancet i: 1122–1225

1019. Traub R (1983) Recent data and hypotheses on Creutzfeldt-Jakob disease. In: Mayeux R, Rosen WG (eds) The Dementias. Raven Press, New York, pp 149–164

1020. Traub R, Gajdusek DC, Gibbs CJ Jr (1977) Transmissible virus dementia: the relation of transmissible spongiform encephalopathy to Creutzfeldt-Jakob disease. In: Kinsbourne M, Smith L (eds) Aging and Dementia. Spectrum Publ Inc, New York, pp 91–172

1021. Tsukamoto T, Diringer H, Ludwig H (1985) Absence of autoantibodies against neurofilament proteins in the sera of scrapie infected mice. Tohoku J Exp Med 146: 483–484

1022. Tully JG, Whitcomb RF, Williamson DL, Clark HF (1976) Suckling mouse cataract agent is a helical wall-free procaryot (spiroplasma) pathogenic for vertebrates. Nature 259: 117–120

1023. Turk E, Teplow DB, Hood LE, Prusiner SB (1988) Purification and properties of the cellular and scrapie hamster prion proteins. Eur J Biochem 176: 21–30

1024. Vallat J-M, Dumas M, Corvisier N, Leboutet M-J, Loubet A, Dumas P, Cathala F (1983) Familial Creutzfeldt-Jakob disease with extensive degeneration of white matter. J Neurol Sci 61: 261–275

1025. Van Rossum (1968) Spastic pseudosclerosis (Creutzfeldt-Jakob disease). In: Vinken P, Bruyn G (eds) Handbook of Clinical Neurology, vol 6. North-Holland, pp 726–760

1026. Verhaart WJC (1940) An unclassified degenerative disease of the central nervous system. Arch Neurol Psychiatr 44: 1262–1270

1027. Westphal KP, Schachenmayr W (1985) Computed tomography during Creutzfeldt-Jakob disease. Neuroradiol 27: 362–264

1028. Vettermann W, Schmeisser S, Gelderblom H, Koch MA (1983) Scrapie related alterations in cells of the lymphoreticular system. In: Court LA, Cathala F (eds) Virus non Conventionnels at Affections du Systeme Nerveaux Central. Masson, Paris, pp 445–452

1029. Vettermann W, Werner HJ, Schmeisser S, Koch MA (1981) Observation on the capping of spleen lymphocytes from scrapie infected mice after treatment with Ig. Med Microbiol Immunol 169: 75–81

1030. Vinters HV, Hudson AJ, Kaufmann JCE (1986) Gerstmann-Sträussler-Scheinker disease: autopsy study of a familial case. Ann Neurol 20: 540–543

1031. Viret J, Dormont D, Molle D, Court L, Feterrier F, Cathala F, Gibbs CJ Jr, Gajdusek DC (1981) Structural modifications of nerve membranes during experimental scrapie evolution in mouse. Biochem Biophys Res Comm 101: 830–836

1032. Vorbrodt AW, Lossinsky AS, Wisniewski HM, Moretz RC, Iwanowski L (1981) Ultrastructural cytochemical studies of cerebral microvasculature in scrapie infected mice. Acta Neuropathol (Berl) 53: 203–211

1033. Walberg F (1966) The fine structure of cuneate nucleus in normal cats and following interruption of afferent fibers. An electron microscopic study with particular reference to findings made in Glees and nauta sections. Exp Brain Res 2: 107–128

1034. Ward RL, Portier DD, Stevens JG (1974) Nature of the scrapie agent: evidence against a viroid. J Virol 14: 1099–1103

1035. Warter JM, Steinmetz G, Heldt N, Rumbach L, Marescaux Ch, Eber AM, Collard M, Rohmer F, Floquet J, Guedenet JC, Gehin P, Weber M (1982) Demence pre-senile familiale syndrome de Gerstmann-Sträussler-Scheinker. Rev Neurol 138: 107–121

1036. Walton JN (1956) Myopathy in sheep. Lancet 271: 841–842

1037. Watanabe I, Bingle GJ (1972) Dysmyelination in "quaking" mouse. J Neuropathol Exp Neurol 31: 352–369

1038. Weissmann C (1989) Sheep disease in human clothing. Nature 338: 298–299

1039. Weissmann C (1991) A "unified theory" of prion propagation. Nature 352: 679–683

1040. Weitgrefe S, Zupanic M, Haase A, Chesebro B, Race R, Frey W II, Rustan T, Friedman RL (1985) Cloning of a gene whose expression is increased in scrapie and in senile plaques in human brain. Science 230: 1177–1179

1041. Wells GAH, Scott AC, Johnson CT, Gunning RF, Hancock RD, Jeffrey M, Dawson M, Bradley R (1987) A novel progressive spongiform encephalopathy in cattle. Vet Rec 121: 419–420

1042. Westaway D, Goodman PA, Mirenda CA, McKinley MP, Carlson GA, Prusiner SB (1987) Distinct prion proteins in short and long scrapie incubation period mice. Cell 51: 651–662

1043. Westaway D, Mirenda C, Foster D, Zebaradjan J, Scott M, Torchia M, Yang S-L, Serban H, DeArmond SJ, Ebeling C, Prusiner SB, Carlson GA (1991) Paradoxical shortening of scrapie incubation times by expression of prion protein transgenes derived from long incubation period mice. Cell 7: 59–68

1044. Westaway D, Prusiner SB (1986) Conservation of the cellular gene encoding the scrapie prion protein. Nucl Acid Res 14: 2035–2044

1045. Westaway D, Prusiner SB (1989) Unravelling prion diseases through molecular genetics. Trend Neurosci 12: 221–227

1046. Wiley CA, Burrola PC, Buchmeier MJ, Wooddell MK, Barry RA, Prusiner SB, Lampert PW (1987) Immuno-gold localization of prion filaments in scrapie-infected hamster brains. Lab Invest 57: 646–655

1047. Wilis PR (1989) Induced frameshifting mechanism of replication for an information-carrying scrapie prion. Microb Pathog 6: 235–249

1048. Will RG, Matthews WB, Smith PG, Hudson C (1986) A retrospective study of Creutzfeldt-Jakob disease in England and Wales 1970–1979. II: epidemiology. J Neurol Neurosurg Psychiatr 49: 749–755

1049. Williamson KA, Sima AF, Curry B, Ludwin SK (1982) Neuroaxonal dystrophy in young adults: a clinicopathological study of two unrelated cases. Ann Neurol 11: 335–343

1050. Wilson DR, Anderson RD, Smith W (1950) Studies in scrapie. J Comp Pathol 60: 267–282

1051. Wisniewski HM, Lossinsky AS, Moretz RG, Vorbrodt AW, Lassmann H, Carp RI (1983) Increased blood-brain barrier permeability in scrapie-infected mice. J Neuropathol Exp Neurol 42: 615–626

1052. Wisniewski HM, Merz GS, Merz PA, Wen GY, Iqbal K (1983) Morphology and biochemistry of neuronal filaments and amyloid fibers in humans and animals. In: Zimmerman HM (ed) Progress in Neuropathology, vol 5. Raven Press, New York, pp 139–150

1053. Wisniewski HM, Moretz RC, Lossinsky AS (1981) Evidence for induction of localized amyloid deposits and neuritic plaques by an infectious agent. Ann Neurol 10: 517–522

1054. Wisniewski HM, Raine CS (1971) An ultrastructural study of experimental demyelination and remyelination. V. Central and peripheral nervous system lesions caused by diphteria toxin. Lab Invest 25: 73–80

1055. Wisniewski HM, Raine CS, Kay WJ (1969) Observation of viral demyelinating encephalomyelitis. Lab Invest 26: 589–599

1056. Wisniewski HM, Sinatra RS, Iqbal K, Grundke-Iqbal I (1981) Neurofibrillary and synaptic pathology in aged brain. In: Johnson JE (ed) Aging and Cell Structures, vol 1. Plenum Press, New York, pp 105–142

1057. Wisniewski HM, Terry RD (1973) Re-examination of the pathogenesis of the senile plaques. In: Zimmerman, HM (ed) Progress in Neuropathology, vol 2. Grune & Stratton, New York, pp 1–26

1058. Wisniewski HM, Vorbrodt AW, Wegiel J, Morys J, Lossinsky AS (1990) Ultrastructure of the cells forming amyloid fibers in Alzheimer disease and scrapie. Am J Med Genetics [Suppl] 7: 287–297

1059. Wisniewski K, Laure-Kamionkowska M, Sher J, Pitter J (1985) Infantile neuroaxonal dystrophyin an albino girl. Acta Neuropathol (Berl) 66: 68–71

1060. Wopdard JC, Collins GH, Hessler JR (1971) Feline hereditary neuroaxonal dystrophy. Am J Pathol 74: 551–556

1061. Wu Y, Brown WT, Robakis N, Dobkin C, Devine-Gage E, Merz PA, Wisniewski HM (1987) A PvuII RFLP detected in the human prion protein (PrP) gene. Nucl Acid Res 15: 3191

1062. Yagashi S, Sima AAF (1986) Neuroaxonal dystrophy in diabetic autonomic neuropathy. Classification and topographic distribution in BB rat. J Neuropathol Exp Neurol 45: 545–565

1063. Yagishita S (1978) Morphological investigations on axonal swellings and spheroids in various human diseases. Virch Arch (Pathol Anat Histol) 378: 182–197

1064. Yagishita S (1981) Creutzfeldt-Jakob disease with kuru-like plaques in Japan. Acta Pathol Jpn 31: 923–942

1065. Yagishita S, Iwabuchi K, Amano N, Yokoi S (1989) Further observation on Japanese Creutzfeldt-Jakob disease with widespread amyloid plaques. J Neurol 236: 145–148

1066. Yaima K, Suzuki K (1979) Neuronal degeneration in the brain of the brindled mouse. I. Chronological studies of the long surviving group. Acta Neuropathol (Berl) 48: 127–132

1067. Yamamoto T, Nagashima K, Oikawa K, Akai J (1985) Familial Creutzfeldt-Jakob disease in Japan. Three cases in a family with white matter involvement. J Neurol Sci 67: 119–130

1068. Yamanouchi H, Budka H, Vass K (1986) Unilateral Creutzfeldt-Jakob disease Neurology 36: 1517–1520

1069. Yoshikawa H, Tarui S, Hashimoto PH (1985) Diminished retrograde transport causes axonal dystrophy in the nucleus gracilis. Acta Neuropathol (Berl) 68: 93–100

1070. Yost CS, Lopez CD, Prusiner SB, Myers RM, Lingappa VR (1990) Non-hydrophobic extracytoplasmic determinant of stop transfer in the prion protein. Nature 343: 669–672

1071. Young PA, Taylor JJ, Yu WHA, Turen LL (1973) Ultrastructural changes in chick cerebellum induced vitamin e-deficiency. Acta Neuropathol (Berl) 25: 149–160

1072. Yu RK, Manuelidis EE (1978) Ganglioside alterations in guinea pigs at end stages of experimental Creutzfeldt-Jakob disease. J Neurol Sci 35: 15–23

1073. Zarranz JJ, Rivera Pomar JM, Salisachs P (1979) Kuru plaques in the brain of two cases of Creutzfeldt-Jakob disease. A common origin of the two diseases? J Neurol Sci 43: 291–300

1074. Zlotnik I (1960) Cerebellar and midbrain lesions in scrapie. Nature 185: 785

1075. Zlotnik I (1963) Experimental transmission of scrapie to golden hamsters. Lancet ii: 1072

1076. Zlotnik I (1962) The pathology of scrapie: a comparative study of lesions in the brain of sheep and goats. Acta Neuropathol (Berl) [Suppl] 1: 61–70

1077. Zlotnik I (1957) Significance of vacuolated neurones in the medulla of sheep infected with scrapie. Nature 180: 393–394

1078. Zlotnik I (1957) Vacuolated neurons in sheep affected with scrapie. Nature 179: 737

1079. Zochodne DW, Young GB, McLachlan RS, Gilbert JJ, Vinters HV, Kaufmann JCE (1988) Creutzfeldt-Jakob disease without periodic sharp wave complexes: a clinical, electroencephalographic, and pathologic study. Neurology 38: 1056–1060

Author index

Subject index

Archives of Virology

Official Journal of the Virology Division of the
International Union of Microbiological Societies

Editorial Board:

- I.W. Halliburton, Leeds
- D.R. Lowy, Bethesda, Md.
- F.A. Murphy, Davis, Calif. (Editor-in-Chief)
- Y. Nagai, Nagoya
- C. Scholtissek, Giessen
- J.H. Strauss, Pasadena, Calif.
- A. Vaheri, Helsinki
- M.H.V. Van Regenmortel, Strasbourg
- D.O. White, Melbourne

Virology Division:

- M.C. Horzinek, Utrecht

Special Issues:

- C.H. Calisher, Fort Collins, Colo.
- H.-D. Klenk, Marburg

Archives of Virology publishes original contributions
from all branches of research on viruses and virus
infections of humans, animals, plants, insects, and
bacteria. Coverage includes the broadest spectrum of
topics, from initial descriptions of newly discovered
viruses, to studies of virus structure, composition, and
genetics, to studies of virus interactions with host cells,
host organisms, and host populations. Multidisciplinary
studies are particularly welcome, as are studies
employing molecular biologic, molecular genetic, and
modern immunologic and epidemiologic approaches. For
example, studies on the molecular pathogenesis,
pathophysiology, and genetics of virus infections in
individual hosts, and studies on the molecular
epidemiology of virus infections in populations, are
encouraged. Studies involving applied research, such as
diagnostic technology development, monoclonal
antibody panel development, vaccine development, and
antiviral drug development, are also encouraged.
However, such studies are often better presented in the
context of a specific application or as they bear upon
general principles of interest to many virologists. In all
cases, it is the quality of the research work, its
significance, and its originality which will decide
acceptability.

Subscription Information:
1993. Vols. 128-133 (4 issues each):
DM 1.944,–, US $ 1.364.00, plus carriage charges
ISSN 0304-8608. Title No. 705

O.W. Barnett (ed.)
Potyvirus Taxonomy
1992. 57 figures. IX, 450 pages. ISBN 3-211-82353-0
Soft cover DM 290,–, öS 2030,–*
(Archives of Virology / Supplementum 5)

C. De Bac, W.H. Gerlich, G. Taliani (eds.)
Chronically Evolving Viral Hepatitis
1992. 72 figures. XIV, 348 pages. ISBN 3-211-82350-6
Soft cover DM 260,–, öS 1820,–*
(Archives of Virology / Supplementum 4)

B. Liess, V. Moennig, J. Pohlenz, G. Trautwein (eds.)
Ruminant Pestivirus Infections
Virology, Pathogenesis, and Perspectives of Prophylaxis
1991. 78 figures. VIII, 271 pages. ISBN 3-211-82279-8
Soft cover DM 220,–, öS 1540,–*
(Archives of Virology / Supplementum 3)

*R.I.B. Francki †, C.M. Fauquet, D.L. Knudson,
F. Brown (eds.)*
Classification and Nomenclature of Viruses
Fifth Report of the International Committee
on Taxonomy of VirusesVirology Division of the
International Union of Microbiological Societies
1991. IV, 450 pages. ISBN 3-211-82286-0
Soft cover DM 110,–, öS 770,–*
(Archives of Virology / Supplementum 2)

C.H. Calisher (ed.)
**Hemorrhagic Fever with Renal Syndrome,
Tick- and Mosquito-Borne Viruses**
1991. 75 figures. VII, 347 pages. ISBN 3-211-82217-8
Soft cover DM 258,–, öS 1800,–*
(Archives of Virology / Supplementum 1)

* 10 % price reduction for subscribers to the journal
"Archives of Virology"

Prices are subject to change without notice

Springer-Verlag Wien New York

E. Kurstak

Measles and Poliomyelitis

1993. 49 figures. Approx. 400 pages.
Cloth approx. DM 200,–
ISBN 3-211-82436-7

Elimination of measles and poliomyelitis diseases from the globe is a priority goal of the World Health Organization. For the first time, in a single volume comprising thirty one well-documented chapters, internationally recognized experts provide a state-of-the-art treatment of these two important viral diseases. The book offers a wide range of new findings and references on the latest advances regarding the measles and poliomyelitis:

- global and molecular genetic epidemiology characteristics and diseases surveillance new strategies
- all available vaccines and research to produce more safe and more potent biotechnology vaccines
- immunization programmes, considering the available vaccines and possibility of vaccinal associations in strategy to eliminate-eradicate these diseases
- immunity to infections and immunogenicity of vaccines
- virus genomes organization and antigenic structures related to vaccine characteristics, stressing their role in immunization strategies
- needs of global cooperation, using all available resources, vaccines and strategies to achieve the global control of diseases.

It is addressed mainly to all public health professionals concerned with measles and poliomyelitis control, especially in hospitals, clinics, governmental health services, international health organizations, centers of infectious diseases, research institutes, medical schools, vaccine producers and experts in immunization strategies and programmes.

E. Kurstak

Viral Hepatitis

Current Status and Issues
In collaboration with Christine Kurstak,
A. Hossain, and A.Al Tuwaijri

1993. 26 figures. Approx. 220 pages.
Soft cover DM 120,–, öS 840,–
ISBN 3-211-82387-5

In the 1990's significant advances in the understanding of viral hepatitis have been observed. In particular, our knowledge of the nature and diversity of viruses causing hepatitis in humans have substantially increased.
"Viral Hepatitis: Current Status and Issues" comprehensively and uniquely presents these valuable information all in a single volume for the utmost benefit of medical practitioners, microbiologists as well as those actively involved in health administration world-wide.
The virological, clinical epidemological, diagnostic, therapeutic, and preventive aspects pertaining to all the types of hepatitis known to date including hepatitis C and E are thoroughly discussed.

Prices are subject to change without notice

Springer-Verlag Wien New York